T0301808

Modeling MEMS and NEMS

Modeling MEMS and NEMS

John A. Pelesko
David H. Bernstein

CRC Press is an imprint of the
Taylor & Francis Group, an **informa** business
A CHAPMAN & HALL BOOK

Chapman & Hall/CRC Press
Taylor & Francis Group
6000 Broken Sound Parkway NW, Suite 300
Boca Raton, FL 33487-2742

ISBN 13: 978-1-58488-306-7 (hbk)

Visit the Taylor & Francis Web site at
http://www.taylorandfrancis.com

and the CRC Press Web site at
http://www.crcpress.com

Library of Congress Card Number 2002031599

Library of Congress Cataloging-in-Publication Data

Pelesko, John A.
Modeling MEMS and NEMS / John A. Pelesko and David H. Bernstein.
p. cm.
Includes bibliographical references and index.
ISBN 1-58488-306-5 (alk. paper)
1. Microelectromechanical systems—Mathematical models. I. Bernstein, David H. II. Title.

TK7875 .P45 2002
621.3—dc21 2002031599

Contents

Preface

Modeling MEMS and NEMS is about the construction, analysis, and interpretation of mathematical models of microelectromechanical and nanoelectromechanical systems (MEMS and NEMS). The study of these models is intended to illuminate microscale and nanoscale phenomena and to aid in the design and optimization of MEMS and NEMS devices. MEMS technology dates to 1964 with the production of the first batch fabricated MEMS device. The resonant gate transistor designed by H.C. Nathanson and his co-workers at Westinghouse, while large by today's standards (1 mm on a side!), incorporated all of the features that characterize modern MEMS and NEMS structures. Right from the start these early MEMS pioneers realized the value of mathematical modeling. In fact, almost half of Nathanson's seminal paper on the resonant gate transistor concerns modeling. This is a reflection of the fact that designing small structures necessitates an understanding of behavior *a priori*. The way to gain such understanding is to construct, analyze, and interpret the proper mathematical model. *Modeling MEMS and NEMS* contains the material needed by the MEMS/NEMS researcher building a mathematical model of their latest design. This text is intended for use by MEMS and NEMS professionals and by students studying science, engineering, or applied mathematicians who hope to become MEMS and NEMS researchers.

Organization

This book is organized around phenomena. After the introductory material, review of continuum mechanics, and study of scaling presented in Chapters 1 to 3, each chapter follows a similar format. A sequence of real devices which share a common feature is described. For example, in Chapter 4, eight devices that rely upon Joule heating for their operation are discussed. The common feature is then abstracted from the sequence of devices and the mathematical tools necessary to model this aspect of the devices are presented. A series of mathematical models increasing in complexity is then constructed, analyzed, and interpreted. At the end of each chapter, we return to one of the devices described, apply the model, and interpret the analysis in the context of the specific device.

Prerequisites

The fields of MEMS and NEMS are truly interdisciplinary. This places a tremendous burden on the MEMS practitioner. An understanding of mechanics, fluid dynamics, electromagnetism, circuit design, materials science, and numerous other subjects is required. Mastery of many of these subjects requires a firm grounding in mathematical methods. Likewise, appreciation of all the material in *Modeling MEMS and NEMS* requires a solid grasp of mathematical methods and continuum mechanics. Nevertheless, we recognize that an undergraduate student studying this text, a beginning graduate student trained in a specific engineering discipline, and even many accomplished MEMS practitioners may not have such a background. Consequently, we have designed the text to be accessible to anyone with a knowledge of calculus and ordinary differential equations. Many of the "back-of-the-envelope" models studied in this book are based on systems familiar to undergraduate engineering and science majors—the mass on a spring, the piston-cylinder, the parallel plate capacitor, etc. These models and their analysis can be appreciated by the typical senior level undergraduate engineering or science major. The reader familiar with the standard partial differential equations of mathematical physics, i.e., the diffusion equation, the wave equation, and Laplace's equation, will appreciate most of the models studied in this text. The reader who has studied continuum mechanics at the beginning graduate level is well equipped to study all of *Modeling MEMS and NEMS*.

Chapter Interdependence

The plot of this text is driven by *phenomena*. The subplot is driven by *mathematical methods*. Both the plot and the subplot proceed from the easy to hard within each chapter and within the text as a whole. This requires a certain amount of chapter interdependence. However, Chapters 4, 6 to 8 and 9 can be covered independently of one another provided the student has a firm grasp on the appropriate material from Chapters 2 and 5. For example, the instructor mainly interested in electrostatic actuation can cover Sections 2.4 and 2.7, Chapter 5, and then Chapter 7. Or, the instructor interested in thermopneumatic actuation can cover Sections 2.3 and 2.4, Chapter 5, and then Sections 6.1, 6.2, and 6.3. The instructor interested in only thermal phenomena can cover Section 2.3 and then Chapter 4. Other such arrangements are possible according to the interest of the instructor. The development and review of various mathematical methods are scattered throughout the text in keeping with our philosophy that methods should be introduced as the models require. Additional background on selected mathematical techniques is provided in Appendix A.

The World Wide Web

The world wide web has created an opportunity for an unprecedented level of communication between authors and readers. To take advantage of this, many texts being published today are accompanied by a companion website. This text is no exception. We have created a website entirely devoted to this book. The web address is

www.modelingmemsandnems.com

On this website you will find updated and new homework problems, MATLAB simulations of models discussed in the text, links to other MEMS and NEMS resources, and other supplementary material designed to complement the text. You will also find a discussion board devoted to *Modeling MEMS and NEMS*. We welcome comments, criticisms, and suggestions from students, instructors, and MEMS professionals. You are invited to submit materials for posting on the website. Please also send us links to your websites for courses based on this text. We look forward to hearing from you!

A Note on Terminology

MEMS, NEMS, microsystems, nanosystems, nanoscience, micromachines, microengineering, micromechanics, the list of terms used to describe the fields of MEMS and NEMS is virtually endless. In this book we shall use almost all such terms and use them nearly interchangeably. As it becomes quite cumbersome to repeatedly say "the fields of MEMS and NEMS," we will sometimes say "MEMS" and at other times say "NEMS" or "nanoscience" or "microsystems." We hope the reader will forgive this lack of precision and will appreciate the enhancement to the exposition.

Acknowledgments

I am indebted to many people whose advice, support, and help during the writing of this book have been invaluable. Thanks to our editor at CRC Press, Sunil Nair, and all of the staff at CRC, especially Jasmin Naim , Helena Redshaw, and Joette Lynch. Thanks to Wenjing Ye, Matthew Realff, and John McCuan, for many stimulating discussions on microsystems, nanosystems, and other topics. Thanks to Guillermo Goldsztein for not convincing me not to write this book. Thanks to my co-author Dave Bernstein, for introducing me to MEMS and pushing this book forward. Thanks to the unwitting students of Math 6514 and Math 6515 at the Georgia Institute of Technology for helping me try out the material of Chapter 7. Thanks to the National Science Foundation for supporting my research in MEMS.

I have been fortunate to have had my career shaped by interactions with many truly wonderful people. I would especially like to acknowledge Gregory A. Kriegsmann, who taught me everything I know about modeling and much more besides, Jonathan Luke for inspiration, and Daljit S. Ahluwalia for creating a fantastic environment in which to learn. Thanks also to Oscar P. Bruno and Donald S. Cohen for their support during and after my time at Caltech. Thanks to my parents for giving me the gift of a love of learning. A special thanks to my wife, Holly, for many things but especially for her support over the past year. Finally, thanks to Julia and John-John for the time I spent writing that I could have spent with you and for continually forcing me to look at the world with fresh eyes.

Newark, Delaware *John A. Pelesko* July 2002

First and foremost, I would like to thank to Mary Ann Maher for her support and encouragement throughout the writing of this book. Many thanks to Barry Dyne at Tanner Research for introducing me to the world of microelectronics and MEMS and for providing a good learning environment during my tenure there. I am grateful to Dubravka Bilic for reading and commenting on a draft of the manuscript. I would also like to thank MEMScAP, Inc., for their support during the initial phase of this project.

I am grateful to have learned from a series of gifted teachers over the years: Bill Sweeney, Jerome Jaffre, Helmut Fritzsche, Robert Gilmore, and Larry Smarr.

The applied mathematics department at Caltech deserves special thanks for their compassionate adoption of a stray computational physicist with a suspicious pedigree. Herb Keller, Mike Holst, Patrick Guidotti, and Eldar Giladi deserve special mention for their particularly open-minded and friendly attitudes. Of course, John Pelesko must get a nod for actually doing most of the work and because *he* was the one who really pushed the project forward. Grateful thanks to Linh Nguyen for many helpful and enlightening discussions. Lastly, my deepest gratitude to Mayya Tokman for her faith and support, despite the lack of any reasonable justification for them.

Berkeley, California *David H. Bernstein* July 2002

List of Figures

List of Tables

Chapter 1

Introduction

Though men now possess the power to dominate and exploit every corner of the natural world, nothing in that fact implies that they have the right or the need to do so.

Edward Abbey

1.1 MEMS and NEMS

These days there is tremendous interest in micro- and nanoelectromechanical systems (MEMS and NEMS) technology. After all, who doesn't get excited at the thought of nanorobots coursing through veins, repairing damaged cells, or smart dust drifting through the air sniffing out pollutants? Who isn't tickled by the idea of microneedles that deliver painless injections or nanostructured water-repellent blue jeans as soft as cotton? Micro- and nanoelectromechanical systems technology promises all of this and much more.

Undoubtedly, both the "gee-whiz" factor and the obvious commercial potential account for much of the recent interest in micro- and nanotechnology. But, perhaps you feel the urge to go beyond merely reading about MEMS and NEMS—perhaps you wish to design and build your own device. If so, more than possibly with any other engineering discipline, you will need to become conversant with the art of *mathematical modeling.* Just as the high cost in time and money forces a new jumbo jet design to be tested in the computer before being tested in the real world, high cost in time and money forces any new MEMS/NEMS device to be modeled before being fabricated. If you want to write down, simplify, or analyze a mathematical model of a micro- or nanoelectromechanical system this book is for you.

Modeling MEMS and NEMS mixes three ingredients: MEMS/NEMS device descriptions, mathematical models, and mathematical methods. Our philosophy is that modeling should be taught by example. The corollary is that modeling of MEMS and NEMS should be taught by examples drawn from MEMS and NEMS. Consequently, after the introductory material of Chapters 1 to 3, each chapter is device driven. At the start of each chapter, we describe a collection of MEMS and NEMS that share a common feature. For instance, in Chapter 4, we describe devices that use resistive heaters, while in Chapter 7 we describe devices that use electrostatic forces for actuation. The common feature is abstracted from the sequence and studied in the remainder of the chapter. This may involve a review of the relevant physics, the introduction of mathematical methods and certainly involves constructing mathematical models. Usually, we construct a sequence of models increasing in complexity. In fact, for the first model in each chapter we strive for simplicity. The idea is that if we can capture the essential features of the system with a simple model, complexity can always be built back in, but the analysis of the complex model will benefit from having studied a simplified version. We emphasize that this includes *numerical analysis* of complex models. Studying the mass-spring parallel plate model of electrostatic actuation yields insight into why a complicated 3-D finite element simulation of an electrostatic actuator takes longer to run near the "pull-in" voltage. Studying Joule heating of a long thin cylinder in limiting situations results in nonlinear partial differential equations involving only one spatial dimension. Numerical methods developed in the context of this simplified model can be scaled up to handle complex geometries.

Our study of the tools needed to model MEMS and NEMS begins in earnest in Section 1.3. But before buckling down we present a brief historical overview of the MEMS and NEMS fields. The related reading section at the end of this chapter provides pointers to more thorough historical treatments.

1.2 A Capsule History of MEMS and NEMS

It's too early to write a complete history of MEMS and NEMS. However, some key landmarks are visible. The roots of microsystem technology clearly lie in the technological developments accompanying World War II. In particular, the development of radar stimulated research in the synthesis of pure semiconducting materials. These materials, especially pure silicon, would become the lifeblood of integrated circuit and modern MEMS technology.

Both MEMS researchers and nanotechnology proponents point to Richard Feynman's famous "There's plenty of room at the bottom" lecture as a seminal event in their respective fields. Given in 1959 at the annual meeting of the American Physical Society, Feynman anticipated much of the next four decades of research in MEMS and NEMS:

> It is a staggeringly small world that is below. In the year 2000, when they look back at this age, they will wonder why it was not until the year 1960 that anybody began to seriously move in this direction.

While Feynman's lecture inspired a few immediate developments, it wasn't

TABLE 1.1: Landmarks in the history of MEMS and NEMS

1940s	Radar drives the development of pure semiconductors.
1959	Richard P. Feynman's famous "There's plenty of room at the bottom" lecture.
1960	Planar batch-fabrication process invented.
1964	H.C. Nathanson and team at Westinghouse produce the resonant gate transistor, the first batch-fabricated MEMS device.
1970	The microprocessor is invented, driving the demand for integrated circuits ever higher.
1979	The first micromachined accelerometer is developed at Stanford University.
1981	K. Eric Drexler's article, *Protein design as a pathway to molecular manufacturing*, is published in the Proceedings of the National Academy of Sciences. This is arguably the first journal article on molecular nanotechnology to appear.
1982	The scanning tunneling microscope is invented.
1984	The polysilicon surface micromachining process is developed at the University of California, Berkeley. MEMS and integrated circuits can be fabricated together for the first time.
1985	The "Buckyball" is discovered.
1986	The atomic force microscope is invented.
1991	The carbon nanotube is discovered.
1996	Richard Smalley develops a technique for producing carbon nanotubes of uniform diameter.
2000s	The number of MEMS devices and applications continually increases. National attention is focused on funding nanotechnology research and education.

until five years later that MEMS technology officially arrived. In 1964, H.C. Nathanson and his colleagues at Westinghouse produced the first batch-fabricated MEMS device. Their resonant gate transistor exhibited all of the features of modern MEMS. Mathematical modeling played a key role in the development of this system.

The invention of the microprocessor in 1970, while technically not a contribution to MEMS or NEMS, sparked an interest in lithographic fabrication techniques that would have a huge impact on MEMS fabrication methods.

In 1979, the first MEMS accelerometer was developed by researchers at Stanford University. The MEMS accelerometer would go on to become the first commercially successful MEMS device. In 1998, 27 million silicon microaccelerometers were shipped.

A good deal of interest in nanotechnology was stimulated by developments of the 1980s and 1990s. The development of the scanning tunneling microscope in 1982, the follow-up development of the atomic force microscope in 1986, the discovery of the carbon nanotube in 1991, and the technique developed by Smalley in 1996 for uniform nanotube production, rank among the most important developments of this period.

At present, the variety of MEMS devices and applications is continually increasing. True nanosystems, while limited in number today, promise to become even more important in the future.

1.3 Dimensional Analysis and Scaling

The vast majority of problems of interest to MEMS and NEMS researchers are *coupled domain* problems. The design of a microscale thermal anemometer couples heat transfer, electrostatics, and fluid dynamics. The design of nanotweezers couples electrostatics and elasticity. And so on. The main stumbling block to successful modeling of such systems is not determining what to put in, but determining what to leave out. How does one decide if tension in an elastic plate is more important than rigidity? Or vice versa? How does one decide if radial variations in temperature in a long thin rod are truly negligible? Or not? The answers to these questions are all the same: Nondimensionalize. Rescale the problem and write your model in terms of quantities without units. Examine dimensionless parameters representing *ratios* of quantities whose relative importance you wish to compare.

Of course, it only makes sense to talk about the size of dimensionless quantities. Is 1μm big? That's 1000nm. Is it big now? The question is meaningless. One micrometer is big compared to the diameter of a carbon nanotube, but small compared to length of a finger in an electrostatic comb drive. It's the relative size of things that matter.

It's worth working through a simple example of scaling. I know of no clearer illustration of the benefits and pitfalls of nondimensionalization than the projectile problem presented in the wonderful book by Lin and Segel [124]. We unabashedly borrow this problem.[1] The reader is urged to consult [124], Chapters 6 and 7, for an even more detailed look at the art of scaling.

So, with an apology to Lin and Segel, let's pretend that you are an engineer working in the early days of rocketry and Von Braun comes to you with the sketch in Figure 1.1. Your task is to answer the question: At what time will the rocket reach its maximum height? Perhaps there is a camera in the rocket and they want it timed to take a picture from as high as possible, or perhaps there is some other type of sensor. Use your imagination.

How to proceed? Let's make a simple model. Recall Newton's second law,

$$\mathbf{F} = m\mathbf{a} \tag{1.1}$$

where the boldface notation denotes vectors. If we apply Newton in the most direct way and use the coordinate system shown in Figure 1.2, we obtain

$$m\frac{d^2x}{dt^2} = -mg \tag{1.2}$$

or

$$\frac{d^2x}{dt^2} = -g, \quad x(0) = 0, \quad \frac{dx}{dt}(0) = V. \tag{1.3}$$

This is easily solved; we find

$$x(t) = -\frac{1}{2}gt^2 + Vt. \tag{1.4}$$

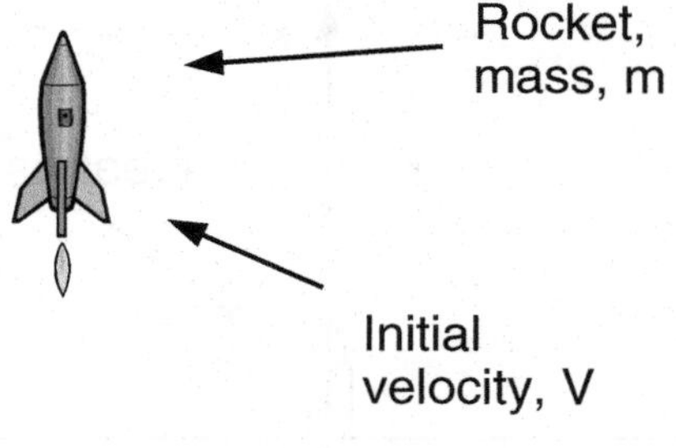

FIGURE 1.1: Von Braun's sketch.

Now, the maximum height occurs when $dx/dt = 0$, i.e., when $-gt + V = 0$ and we find

$$t_{\text{max height}} = \frac{V}{g}. \tag{1.5}$$

So, you take the rest of the week off and the next week Von Braun asks you to present your results in the weekly group meeting. Well, you go ahead and in the most obtuse way you can think of present the results above. At the end of your presentation, a little old guy stands up in the back of the room and says "That's all very nice, but you've forgotten one thing. The earth is round!" The audience laughs, but the old guy is holding up a picture, Figure 1.3, and you realize that he has a real point. The earth is in fact curved! But, we've treated the earth as a point mass. Why should this matter? Does it matter? Perhaps you want to argue that the rocket is not going up so high that it does matter? How do we justify such an argument? Let's make a better model and see what we can do. Newton's law of gravity tells us that the gravitational force between a pair of point masses is inversely proportional to the square of the distance between them. That is,

$$F_{\text{gravity}} = \frac{GM_e m}{d^2}. \tag{1.6}$$

Look at Figure 1.4. In our previous coordinates, $d = x + R$. So, we should really compute the force from

$$F_{\text{gravity}} = \frac{GM_e m}{(x+R)^2}. \tag{1.7}$$

To be consistent with our previous model, we note that

$$F_{\text{gravity}}(x = 0) = \frac{GM_e m}{R^2} = mg, \tag{1.8}$$

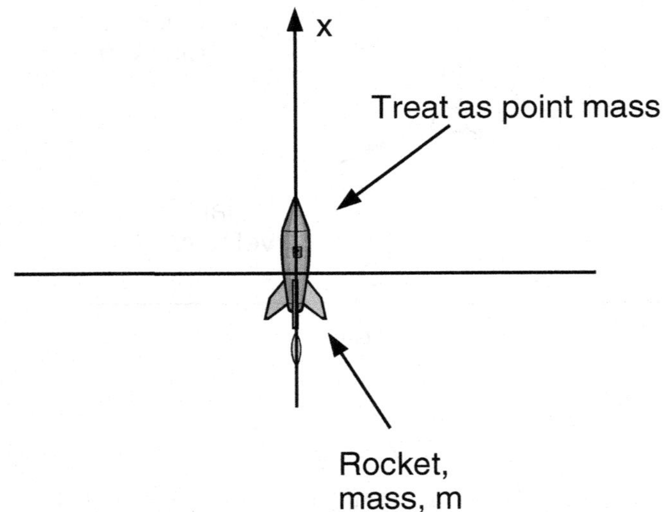

FIGURE 1.2: Coordinate system for our first model.

which implies that $g = GM_e/R^2$ or $GM_e = gR^2$. Rewriting the force in terms of g so we can compare to our previous result yields

$$F_{\text{gravity}} = \frac{mgR^2}{(x+R)^2}. \tag{1.9}$$

So, we have a new formulation of our problem

$$\frac{d^2x}{dt^2} = -\frac{gR^2}{(x+R)^2}, \quad x(0) = 0, \quad \frac{dx}{dt}(0) = V. \tag{1.10}$$

But now, our model is nonlinear! Ugh! Our intuition (or fervent wish) is that small V should mean that curvature effects are negligible. That is, we expect $x \ll R$ and therefore wish to claim that our first model is a really good approximation and we don't need to worry about this more complicated model. But how do we make this precise? And what if we want to know how big these negligible effects really are? The answer is to *nondimensionalize.*

The parameters and variables in our problem appear in Tables 1.2 and 1.3. Using these, let's form dimensionless variables

$$y = \frac{x}{R}, \quad \tau = \frac{Vt}{R}. \tag{1.11}$$

We want to substitute these into our model. First we need to compute how derivatives change. Recall the chain rule,

$$\frac{d^2x}{dt^2} = \frac{d^2}{dt^2}(Ry) = R\frac{d}{dt}\frac{dy}{dt} = R\frac{d}{dt}\left(\frac{dy}{d\tau}\frac{d\tau}{dt}\right)$$

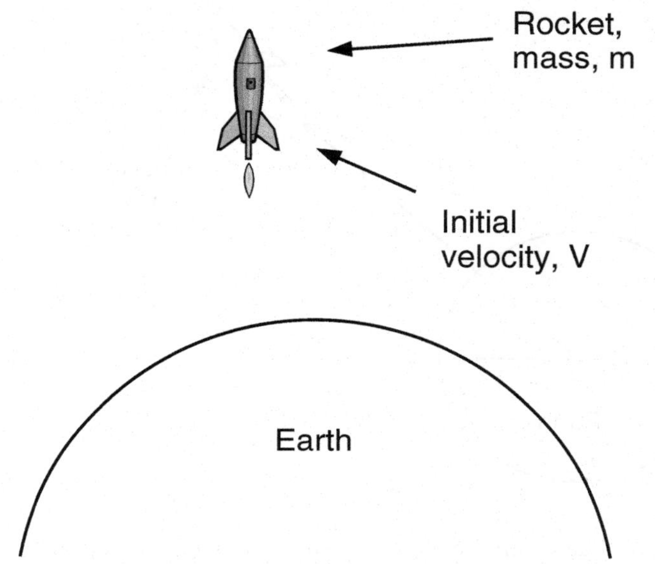

FIGURE 1.3: Old guy's sketch.

TABLE 1.2: The variables and units for the projectile problem

Variables in the problem	Associated units
x	Length
t	Time

$$= V \frac{d}{dt} \frac{dy}{d\tau} = V \frac{d}{d\tau} \frac{dy}{d\tau} = \frac{d\tau}{dt} = \frac{V^2}{R} \frac{d^2y}{d\tau^2}. \tag{1.12}$$

Whew! That's the last time we'll do it the hard way; from now on, we'll just do it in our head directly as

$$\frac{d^2x}{dt^2} \longrightarrow V^2 \frac{R}{R^2} \frac{d^2y}{d\tau^2} = \frac{V^2}{R} \frac{d^2y}{d\tau^2}. \tag{1.13}$$

Okay, now, let's use this result to scale our model. Introducing the dimensionless variables into our model yields

$$\frac{V^2}{R} \frac{d^2y}{d\tau^2} = -\frac{gR^2}{(Ry + R)^2}, \tag{1.14}$$

which simplifies to

$$\frac{V^2}{gR} \frac{d^2y}{d\tau^2} = -\frac{1}{(1 + y)^2}. \tag{1.15}$$

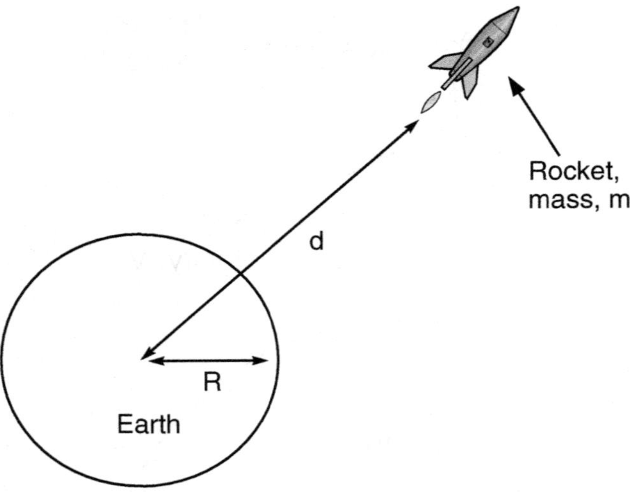

FIGURE 1.4: The setup for a more accurate model.

TABLE 1.3: The parameters and units for the projectile problem

Parameters in the problem	Associated units
R	Length
V	Length/time
g	Length/time^2

The grouping multiplying the derivative is dimensionless. Let's name it ϵ. That is,

$$\epsilon = \frac{V^2}{gR}. \tag{1.16}$$

Note if it isn't dimensionless, then we've made a mistake! Also notice that in dimensionless variables our model

$$\epsilon\frac{d^2y}{d\tau^2} = -\frac{1}{(1+y)^2}, \quad y(0) = 0, \quad \frac{dy}{d\tau}(0) = 1 \tag{1.17}$$

contains only one parameter rather than three! Hopefully, you see one of the benefits of nondimensionalizing. If you wanted to do a parameter study, parameter space has collapsed from three dimensions to one. Let's think about ϵ a bit more closely. We wanted to claim that if velocities were sufficiently small, then we could ignore curvature. Well, ϵ provides us with just such a measure. That is, ϵ compares rocket velocity to a reference velocity based upon gravity and earth radius. So, for our experiment we could argue that $\epsilon \ll 1$. So, let's just set $\epsilon = 0$.[2] What happens to our model? There's no solution! What went wrong? Let's rethink what we just did. What we want to claim is if V is small, then in some sense x will be small. We scaled

$$y = \frac{x}{R}. \tag{1.18}$$

That is, we're measuring x in units of R. This is not so good. We don't know that the term that ϵ multiplies is of "size" one. That is, one of the points of scaling is to get our model in a form where parameters measure the size of terms. In order for this to be true, the terms multiplied by parameters must be "about" equal to one. But according to our scaling, y will be very small and who knows what its acceleration will be. Observe that we have another length scale in the problem

$$\left[\frac{V^2}{g}\right] = \text{length}. \tag{1.19}$$

(A bit of notation, $[x]$ means the units of x.) Is this length scale more appropriate? Recall the result from our initial model was that

$$t_{\text{max height}} = \frac{V}{g}. \tag{1.20}$$

At this time, x attains the value

$$x_{\text{max}} = \frac{V^2}{2g}, \tag{1.21}$$

which suggests that this is indeed a more appropriate length scale to use.

So, let's try again. We introduce the dimensionless variables,

$$y = \frac{gx}{V^2}, \quad \tau = \frac{t}{T}, \tag{1.22}$$

where T is an unknown time scale. Substitution into our model yields

$$\frac{V^2}{gT^2}\frac{d^2y}{d\tau^2} = -\frac{gR^2}{\left(\frac{V^2}{g}y + R\right)^2} \tag{1.23}$$

or

$$\frac{d^2y}{d\tau^2} = -\frac{g^2R^2T^2}{V^2R^2}\left(1 + \frac{V^2y}{gR^2}\right)^{-2}. \tag{1.24}$$

This suggests we choose $T = V/g$. Then our model reduces to

$$\frac{d^2y}{d\tau^2} = -\frac{1}{(1+\epsilon y)^2}, \quad y(0) = 0, \quad \frac{dy}{d\tau}(0) = 1, \tag{1.25}$$

which looks much better. There is an important lesson here. *The proper choice of scale depends upon the parameter range under consideration.* Now, if we again argue that $\epsilon \to 0$, our model reduces to

$$\frac{d^2y}{d\tau^2} = -1, \tag{1.26}$$

which, rewritten in terms of dimensional variables, is just

$$\frac{d^2x}{dt^2} = -g. \tag{1.27}$$

This, of course, is our original approximation. So, in the limit of "small velocities" the uniform acceleration approximation is good.

Before we conclude let's notice one more benefit of dimensional analysis. What we've shown is that

$$y = y(\tau; \epsilon). \tag{1.28}$$

We're interested in the time to maximum height, say τ_m. This occurs when $dy/d\tau = 0$, which implies that

$$\tau_m = f(\epsilon). \tag{1.29}$$

That means, once we find f, we can plot τ_m as a function of ϵ and combine data. We can get the result for Mars or Venus with no additional work!

As we've seen in this example, scaling a model to clearly bring out the relative importance of the various terms is where the art of nondimensionalization comes into play. As a general rule, we try and choose the scales to make the largest dimensionless parameters in the system about equal to one. The only real way to learn this skill is by studying examples and by practice. Throughout the remainder of the text, we will practice this skill by scaling our models of MEMS and NEMS systems. We'll use our scaling to simplify our models in a rational way. We again emphasize that even if you do not wish to simplify a model, scaling reduces parameter space. This has an obvious benefit when doing a numerical parameter study.

1.4 Chapter Highlights

- The design of MEMS and NEMS devices relies on mathematical modeling. If you want to write down, simplify, or analyze a mathematical model of a MEMS or NEMS device, this text is for you.

- The simplification of mathematical models relies upon the process of scaling. Scaling reveals the relative importance of various terms in a model.

- The proper way to scale a model depends on the parameter range of interest. There is an art to the process of scaling.
- Scaling reduces parameter space.

1.5 Exercises

Section 1.3

1. Consider the damped mass-spring oscillator,

$$m\frac{d^2u}{dt'^2} + a\frac{du}{dt'} + ku = 0.$$

Here a is a damping coefficient. What are the units of a? Scale this problem. How many dimensionless parameters remain in the equation after you have scaled? Give a physical interpretation of any dimensionless parameters in the scaled system.

2. The motion of a bead of mass m sliding on a wire hoop of radius r rotating about a diameter is governed by

$$mr\frac{d^2\phi}{dt'^2} + b\frac{d\phi}{dt'} + mg\sin(\phi) - mr\omega^2\sin(\phi)\cos(\phi) = 0,$$

where ω is the angular velocity of the hoop, b is the damping coefficient, g is the gravitational acceleration, and ϕ is the angular position of the bead. Nondimensionalize this equation in such a way that two dimensionless parameters arise and one dimensionless parameter sits in front of the second derivative term. The dimensionless parameter you should find is $\epsilon = m^2gr/b^2$. Give a physical interpretation of this parameter. When might ϵ be small?

3. Consider the motion of a projectile near the surface of the earth subjected to air resistance. That is, consider

$$m\frac{d^2u}{dt'^2} + k\frac{du}{dt'} = -mg,$$

where m is mass, k is the damping constant, g is the gravitational acceleration, and u is the projectile position. Suppose you have initial conditions

$$u(0) = 0, \quad \frac{du}{dt'}(0) = V.$$

Scale this problem under the assumption that the initial velocity V is small. Show that only one dimensionless parameter arises, namely $\beta = kV/(mg)$, and give a physical interpretation of this parameter. How do you expect the projectile to behave when this parameter is small? Large?

1.6 Related Reading

New books on MEMS and NEMS are appearing almost as quickly as new MEMS and NEMS devices. Nevertheless, some of them are quite good. The text by Madou on fabrication is one of the best.

M.J. Madou, *Fundamentals of Microfabrication: The Science of Miniaturization, Second Edition*, CRC Press, 2002.

The book by Maluf is a wonderful readable introduction to MEMS.

N. Maluf, *An Introduction to Microelectromechanical Systems Engineering*, Artech House, 1999.

Trimmer's collection of classic and seminal papers belongs on the shelf of anyone involved in MEMS and NEMS.

W. Trimmer (Editor), *Micromechanics and MEMS: Classic and Seminal Papers to 1990*, IEEE, 1997.

Two textbooks on MEMS worth owning are the texts by Elwenspoek and Senturia.

M. Elwenspoek and R. Wiegerink, *Mechanical Microsensors*, Springer, 2001.

S.D. Senturia, *Microsystem Design*, Kluwer Academic Publishers, 2001.

A few papers that appear in the Trimmer collection but are worth highlighting are the articles by Nathanson and Feynman.

H.C. Nathanson, *The Resonant Gate Transistor*, IEEE Trans. on Elec. Dev., ED-14 (1967), pp. 117-133.

R.P. Feynman, *There's Plenty of Room at the Bottom*, J. Microelectromechanical Sys., 1 (1992), pp. 60-66.

A recent review article by Judy is worth a look.

J.W. Judy, *Microelectromechanical Systems (MEMS): Fabrication, Design and Applications*, Smart Materials and Structures, 10 (2001), pp. 1115-1134.

Two entertaining popular treatments of nanotechnology are the books by Regis and Gross.

E. Regis, *Nano: The Emerging Science of Nanotechnology*, Little, Brown and Company, 1995.

M. Gross, *Travels to the Nanoworld*, Plenum Trade, 1995.

An excellent discussion of scaling and dimensional analysis can be found in the book by Lin and Segel.

C.C. Lin and L.A. Segel, *Mathematics Applied to Deterministic Problems in the Natural Sciences*, Philadelphia: SIAM, 1988.

1.7 Notes

1. We're not the only ones. This problem is also discussed in [83].

2. What we're really doing here is computing the leading-order term in a perturbation theory. We'll explore perturbation theory in more detail in later chapters.

Chapter 2

A Refresher on Continuum Mechanics

Only a fool would leave the enjoyment of rainbows to the opticians. Or give the science of optics the last word on the matter.

Edward Abbey

2.1 Introduction

Ultimately this is a book about continuum mechanics. The majority of the models we will study are derived from one or more of the basic continuum theories: heat transfer, elasticity, fluid dynamics, and electromagnetism. In this chapter we review the basis of continuum mechanics and collect together the equations comprising these continuum theories.

We begin in the next section by discussing the continuum hypothesis. Essentially, continuum theory requires that variables such as temperature, density, pressure, and velocity be defined by some averaging process and be determined as the solution to some system of equations. One of the most interesting and challenging aspects of micro- and nanoscale research is that we are pushing continuum mechanics to its limits. As the length scales of interest become microns and nanometers, quantities such as density and temperature begin to lose meaning. It is important that we recognize where continuum mechanics loses validity. In order to do so we must keep the underlying assumptions in mind.

In Section 2.3 we examine the simplest continuum theory: the theory of heat conduction. We derive the heat equation via the integral method by applying conservation of energy to a control volume and making the constitutive assumption that heat flow is proportional to the local gradient of temperature. Each of the remaining continuum theories may be derived in a similar manner. While we do not derive the equations of elasticity, fluid dynamics, or electromagnetism, we point out the constitutive assumptions underlying each theory.

In Section 2.4 we present the equations of linear elasticity. In contrast with the single scalar equation governing heat transfer, this time we are faced with a coupled system of three partial differential equations. In a pair of subsections we consider simplifications of this system that arise when studying strings and membranes. We will make use of these simplified theories when studying thermopneumatic actuators, electrostatic actuators, and other MEMS and NEMS systems. We also present the equations governing the behavior of elastic beams and plates. While in principle these follow from the Navier equations of elasticity, in practice obtaining them from the Navier equations is cumbersome. We present a brief introduction to the calculus of variations as an alternate method for deriving the equations governing beams and plates. In addition to being useful in the theory of elasticity, the calculus of variations is a generally useful modeling tool. Beam and plate-like structures will appear in various systems studied throughout this text.

In Section 2.5 we present the equations of linear thermoelasticity. This requires a modification of the constitutive law for linear elasticity to allow for stresses created by thermal gradients. This modified constitutive law couples the equation of heat conduction to the equations of linear elasticity. The coupling occurs in two ways. First, thermal gradients induce stresses and second elastic energy dissipates as heat. For most engineering systems, including microsystems[1], the change in temperature due to elastic energy dissipation may be ignored. Also, if the temperature rise is sufficiently slow, the inertial terms in the equations governing elastic displacement may be ignored. These two approximations lead to the uncoupled quasistatic thermoelastic model that we present in a subsection. We also discuss the behavior of a heated elastic rod. This system will be central to the discussion of thermally actuated devices in Chapter 6.

In Section 2.6 we turn our attention to the behavior of fluids. We present the Navier-Stokes equations that govern the motion of a Newtonian fluid. This system is even more complex than the Navier equations. This time we are faced with a coupled system of *nonlinear* partial differential equations. However, as with elasticity, certain simplifying assumptions are possible. If the fluid is assumed incompressible, the Navier-Stokes equations simplify to yield the incompressible Navier-Stokes equations. Additionally, if the fluid is assumed to be inviscid, inertial terms may be ignored. This results in the Euler equations. If the Reynolds number for the flow is small, the quadratic nonlinearity in the Navier-Stokes equations may be neglected. This results in the Stokes equations. We discuss simple flows that are important in microfluidics.

In Section 2.7 we present Maxwell's electromagnetic equations. While this system is linear, the unknown field quantities are vectors. Consequently, in general we are faced with solving a coupled system of eight partial differential equations for the components of the electric and magnetic fields. Further, the constitutive laws governing the electromagnetic behavior of many materials are often nonlinear and further complicate the solution of Maxwell's equations. As we shall see in Chapter 7, electrostatics plays an important role in

MEMS and NEMS. We show how to simplify Maxwell's equations in the electrostatic case. Also, we consider the magnetostatic case, which is important for understanding magnetically actuated devices. We will need the equations of magnetostatics when we study magnetically actuated devices in Chapter 8.

In the final section of this chapter, we discuss numerical methods commonly used to solve the equations of continuum mechanics. Most micro- and nanosystems involve more than one of the continuum domains discussed in this chapter. For example, the mass-flow sensor in Chapter 4 couples electrostatics, heat transfer, and fluid flow, while the electrostatic actuators discussed in Chapter 7 couple electrostatics with elasticity. Hence mathematical models of MEMS and NEMS devices often consist of coupled systems of nonlinear partial differential equations. Except in certain special cases, we have no hope of obtaining exact solutions. Often we resort to computation. In Section 2.8 we will briefly review finite difference, finite element, and spectral methods, three techniques in the toolbox of every numerical analyst. The finite difference method is also discussed in Chapter 8.

2.2 The Continuum Hypothesis

The most striking aspect of micro- and nanoscale science is the size of the systems under consideration. The average human hair is one-tenth of a millimeter in diameter with a cross-sectional area of approximately $1.6 \times 10^{-8}\text{m}^2$. Small. Yet, microbeams of diameter 20μm are routinely fabricated and used in MEMS devices. Such a beam has a cross-sectional area of only $3.8 \times 10^{-16}\text{m}^2$. More than 40 million such beams could be abutted against the end of a human hair at one time! Many many more carbon nanotubes with a typical radius on the order of 1nm, i.e., 10^{-9}m could be so abutted. Yet, we think of MEMS and NEMS devices as being comprised of mechanical beams, plates, and shells, we imagine carbon nanotubes as being beams of cylindrical cross-section, we think of micropumps as pumping fluid with continuous properties of velocity, temperature, and density. The literature is filled with articles where MEMS and NEMS devices are treated in the same way we treat macroscopic objects. Molecular details are ignored and components are treated as a *continuum.* Functions such as density, temperature, velocity, and displacement are assumed to be smoothly varying continuous functions of position. Is this justified? From a pragmatic point of view the answer is clearly yes. It is justified because it works. Mathematical models of micro- and nanoscale systems based on continuum theory return answers in good agreement with experiments. Most of the time. In this text we'll try and point out the places where such agreement does not occur and continuum theory needs to be patched up or scratched all together. Yet, *most* of the time is a *lot* of the time! Why does continuum theory work so well at these scales? When should we expect the breakdown of continuum theory? Answers to these questions are not easy and are the subject of a good deal of modern research. However we can give an answer that at the very least closely approximates truth; continuum theory is good provided the length scales of interest are much larger than the length scales of molecular variation in the system under consideration. An elegant illustration of this idea is provided by Lin and Segel in their wonderful text on mathematical modeling [124]. We borrow their thought experiment here. Imagine we are interesting in defining the density of a fluid at a point in space and a moment in time, say $\rho(x, y, z, t)$. In order to define this density, we imagine drawing a little ball of radius r around the point (x, y, z), measuring the mass M inside the ball, and defining ρ by dividing M by the volume, that is, $3M/(4\pi r^3)$. How does the answer depend on r? A little thought reveals that ρ computed in this manner will vary as in Figure 2.1. When

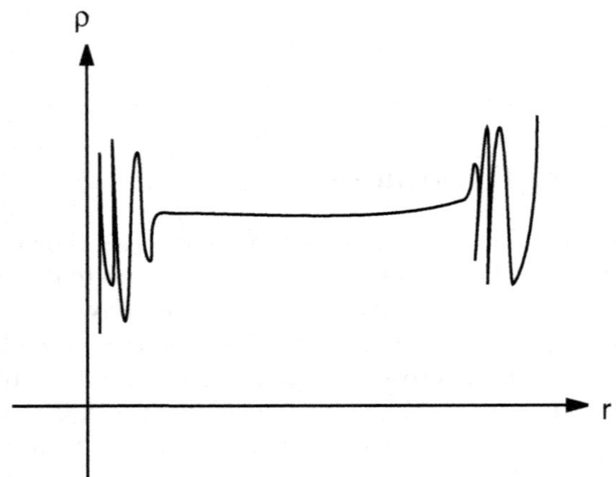

FIGURE 2.1: Variations in computed density with sample radius.

the ball is very, very small, a tiny change in radius may include very few or very many additional molecules of fluid. This results in a big change in the value of M and the wiggly region on the left of the plot in Figure 2.1. As the radius of the ball becomes bigger, this phenomenon starts to smooth itself out. Now increases in r lead to the same increases in included mass, and the value obtained for ρ is nearly the same. As r continues to increase, the wiggling phenomenon occurs again. Now we should think of the ball as being several miles in diameter and situated in the ocean. Changes in r can suddenly include large pockets of warmer, less dense water or cooler, more dense water and ρ wiggles again. Continuum theory is good where ρ doesn't wiggle. A further discussion of the limits of continuum theory may be found in Chapter 10.

2.3 Heat Conduction

Problems involving heat transfer abound at the microscale. Everyone who has ever looked inside a personal computer has undoubtedly noticed the multi-finned heat sink covering the CPU; large collections of microscale devices generate significant quantities of heat. Consequently, every MEMS designer must pay attention to thermal management issues. Additionally, numerous MEMS and NEMS devices use thermal phenomena in their operation. In Chapter 4 we will study a mass flow sensor that works by correlating fluid velocity with temperature distribution. Understanding the flow of heat is an essential part of designing this thermal anemometer. Other thermally based devices such as the thermopneumatic valves, thermoelastic actuators, and micro heat pumps are discussed in the literature [87, 94, 141, 152, 163]. Thermally driven MEMS devices are even common on the desktop; many commercially available ink jet printers use a thermal ink jet print head, which uses resistive heating to produce tiny ink droplets. We begin our review of continuum mechanics by deriving the heat equation.

2.3.1 Derivation of the Heat Equation

Consider an arbitrary control volume inside a material of interest. For simplicity, assume no internal sources of heat. Conservation of energy says that for this volume V,

$$\text{Change in energy} = \text{Flux of energy through surface.} \tag{2.1}$$

By the change in energy we mean the derivative with respect to time of the thermal energy in V. At a point, the thermal energy is given by $\rho c_p T$ where ρ denotes the density of the material, c_p its specific heat at constant pressure, and T its temperature. So

$$\text{Change in energy} = \frac{d}{dt}\int_V \rho c_p T \, dV. \tag{2.2}$$

The integral is a volume integral over the volume V. Now, assume the flux or flow of heat *into* the body is described by the vector $\mathbf{Q}$. Throughout this text we will use the boldface notation to denote vectors. Then the flux through the surface is given by the surface integral over the normal component of $\mathbf{Q}$, i.e.,

$$\text{Flux through surface} = -\int_S \mathbf{Q} \cdot \mathbf{n} \, dS. \tag{2.3}$$

Hence we have

$$\frac{d}{dt} \int_V \rho c_p T \, dV = -\int_S \mathbf{Q} \cdot \mathbf{n} \, dS. \tag{2.4}$$

Via the divergence theorem[2] the surface integral on the right-hand side may be rewritten as a volume integral,

$$\frac{d}{dt} \int_V \rho c_p T dV = -\int_V \nabla \cdot \mathbf{Q} \, dV, \tag{2.5}$$

and the order of integration and differentiation exchanged on the left to obtain

$$\int_V \left(\rho c_p \frac{\partial T}{\partial t} + \nabla \cdot \mathbf{Q} \right) dV = 0. \tag{2.6}$$

Now, since the choice of volume V is arbitrary the integrand must be identically zero[3]:

$$\rho c_p \frac{\partial T}{\partial t} + \nabla \cdot \mathbf{Q} = 0. \tag{2.7}$$

Finally, to complete the derivation we must make a constitutive assumption. Following Fourier we assume that the heat flux is proportional to the temperature gradient. That is,

$$\mathbf{Q} = -k \nabla T, \tag{2.8}$$

where k is called the thermal conductivity of the medium. In general, k is a tensor. Here we assume k is a scalar quantity. Note that this is an assumption about the behavior of a particular material. Contrast this with the principle of conservation of energy, which is an assumption about the behavior of the universe. Inserting equation (2.8) into equation (2.7) we obtain the heat equation,

$$\rho c_p \frac{\partial T}{\partial t} = \nabla \cdot k \nabla T. \tag{2.9}$$

Equation (2.9) is valid for an anisotropic inhomogeneous medium. Anisotropy may be modeled by treating k as a tensor, the heat flux may vary with direction in the material. The inhomogeneous nature of the material is reflected in the fact that ρ, c_p, and k are in general functions of position—material properties may vary from point to point within the material. In an isotropic homogeneous material equation (2.9) simplifies to

$$\frac{\partial T}{\partial t} = \frac{k}{\rho c_p} \nabla^2 T. \tag{2.10}$$

The ratio $k/(\rho c_p)$ is often denoted by κ and called the thermal diffusivity.

2.3.2 Initial and Boundary Conditions

In order to complete the specification of a problem in heat transfer, it is necessary to specify initial and boundary conditions in addition to equation (2.9). Initial conditions are easy, we simply specify the temperature of the body at every point in space at time $t = 0$,

$$T(x, y, z, 0) = f(x, y, z). \tag{2.11}$$

The choice of boundary conditions is wider. Perhaps the simplest is the Dirichlet boundary condition where the temperature is specified on the boundary of the body,

$$T|_S = T_A. \tag{2.12}$$

Physically, this boundary condition assumes that the body is in contact with a constant temperature bath. A second alternative is the Neumann boundary condition which specifies the flux of heat through the boundary

$$k\nabla T \cdot \mathbf{n}|_S = q. \tag{2.13}$$

The special case where $q = 0$ models a perfectly insulated body, i.e., heat does not flow through the surface,

$$k\nabla T \cdot \mathbf{n}|_S = 0. \tag{2.14}$$

Another alternative is to employ Newton's Law of Cooling,

$$k\nabla T \cdot \mathbf{n}|_S = -h(T - T_A)|_S, \tag{2.15}$$

where h is known as the heat transfer coefficient and T_A is the temperature of the surrounding ambient environment. This boundary condition models the situation where the heat flux through the surface depends on the difference in temperature between the surface and the ambient environment. Since we will make use of the Newton cooling boundary condition repeatedly throughout this text, it warrants closer inspection. In Chapter 3 we will see that for microsystems the dominant mode of heat transfer between solids and liquids is diffusion. That is, neither free nor forced convection contribute significantly to the exchange of heat between a microsystem and its environment. Often, we think of the Newton cooling condition as merely being a model of forced convection. After all, this was the context in which Newton first introduced the condition. However, its utility extends beyond modeling forced convection. The heat transfer coefficient h may be thought of as a "fudge factor" obtained from experiment or lower level modeling and the Newton cooling condition may be used to model heat exchange by forced convection, free convection, boiling, condensation, or even diffusion. The simplicity of the Newton cooling condition, i.e., heat flux is proportional to temperature difference, makes this an attractive model from the mathematical point of view. Whenever we use the Newton cooling condition we should keep in mind the distinction between the model and the true mode of heat transfer present in our system.

When radiant heat transfer between the body and the surrounding medium is important, the nonlinear boundary condition

$$k\nabla T \cdot \mathbf{n}|_S = -\sigma\epsilon(T^4 - T_A^4)|_S \tag{2.16}$$

is often used. Here, σ is the Stefan-Boltzmann constant and ϵ is the emissivity of the surface. The emissivity measures how closely the body approximates a black-body and hence is a dimensionless number between zero and one.

Other boundary conditions are possible. If we were studying a large body coated with a thin layer of a different material, the thin layer could be modeled by introducing the appropriate boundary condition. The reader is referred to the vast heat transfer literature and related reading given at the end of this chapter for further information on thermal boundary conditions.

2.3.3 The Cooling Sphere

In Chapter 3, when we discuss the science of scale and how micro- and nanoscale systems differ from their macroscopic counterparts, we shall use the cooling of a solid sphere as an example. Here, as a refresher, we set up the equations governing the temperature of a cooling sphere and present the solution to these equations.

Consider an isotropic homogeneous sphere of radius a initially at temperature T_0 immersed in a constant temperature bath of temperature T_A. The system is spherically symmetric so the temperature will only be a function of the radial coordinate r and time t. This temperature satisfies the heat equation,

$$\rho c_p \frac{\partial T}{\partial t} = \frac{k}{r^2}\frac{\partial}{\partial r}\left(r^2 \frac{\partial T}{\partial r}\right), \tag{2.17}$$

with initial condition,

$$T(r,0) = T_0, \tag{2.18}$$

and boundary condition at the outer surface,

$$T(a,t) = T_A. \tag{2.19}$$

Mathematically, to complete the specification of the problem we need to impose a second boundary condition at $r = 0$. It is sufficient to require that T be bounded[4] at $r = 0$, that is,

$$|T(0,t)| < \infty. \tag{2.20}$$

Now, to solve the system we begin by introducing the change of variables,

$$u(r,t) = rT(r,t), \tag{2.21}$$

into equations (2.17)–(2.20). The effect of this change of variables is to transform the problem into the simpler problem governing heat flow in a slab of thickness a,

$$\rho c_p \frac{\partial u}{\partial t} = k \frac{\partial^2 u}{\partial r^2} \tag{2.22}$$

$$u(r,0) = T_0 \tag{2.23}$$

$$u(a,t) = aT_A \tag{2.24}$$

$$u(0,t) = 0. \tag{2.25}$$

This problem for $u(r,t)$ is easily solved via separation of variables[5] and the solution for $T(r,t)$ thus obtained,

$$T(r,t) = T_A - \frac{2}{\pi r}\sum_{n=1}^{\infty}\frac{1}{n}\left(aT_A(-1)^n + T_0((-1)^n - 1)\right)e^{-n^2\pi^2 a^{-2}\kappa t}\sin\left(\frac{n\pi r}{a}\right). \tag{2.26}$$

Here κ is the thermal diffusivity as defined above.

2.4 Elasticity

The "mechanical" in "microelectromechanical" and "nanoelectromechanical" usually refers to the elastic behavior of some system. Pumps flex, tweezers bend, optical switches twist, and pressure sensors deflect. In order to understand the behavior and optimize the design of any device that involves an elastic deflection, we must be familiar with the equations of elasticity. In general, when allowing for large deflections, the equations of elasticity are nonlinear. However, in many engineering applications treatment via the linearized Navier equations is sufficient.

2.4.1 The Navier Equations

The elastic behavior of a solid body is characterized in terms of *stress* and *strain.* The stress in a solid body is expressed in terms of the *stress tensor*, σ_{ij}. If you are unfamiliar with tensors, you can think of σ_{ij} as simply being a 3×3 matrix:

$$\begin{pmatrix} \sigma_{11} & \sigma_{12} & \sigma_{13} \\ \sigma_{21} & \sigma_{22} & \sigma_{23} \\ \sigma_{31} & \sigma_{32} & \sigma_{33} \end{pmatrix}. \tag{2.27}$$

The ijth component of this matrix, σ_{ij}, denotes the force per unit area in the direction i exerted on a surface element with normal in the j direction.

Balance of angular momentum within an elastic body requires that the stress tensor be symmetric, i.e.,

$$\sigma_{ij} = \sigma_{ji}. \tag{2.28}$$

Or in matrix notation,

$$\sigma_{ij} = \begin{pmatrix} \sigma_{11} & \sigma_{12} & \sigma_{13} \\ \sigma_{12} & \sigma_{22} & \sigma_{23} \\ \sigma_{13} & \sigma_{23} & \sigma_{33} \end{pmatrix} \tag{2.29}$$

so that we have only six independent components of the stress tensor. The strain in a solid body is also expressed in terms of a tensor, called the *strain tensor*, and is denoted ϵ_{ij}. For small displacements, the strain tensor is related to displacements of the elastic body by

$$\epsilon_{ij} = \frac{1}{2}\left(\frac{\partial u_i}{\partial x_j} + \frac{\partial u_j}{\partial x_i}\right). \tag{2.30}$$

Here, u_i is the ith component of the displacement vector and the partial derivatives are taken with respect to the directions x_1, x_2, and x_3 in a Cartesian coordinate system. Note that equation (2.30) is only valid for small displacements. This equation has been linearized by ignoring quadratic terms. In general, the strain tensor is

$$\epsilon_{ij} = \frac{1}{2}\left(\frac{\partial u_i}{\partial x_j} + \frac{\partial u_j}{\partial x_i} + \frac{\partial u_k}{\partial x_i}\frac{\partial u_k}{\partial x_j}\right). \tag{2.31}$$

We have assumed the Einstein summation convention. This simply says that when you see a repeated index, it denotes summation over that index. Above, the repeated index of k indicates summation over k. The approximation of small displacements allows us to use equation (2.30) in place of equation (2.31). Just as the basic equation of heat transfer arose from applying conservation of energy, the basic equation of elasticity comes from applying conservation of momentum. That is, we apply Newton's second law to an appropriate control volume and obtain

$$\rho\frac{\partial^2 u_i}{\partial t^2} = \frac{\partial \sigma_{ij}}{\partial x_j} + f_i, \tag{2.32}$$

where ρ is the density of the body and f_i is the body force per unit volume. For instance, in modeling an electrostatically actuated device, f_i would be computed from electrostatic forces while in a magnetic device f_i would be computed from magnetic forces. Note that equation (2.32) is really three equations, one for each of the the components of the displacement vector. Again we have assumed the Einstein summation convention. For clarity we write out the three equations in (2.32):

$$\rho\frac{\partial^2 u_1}{\partial t^2} = \frac{\partial \sigma_{11}}{\partial x_1} + \frac{\partial \sigma_{21}}{\partial x_2} + \frac{\partial \sigma_{31}}{\partial x_3} + f_1, \tag{2.33}$$

$$\rho \frac{\partial^2 u_2}{\partial t^2} = \frac{\partial \sigma_{12}}{\partial x_1} + \frac{\partial \sigma_{22}}{\partial x_2} + \frac{\partial \sigma_{32}}{\partial x_3} + f_2, \tag{2.34}$$

$$\rho \frac{\partial^2 u_3}{\partial t^2} = \frac{\partial \sigma_{13}}{\partial x_1} + \frac{\partial \sigma_{23}}{\partial x_2} + \frac{\partial \sigma_{33}}{\partial x_3} + f_3. \tag{2.35}$$

Also notice that these equations really are the familiar $\mathbf{F} = m\mathbf{a}$. The left-hand side is just ρ, that's the m, times the second derivative with respect to time of the displacement, that's the $\mathbf{a}$, while the right-hand side contains all the forces.

To close the system we must make a constitutive assumption. We assume a generalized Hooke's law,

$$\sigma_{ij} = 2\mu\epsilon_{ij} + \lambda\epsilon_{kk}\delta_{ij}. \tag{2.36}$$

Just as with the Hooke's law used to find the motion of a mass on a spring, equation (2.36) assumes that stress and strain are linearly related. Here μ and λ are the *Lame constants* while δ_{ij} is called the *Kronecker delta.* In matrix terms, δ_{ij} is simply the 3×3 identity matrix. The Lame constants are related to the more familiar Young's modulus,

$$E = \frac{\mu(2\mu + 3\lambda)}{\lambda + \mu} \tag{2.37}$$

and Poisson ratio

$$\nu = \frac{\lambda}{2(\lambda + \mu)}. \tag{2.38}$$

Hence, equation (2.36) may be rewritten as

$$E\epsilon_{ij} = (1 + \nu)\sigma_{ij} - \nu\sigma_{kk}\delta_{ij}. \tag{2.39}$$

Now, substituting equation (2.36) into equations (2.32), we obtain the Navier equations of elasticity,

$$\rho \frac{\partial^2 u_i}{\partial t^2} = \frac{\partial}{\partial x_j}(2\mu\epsilon_{ij} + \lambda\epsilon_{kk}\delta_{ij}). \tag{2.40}$$

Eliminating the strain tensor in favor of the displacement vector and using vector notation yields

$$\rho \frac{\partial^2 \mathbf{u}}{\partial t^2} = \mu\nabla^2\mathbf{u} + (\mu + \lambda)\nabla(\nabla \cdot \mathbf{u}) + \mathbf{f}. \tag{2.41}$$

Equations (2.41) are called the Navier equations of linear elasticity.

2.4.2 Initial and Boundary Conditions

In order to complete the specification of a problem in linear elasticity, it is necessary to specify initial and boundary conditions in addition to equations (2.41). It is typical to specify the initial displacement and velocity and the prescribed displacements and stresses on the boundary. The particular realizations of such conditions are considerably wider than in the case of heat transfer and varies according to assumed approximations. We will not attempt to catalog all such possible choices here. Rather, we will consider a few examples that we will draw on later in the book and leave a full discussion to the texts on elasticity mentioned at the end of this chapter.

2.4.3 The Vibrating String

Consider a thin elastic string suspended between two rigid walls as shown in Figure 2.2. Assume no body forces are acting. If we assume the string

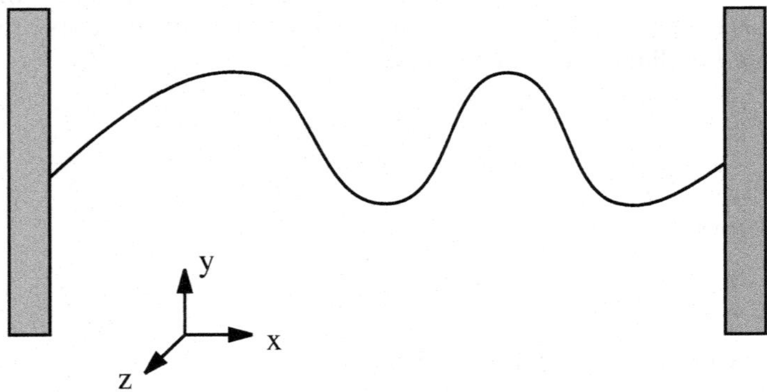

FIGURE 2.2: A thin vibrating elastic string.

is only allowed to move in the x_2 or y direction, our displacement vector $\mathbf{u}$ reduces to

$$\begin{pmatrix} 0 \\ u_2 \\ 0 \end{pmatrix}. \tag{2.42}$$

Further, the assumption that the string is thin implies that u_2 is a function of x and t alone, i.e., $u_2 = u_2(x,t)$. Hence, the Navier equations, equations (2.41), reduce to the single equation for u_2,

$$\rho \frac{\partial^2 u_2}{\partial t^2} = \mu \frac{\partial^2 u_2}{\partial x^2}. \tag{2.43}$$

To complete the specification of this problem, we need initial and boundary conditions. Since the equation is second order in time, we need a pair of initial conditions. Specifying the initial displacement and velocity is common,

$$u_2(x,0) = f(x) \tag{2.44}$$

$$\frac{\partial u_2}{\partial t}(x,0) = g(x). \tag{2.45}$$

If the string is attached to the two walls as shown in Figure 2.2, we can specify fixed boundary conditions,

$$u_2(0,t) = 0 \tag{2.46}$$

$$u_2(L,t) = 0. \tag{2.47}$$

Note, we have assumed the string has length L and occupies the interval $[0, L]$ in the x direction.

2.4.4 The Vibrating Membrane

Consider the thin elastic membrane shown in Figure 2.3. Assume no body forces are acting. Assume that the membrane is only allowed to deflect in the

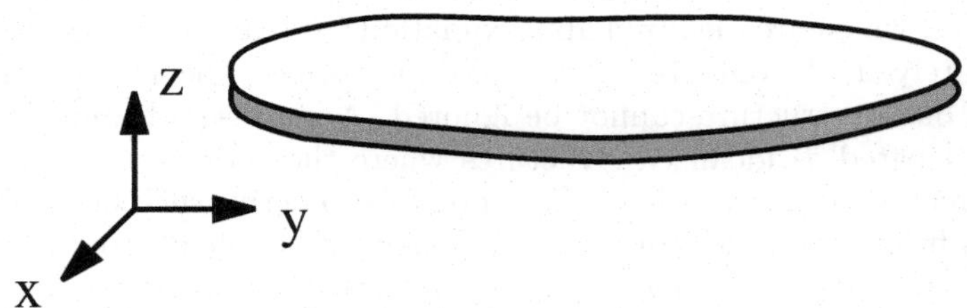

FIGURE 2.3: A thin vibrating elastic membrane.

x_3 or z direction. Then, our displacement vector reduces to

$$\begin{pmatrix} 0 \\ 0 \\ u_3 \end{pmatrix}. \tag{2.48}$$

Assuming that the membrane is thin allows us to further assume that u_3 is a function of x, y and t only, i.e., $u_3 = u_3(x, y, t)$. Hence, the Navier equations, equations (2.41), reduce to the single equation for u_3,

$$\rho \frac{\partial^2 u_3}{\partial t^2} = \mu \nabla^2 u_3, \tag{2.49}$$

where the Laplace operator, ∇^2, acts in the x and y directions only; that is,

$$\nabla^2 = \frac{\partial^2}{\partial x^2} + \frac{\partial^2}{\partial y^2}. \tag{2.50}$$

As with the vibrating string, to complete the specification of this problem, we need initial and boundary conditions. This equation is second order in time, and we need a pair of initial conditions. Specifying the initial displacement and velocity,

$$u_3(x, y, 0) = f(x, y) \tag{2.51}$$

$$\frac{\partial u_3}{\partial t}(x, y, 0) = g(x, y). \tag{2.52}$$

A variety of conditions may be applied on the boundary. Typical would be the fixed Dirichlet type boundary condition,

$$u_3(x, y, t) = 0 \quad \text{on boundary.} \tag{2.53}$$

2.4.5 Beams, Plates, and Energy Minimization

In addition to strings and membranes, the idealized structures of beams and plates play a central role in engineering elasticity. In contrast to strings and membranes, which are idealized approximations to real structures with thickness infinitely small compared to lateral dimensions, in beams and plates the thickness of the structure cannot be ignored. Nonetheless beams and plates are often treated as idealized structures where the ratio of the thickness to lateral dimensions is small. While it is possible to derive such beam and plate equations from the Navier equations, in general it is difficult to do so. In this section we present an alternate approach to the derivation of these equations based on the principle of energy minimization via the calculus of variations. The calculus of variations has wide-ranging applicability beyond the theory of elasticity. In many cases, especially when the proper modeling approach is unclear, a derivation via the variational calculus is often illuminating. Further references on the calculus of variations are provided at the end of the chapter.

2.4.6 The String Revisited

To introduce the calculus of variations, we revisit the one-dimensional elastic string of Section 2.4.3. To further simplify matters, we consider the steady-state behavior of such a string and assume that gravity is the only relevant body force. The reader may refer to Figure 2.2, keeping in mind the steady-state and gravitational assumptions. The energy, E, of this system consists of two parts and may be written

$$\text{E} = \text{Elastic energy} + \text{Gravitational energy}. \tag{2.54}$$

It is reasonable to assume that the elastic energy in the system is proportional to the change in length of the string from its un-stretched configuration. If we assume the string is held fixed at its two ends, that these ends occur at $x = 0$ and $x = L$, and if we denote the deflection of the string at position x from the horizontal by $y(x)$, we may write the elastic energy as

$$\text{Elastic energy} = T\left(\int_0^L \sqrt{1+y'^2}\,dx - L\right). \tag{2.55}$$

The constant of proportionality, T, is simply the tension in the string and the prime denotes differentiation with respect to x. The expression in parentheses is the change in length from length L. The gravitational energy may also be written in terms of $y(x)$. If σ is the mass per unit length of the string and g the gravitational acceleration, then

$$\text{Gravitational energy} = \sigma g \int_0^L y(x)\,dx. \tag{2.56}$$

Hence the total energy is

$$\mathrm{E} = T\left(\int_0^L \sqrt{1+y'^2}\,dx - L\right) + \sigma g \int_0^L y(x)\,dx. \tag{2.57}$$

Since our goal is to compare with the *linear* theory of Section 2.4.3, we should introduce the same linearizing assumptions here as earlier. That is, we should assume small deflections and expand the square root assuming y'^2 is small as,

$$\sqrt{1+y'^2} \sim 1 + \frac{1}{2}y'^2 + \cdots. \tag{2.58}$$

Notice this is simply the Taylor series expansion of $\sqrt{1+x}$ about the point $x = 0$. Then, we ignore all but the first two terms in this expansion to obtain the linearized energy,

$$\mathrm{E} = \frac{T}{2}\int_0^L y'^2\,dx + \sigma g \int_0^L y(x)\,dx. \tag{2.59}$$

Notice that the first term in the expansion of the square root has canceled the constant length in our energy. This is why we retained the first two terms in the expansion of the square root. Our energy may be written as a single integral,

$$\mathrm{E} = \int_0^L \left(\frac{T}{2}y'^2 + \sigma g y\right)\,dx. \tag{2.60}$$

Now, the idea is that the deflection that the string actually assumes is one for which the energy, E, is a *minimum*. Hence, our goal is to minimize equation (2.60) over all possible string deflections $y(x)$. In the exercises the reader is asked to formulate an isoperimetric problem and the problem of a hanging rope in a similar manner. All of these problems (and many others) lead to the problem of minimizing an integral I of the form

$$I = \int_{x_1}^{x_2} F(x, y, y')\, dx \tag{2.61}$$

over some set of functions $y(x)$. Rather than working with the special problem of minimizing E, we'll figure out how to minimize I and then apply this to our string. *The basic idea behind the calculus of variations is to turn the problem of minimizing an integral over a set of functions into the problem of minimizing a function of a single real variable.* Performing a minimization of a function of a single real variable is then a simple calculus exercise. To accomplish this, we first imagine that we know the actual minimizer $y(x)$. Assume that $y(x_1) = y_1$ and $y(x_2) = y_2$. Consider the function

$$y(x) + \epsilon\eta(x), \tag{2.62}$$

where we impose the conditions $\eta(x_1) = \eta(x_2) = 0$. This function is ϵ away from the exact solution $y(x)$ and agrees with the exact solution at the boundary points x_1 and x_2. See Figure 2.4. Plug this function into the integral to

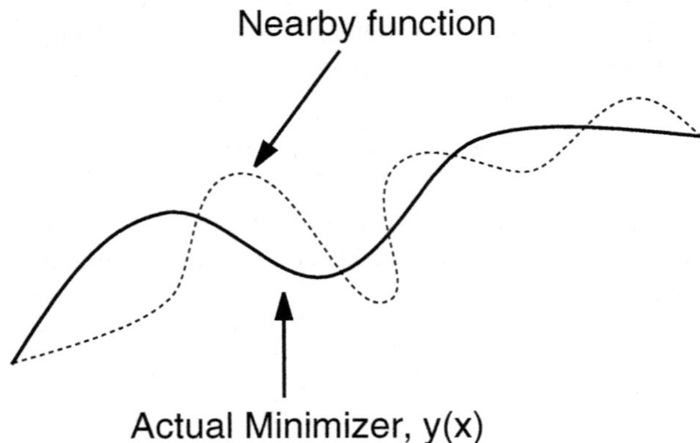

FIGURE 2.4: A variation from the actual minimizer of the energy. The minimizer, $y(x)$, and the variation, $y(x) + \epsilon\eta(x)$, agree at the endpoints.

be minimized to obtain

$$I(\epsilon) = \int_{x_1}^{x_2} F(x, y + \epsilon\eta, y' + \epsilon\eta')\, dx. \tag{2.63}$$

Now, notice that $I(\epsilon)$ is a function of the single real variable ϵ and that $I(0)$ must be a minimum since $y(x)$ was assumed to be the actual minimizer for I! But, $I(\epsilon)$ viewed as a function of the real variable ϵ having a minimum at $\epsilon = 0$ implies that

$$\frac{dI}{d\epsilon}(0) = 0. \tag{2.64}$$

So, let us compute the derivative of equation (2.63) with respect to epsilon, evaluate at $\epsilon = 0$ and set the result equal to zero as required by equation (2.64). We note that

$$\left.\frac{\partial F}{\partial \epsilon}\right|_{\epsilon=0} = \frac{\partial F}{\partial y}\eta + \frac{\partial F}{\partial y'}\eta'. \tag{2.65}$$

hence

$$I'(0) = \int_{x_1}^{x_2} \left(\frac{\partial F}{\partial y}\eta + \frac{\partial F}{\partial y'}\eta'\right) dx \tag{2.66}$$

and equation (2.64) requires

$$\int_{x_1}^{x_2} \left(\frac{\partial F}{\partial y}\eta + \frac{\partial F}{\partial y'}\eta'\right) dx = 0. \tag{2.67}$$

Now, integrate the second term in the integral by parts one time to obtain

$$\int_{x_1}^{x_2} \frac{\partial F}{\partial y}\eta\, dx + \left.\frac{\partial F}{\partial y'}\eta\right|_{x_1}^{x_2} - \int_{x_1}^{x_2} \frac{d}{dx}\frac{\partial F}{\partial y'}\eta\, dx = 0. \tag{2.68}$$

But, η vanishes at x_1 and x_2 and hence this reduces to

$$\int_{x_1}^{x_2} \frac{\partial F}{\partial y}\eta\, dx - \int_{x_1}^{x_2} \frac{d}{dx}\frac{\partial F}{\partial y'}\eta\, dx = 0 \tag{2.69}$$

or

$$\int_{x_1}^{x_2} \left(\frac{\partial F}{\partial y} - \frac{d}{dx}\frac{\partial F}{\partial y'}\right) \eta\, dx = 0. \tag{2.70}$$

Now, this must be true for any η satisfying $\eta(x_1) = 0$ and $\eta(x_2) = 0$. Hence the expression in the integrand multiplying η must be identically zero[6]. That is,

$$\frac{\partial F}{\partial y} - \frac{d}{dx}\frac{\partial F}{\partial y'} = 0. \tag{2.71}$$

Equation (2.71) is known as the Euler-Lagrange equation for I. Notice that this is an ordinary differential equation for the unknown function y. That is, we have turned the problem of minimizing I over a set of functions into the more familiar and tractable problem of solving an ordinary differential equation defining the minimizer. This process of turning a minimization problem into a differential equation is often referred to as an *indirect method.*[7]

Finally, let's apply the Euler-Lagrange equation to our gravitationally loaded string. From equation (2.60) we identify

$$F(x, y, y') = \frac{T}{2} y'^2 + \sigma g y, \tag{2.72}$$

and inserting into equation (2.71) we find the string's deflection must satisfy

$$y'' = \frac{\sigma g}{T}. \tag{2.73}$$

This ordinary differential equation is easily integrated. We leave it as an exercise for the reader to compute the parabolic profile of the string.

2.4.7 Elastic Beams

The machinery of the variational calculus was introduced in order to simplify the derivation of the equations governing elastic beams and plates. As an example of an elastic beam, consider a thin strip of paper. If allowed to fall freely, a thin strip of paper will generally lie flat. Now, bend the paper slightly and let it go. It returns to its flat configuration; that is, the paper, or any elastic beam, resists bending. This is a different phenomenon than discussed above for the elastic string that resisted changes in its length. If we consider an unloaded static beam, the energy of the system, denoted here by Σ, may be written

$$\Sigma = \text{Elastic energy} = \text{Stretching energy} + \text{Bending energy}, \tag{2.74}$$

and from our study of strings we know

$$\text{Stretching energy} = T\left(\int_0^L \sqrt{1+y'^2}\, dx - L\right). \tag{2.75}$$

Here T is tension, L is the length of the beam, and $y(x)$ describes the deflection of the beam from $y = 0$ at location x. We linearize this expression as in the previous section to obtain

$$\text{Stretching energy} = \frac{T}{2}\int_0^L y'^2\, dx. \tag{2.76}$$

The bending energy for a beam is assumed to be proportional to the linearized curvature of the beam. That is,

$$\text{Bending energy} = \frac{EI}{2}\int_0^L y''^2\, dx. \tag{2.77}$$

Here, E is the Young's modulus of the beam, and I is the cross-sectional area moment of inertia. The quantity EI is called the *flexural rigidity* of the beam.

For the total energy we have

$$\Sigma = \frac{T}{2}\int_0^L y'^2\,dx + \frac{1}{2}EI\int_0^L y''^2\,dx \tag{2.78}$$

or

$$\Sigma = \int_0^L \left(\frac{T}{2}y'^2 + \frac{EI}{2}y''^2\right)dx. \tag{2.79}$$

Notice that this integral is of a slightly different type than the one minimized earlier. Here, in addition to y', our integral contains a second derivative term, y''. Finding the Euler-Lagrange equation for this type of integral is straightforward and is left as an exercise. The reader who completes the exercises will show that the Euler-Lagrange equation for integrals of the form

$$\int_{x_1}^{x_2} F(x, y, y', y'')\,dx \tag{2.80}$$

is

$$\frac{\partial F}{\partial y} - \frac{d}{dx}\frac{\partial F}{\partial y'} + \frac{d^2}{dx^2}\frac{\partial F}{\partial y''} = 0. \tag{2.81}$$

For our beam we identify

$$F(x, y, y', y'') = \frac{T}{2}y'^2 + \frac{EI}{2}y''^2 \tag{2.82}$$

and apply the Euler-Lagrange equation to obtain

$$EIy'''' - Ty'' = 0 \tag{2.83}$$

as the equation of a one-dimensional elastic beam with no body forces but containing both tension and bending rigidity.

Thus far we have discussed static deflections of an elastic beam. The equation governing beam dynamics may also be derived using the variational calculus. In contrast to the static case, where the energy of the system was minimized, in the dynamic case we apply Hamilton's principle and minimize the *action* of the system. This is referred to as the principle of least action. Here, the action is defined as the difference between kinetic and potential energy in the system. The pointwise difference of kinetic and potential energy is the *Lagrangian* for the system. Denoting the Lagrangian by $\mathcal{L}$ we have

$$\mathcal{L} = \text{Kinetic Energy} - \text{Potential Energy}. \tag{2.84}$$

The Lagrangian is typically a function of y, its spatial derivatives, and its temporal derivatives. That is, $\mathcal{L} = \mathcal{L}(x, t, y, y', y'', \cdots, \dot{y}, \ddot{y}, \cdots)$. Hamilton's principle is to minimize the double integral,

$$\int_{t_1}^{t_2}\int_0^L \mathcal{L}(x, t, y, y', y'', \cdots, \dot{y}, \ddot{y}, \cdots)\,dx\,dt. \tag{2.85}$$

Deriving the Euler-Lagrange equation here is again a straightforward generalization of our previous analysis. The reader is urged to consult the references for details, in particular the treatment of Weinstock [204]. Here we will restrict our attention to the vibrating beam. In our static study above, we computed the potential energy of the system. The kinetic energy at any instant in time is

$$\frac{\rho A}{2}\int_0^L \dot{y}^2\,dx, \tag{2.86}$$

where ρ is the mass density per unit volume of the beam and A is the cross-sectional area. This expression is the familiar one-half mass times velocity squared expression for kinetic energy. Then, according to Hamilton's principle, we must minimize

$$\int_{t_1}^{t_2}\int_0^L \left(\frac{\rho A}{2}\dot{y}^2 - \frac{T}{2}y'^2 - \frac{EI}{2}y''^2\right) dx\,dt. \tag{2.87}$$

Our Lagrangian is

$$\mathcal{L} = \frac{\rho A}{2}\dot{y}^2 - \frac{T}{2}y'^2 - \frac{EI}{2}y''^2, \tag{2.88}$$

and as shown in the exercises the Euler-Lagrange equation in terms of $\mathcal{L}$ is

$$\frac{\partial}{\partial t}\frac{\partial \mathcal{L}}{\partial \dot{y}} + \frac{d}{dx}\frac{\partial \mathcal{L}}{\partial y'} - \frac{d^2}{dx^2}\frac{\partial \mathcal{L}}{\partial y''} = 0. \tag{2.89}$$

Inserting equation (2.88) into equation (2.89) we obtain

$$\rho A\ddot{y} - Ty'' + EIy'''' = 0 \tag{2.90}$$

as the equation of a vibrating elastic beam.

As compared to the string equation, the beam equation is fourth order in space, i.e., the highest spatial derivative on y is a fourth derivative. This implies that for a boundary value problem concerning beams, we will need four boundary conditions. Generally speaking we impose two at each endpoint. We give seven standard sets of conditions that model various endpoint constraints. If the end of the beam is free, as in the right end of the beam in Figure 2.5, the second and third derivatives are set to zero at that point. That is,

$$y''(L,t) = y'''(L,t) = 0. \tag{2.91}$$

If the end of the beam is held fixed, as in the left end of the beam in Figure 2.5, the function and its first derivative are set to zero at that point. That is,

$$y(0,t) = y'(0,t) = 0. \tag{2.92}$$

If the end is simply supported or pinned, as in the left of the beam in Figure 2.6, the function and its second derivative are set to zero at that point. That is,

$$y(0,t) = y''(0,t) = 0. \tag{2.93}$$

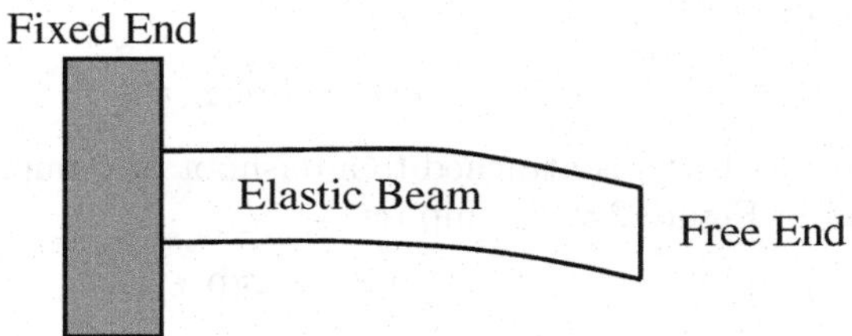

FIGURE 2.5: Elastic beam with free and fixed ends.

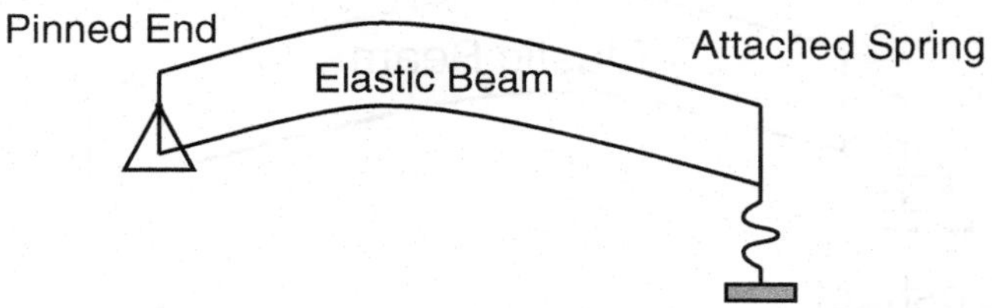

FIGURE 2.6: Elastic beam with pinned end and attached elastic spring.

If the end of the beam is attached to a spring with spring constant k, as in the right end of the beam in Figure 2.6, we impose

$$y''(L,t) = 0, \quad -EIy'''(L,t) = ky(L,t). \tag{2.94}$$

If the end of the beam is attached to a torsion spring with spring constant

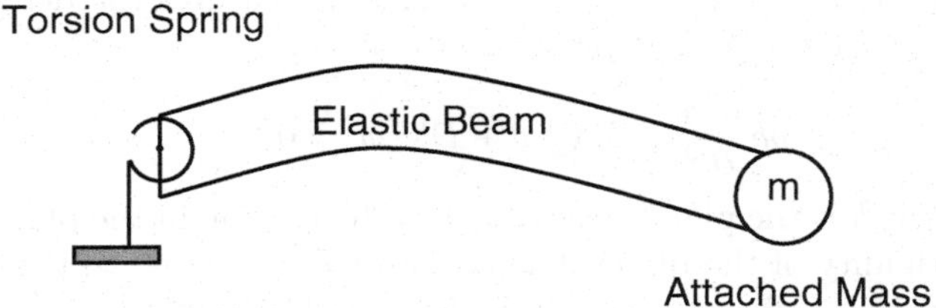

FIGURE 2.7: Elastic beam with torsion spring and attached mass.

k, as the left end of the beam in Figure 2.7, we impose

$$y(0,t) = 0, \quad EIy''(0,t) = ky'(0,t). \tag{2.95}$$

If the end of the beam is attached to a point mass m, as in the right end of the beam in Figure 2.7, we impose

$$y''(L,t) = 0, \quad -EIy'''(L,t) = m\ddot{y}(L,t). \tag{2.96}$$

Finally, if the end of the beam is attached to a dashpot or damper, as in the left end of the beam in Figure 2.8, we impose

$$y''(0,t) = 0, \quad -EIy'''(0,t) = c\dot{y}(0,t) \tag{2.97}$$

where c is the damping constant of the dashpot.

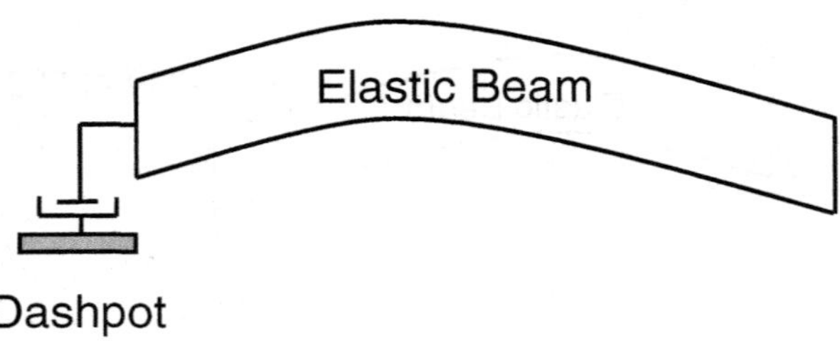

FIGURE 2.8: Elastic beam with dashpot attached.

2.4.8 Elastic Plates

The equation governing the motion of a thin elastic plate may be derived in the same manner as was the beam equation. The key difference between the beam and plate is that the plate's deflection, say w, is a function of *two* space variables, say x and y. Hence $w = w(x, y, t)$ measures the deflection of the plate in the z direction. The plate equation[8] is

$$\rho h \frac{\partial^2 w}{\partial t^2} - T\nabla^2 w + D\nabla^4 w = 0, \tag{2.98}$$

where ρ is density, h is the plate thickness, T is the tension in the plate, and D is the flexural rigidity of the plate. The constant D is related to the Young's modulus, E, Poisson ratio, ν, and plate thickness, h, through

$$D = \frac{2h^3 E}{3(1-\nu^2)}. \tag{2.99}$$

As with beams, a variety of boundary conditions are possible according to the type of support. The situation is more complicated than for beams due to the added spatial dimension. The reader is referred to the texts on elasticity mentioned at the end of the chapter for a full discussion of these boundary conditions. Further study of plate boundary conditions will be deferred to Chapter 5.

2.5 Linear Thermoelasticity

Elastic waves dissipate in the form of heat. Conversely, thermal gradients in a material induce stresses.[9] Devices such as thermoelastic actuators, shape-memory actuators, and bimetallic valves, exploit these facts to create a source of force in MEMS and NEMS devices. To model devices such as these, we must modify and combine the theories of heat conduction and linear elasticity developed in the previous two sections. The resulting theory is known as linear thermoelasticity.

2.5.1 The Equations of Linear Thermoelasticity

The first modification of linear thermoelasticity is to equation (2.9). In thermoelasticity, we incorporate a term that reflects the fact that elastic energy is converted into thermal energy. The modified heat equation is

$$\rho c_p \frac{\partial T}{\partial t} = \nabla \cdot k \nabla T - \alpha(3\lambda + 2\mu) T_0 \frac{\partial \epsilon_{kk}}{\partial t}. \tag{2.100}$$

Here, α is called the coefficient of thermal expansion and T_0 is a suitably defined reference temperature. The definition of the strain, ϵ_{ij}, remains unchanged from linear elasticity,

$$\epsilon_{ij} = \frac{1}{2}\left(\frac{\partial u_i}{\partial x_j} + \frac{\partial u_j}{\partial x_i}\right), \tag{2.101}$$

as does the balance of linear momentum,

$$\rho \frac{\partial^2 u_i}{\partial t^2} = \frac{\partial \sigma_{ij}}{\partial x_j} + f_i. \tag{2.102}$$

Also, as in linear elasticity, the stress tensor is symmetric $\sigma_{ij} = \sigma_{ji}$, reflecting the balance of angular momentum. The major modification to the linear elastic theory is to the constitutive law. A modified Hooke's law known as the Duhamel-Neumann relation is introduced. This modified Hooke's law contains a term expressing the fact that a rise in temperature creates a stress within a material,

$$\sigma_{ij} = 2\mu\epsilon_{ij} + \lambda\delta_{ij}\epsilon_{kk} - \delta_{ij}\alpha(3\lambda + 2\mu)(T - T_0). \tag{2.103}$$

As in the previous section we have assumed the summation convention. Notice that this is the same stress-strain relation introduced in Section 2.4 with the addition of a term depending on the change in temperature above some reference point. Notice that if all displacements in the body are zero, so that the strain tensor is identically zero, a rise in temperature still creates stresses. Equations (2.100)–(2.103) are known as the equations of linear thermoelasticity.

2.5.2 Uncoupled Quasistatic Thermoelasticity

The equations of linear thermoelasticity, equations (2.100)–(2.103), while *linear* still form a *coupled* set of partial differential equations. Fortunately, in most engineering applications, several simplifications are permissible. The first approximation is to neglect the coupling term in the heat equation, equation (2.100). That is, return to the regular equation of heat conduction,

$$\rho c_p \frac{\partial T}{\partial t} = \nabla \cdot k \nabla T. \tag{2.104}$$

This approximation is reasonable provided the elastic energy dissipated in the form of heat is small compared to other sources of heat in the system being studied. For example, in a thermal actuator where stresses in elastic materials are created through Joule heating, the temperature changes due to Joule heating will outweigh any temperature changes due to elastic dissipation. The second approximation is to neglect the inertial terms in equation (2.102). That is, replace equation (2.102) with

$$\frac{\partial \sigma_{ij}}{\partial x_j} + f_i = 0. \tag{2.105}$$

This approximation is reasonable if the time history of the stresses and strains closely follows the time history of the temperature.[10] Again in a thermal actuator where one expects the system to evolve on a time scale dictated by thermal events, this is a reasonable approximation. Scaling issues directly related to micro- and nanosystems will be discussed in Chapter 3. Equations (2.104) and (2.105) together with equation (2.101) are referred to as the equations of uncoupled quasistatic thermoelasticity.

2.5.3 Thermal Expansion of a Rod

Consider the thin elastic rod shown in Figure 2.9. Imagine that a source of internal heat generation is present; perhaps a current runs through the rod providing a source of Joule heating. In a MEMS system, computing the thermally induced displacement of the free end of the rod is of interest. As an exercise in thermoelasticity, let's compute this displacement for a simplified one-dimensional system. We assume the rod is anchored at its left end, is free at its right end, has length L, and is only free to deform in the x direction. Then, our displacement vector $\mathbf{u}$ reduces to

$$\begin{pmatrix} u_1 \\ 0 \\ 0 \end{pmatrix}. \tag{2.106}$$

The assumption that the rod is thin implies that u_1 may be assumed a function of x and t alone. That is, $u_1 = u_1(x,t)$. We also assume that the sidewalls of

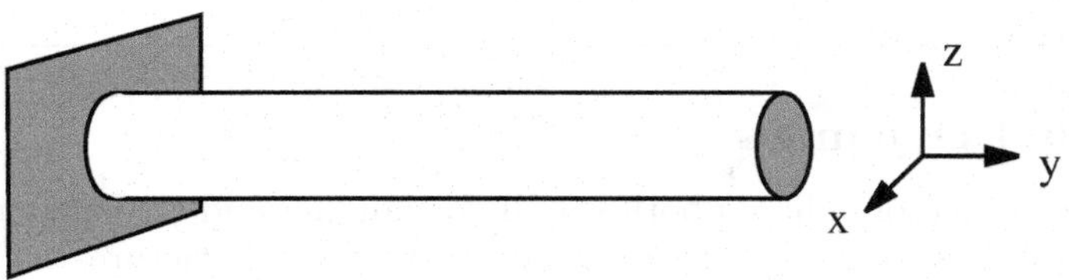

FIGURE 2.9: The thermoelastic rod.

the rod are perfectly thermally insulated and that temperature gradients in the y and z directions may be neglected. That is, the temperature is also a function of x and t alone, $T = T(x,t)$. We assume a spatially and temporally uniform heat source acts within the rod, producing heat at rate q. Employing the uncoupled quasistatic theory and eliminating the stress in favor of the displacement yields

$$\rho c_p \frac{\partial T}{\partial t} = k\frac{\partial^2 T}{\partial x^2} + q, \tag{2.107}$$

$$\frac{\partial^2 u_1}{\partial x^2} = \alpha\frac{3\lambda + 2\mu}{\lambda + 2\mu}\frac{\partial T}{\partial x}. \tag{2.108}$$

Assuming the rod is fixed at the left end and free at the right end implies the boundary conditions,

$$u_1(0,t) = 0, \quad \frac{\partial u_1}{\partial x}(L,t) = 0, \tag{2.109}$$

while if we assume the temperature of the wall and the ambient environment are held fixed at T_0 we may impose the thermal boundary conditions,

$$T(0,t) = T_0, \quad T(L,t) = T_0. \tag{2.110}$$

Finally, if we assume that initially the rod is at temperature T_0 we should impose the initial condition,

$$T(x,0) = T_0. \tag{2.111}$$

The advantage of using the uncoupled thermoelastic theory is that the equations for the displacement and temperature may be solved independently. In particular, we may integrate equation (2.108) twice and use the boundary conditions to find

$$u_1(x,t) = \alpha\frac{3\lambda + 2\mu}{\lambda + 2\mu}\int_0^x T(z,t)\,dz - \alpha\frac{3\lambda + 2\mu}{\lambda + 2\mu}T_0 x. \tag{2.112}$$

We may evaluate equation (2.112) at $x = L$ to compute the displacement at the right end of the rod in terms of the temperature. We find

$$u_1(L,t) = \alpha\frac{3\lambda + 2\mu}{\lambda + 2\mu}\int_0^L T(z,t)\,dz - \alpha\frac{3\lambda + 2\mu}{\lambda + 2\mu}T_0 L. \tag{2.113}$$

The solution of the thermal problem is left as an exercise for the reader.

2.6 Fluid Dynamics

The field of microfluidics is both rapidly expanding and challenging. From passive structures such as ink jet print nozzles to active structures such as micropumps designed for miniaturized drug-delivery systems, MEMS and NEMS devices are interacting with fluids in an increasing number of ways. Fluids are described by a set of nonlinear partial differential equations known as the Navier-Stokes equations.

2.6.1 The Navier-Stokes Equations

As with the Navier equations describing the motion of elastic bodies, the behavior of a fluid is characterized in terms of *stress* and *strain*. In fluids, however, in addition to conservation of momentum, we also have an equation derived from the principle of conservation of mass. If ρ denotes the density of fluid and u_i is a vector of fluid velocities whose ith component is fluid velocity in direction i, then conservation of mass says

$$\frac{\partial \rho}{\partial t} + u_j \frac{\partial \rho}{\partial x_j} + \rho \frac{\partial u_i}{\partial x_i} = 0. \tag{2.114}$$

As in the sections on elasticity, the summation convention is assumed. In contrast with elasticity, the reader should keep in mind that here we are working with *velocities* rather than *displacements*. This implies that it is convenient to define the strain rate tensor as opposed to the strain tensor. In particular we define

$$\frac{\partial \epsilon_{ij}}{\partial t} = \frac{1}{2}\left(\frac{\partial u_i}{\partial x_j} + \frac{\partial u_j}{\partial x_i}\right). \tag{2.115}$$

Stress in a fluid is expressed in terms of the stress tensor σ_{ij}, which as with elasticity, may be thought of as the 3×3 matrix,

$$\begin{pmatrix} \sigma_{11} & \sigma_{12} & \sigma_{13} \\ \sigma_{21} & \sigma_{22} & \sigma_{23} \\ \sigma_{31} & \sigma_{32} & \sigma_{33} \end{pmatrix}. \tag{2.116}$$

The ijth component of this matrix, σ_{ij} again, denotes the force per unit area in the direction i exerted on a surface element with normal in the j direction. The stress tensor is related to the rate of strain tensor through

$$\sigma_{ij} = -p\delta_{ij} + 2\mu\dot{\epsilon}_{ij} + \lambda\dot{\epsilon}_{kk}\delta_{ij}. \tag{2.117}$$

Here the dot denotes differentiation with respect to time, p is the pressure in the fluid, μ is the dynamic fluid viscosity, and λ is a second viscosity coefficient. Recall that repeated indices indicate summation and that δ_{ij} denotes the

Kroneker delta or 3×3 identity matrix. The equation of conservation of momentum can now be written as

$$\rho \frac{\partial u_i}{\partial t} + \rho u_j \frac{\partial u_i}{\partial x_j} = \rho F_i + \frac{\partial \sigma_{ij}}{\partial x_j}. \tag{2.118}$$

As with elasticity, this is nothing more than a statement of Newton's second law, $\mathbf{F} = m\mathbf{a}$. The left-hand side is the mass times the acceleration, while the right-hand side is the sum of the forces. The vector F_i represents body forces, while the internal stresses are captured in the stress tensor. Using equation (2.117) in equation (2.118), we obtain

$$\rho \frac{\partial u_i}{\partial t} + \rho u_j \frac{\partial u_i}{\partial x_j} = \rho F_i - \frac{\partial p}{\partial x_i} + \frac{\partial}{\partial x_j}(2\mu \dot{\epsilon}_{ij} + \lambda \dot{\epsilon}_{kk}\delta_{ij}). \tag{2.119}$$

Equation (2.119) is usually called the Navier-Stokes equation of motion and equations (2.114) and (2.119) are called the Navier-Stokes equations. We may rewrite the Navier-Stokes equations in vector form,

$$\frac{\partial \rho}{\partial t} + \nabla \cdot (\rho \mathbf{u}) = 0 \tag{2.120}$$

$$\rho \frac{\partial \mathbf{u}}{\partial t} + \rho(\mathbf{u} \cdot \nabla)\mathbf{u} + \nabla p - \mu \nabla^2 \mathbf{u} - (\lambda + \mu)\nabla(\nabla \cdot \mathbf{u}) = \mathbf{f}. \tag{2.121}$$

In equation (2.121) we have eliminated the strain rate tensor in favor of the velocity vector and relabeled the body force vector as $\mathbf{f}$.

Equations (2.120) and (2.121), the Navier-Stokes equations for a viscous compressible fluid, are a system of four coupled nonlinear partial differential equations. Notice, however, that the system contains *five* unknown functions. The pressure, the density, and the three components of the velocity vector field are all unknown. To close this system one usually appends the equation of conservation of energy to the Navier-Stokes equations. This introduces one more equation and one more unknown function, the temperature T. The system is then completed by introducing an equation of state which relates ρ, p, and T.

2.6.2 Incompressible Flow

In the special case where a fluid may be assumed to be incompressible, the Navier-Stokes equations may be simplified. We define the notion of incompressibility to mean that the density, ρ, is a constant. Then, equations (2.120) and (2.121) reduce to

$$\nabla \cdot \mathbf{u} = 0 \tag{2.122}$$

$$\frac{\partial \mathbf{u}}{\partial t} + (\mathbf{u} \cdot \nabla)\mathbf{u} = -\frac{1}{\rho}\nabla p + \nu \nabla^2 \mathbf{u}, \tag{2.123}$$

where $\nu = \mu/\rho$ is called the kinematic viscosity. Notice that the assumption of constant density is an equation of state and the incompressible Navier-Stokes equations is a system of four equations in four unknowns.

2.6.3 The Euler Equations

Under the additional assumption that the fluid is inviscid as well as incompressible, the Navier-Stokes equations may be further simplified to obtain the Euler equations. The assumption that the fluid is inviscid means that $\nu = 0$ and hence equations (2.122) and (2.123) reduce to

$$\nabla \cdot \mathbf{u} = 0 \tag{2.124}$$

$$\frac{\partial \mathbf{u}}{\partial t} + (\mathbf{u} \cdot \nabla)\mathbf{u} = -\frac{1}{\rho}\nabla p. \tag{2.125}$$

Equations (2.124) and (2.125) are also called the incompressible inviscid Navier-Stokes equations.

2.6.4 The Stokes Equations

Consider the incompressible Navier-Stokes equations, (2.122) and (2.123), for a flow with characteristic velocity U, in a spatial region with characteristic length l. If we scale equations (2.122) and (2.123) with U and l by introducing

$$\mathbf{u}^* = \frac{\mathbf{u}}{U}, \quad x^* = \frac{x}{l}, \quad y^* = \frac{y}{l}, \quad z^* = \frac{z}{l}, \quad t^* = \frac{Ut}{l}, \quad p^* = \frac{p - p_\infty}{\rho U^2}, \tag{2.126}$$

we obtain the dimensionless system

$$\nabla \cdot \mathbf{u}^* = 0 \tag{2.127}$$

$$\frac{\partial \mathbf{u}^*}{\partial t^*} + (\mathbf{u}^* \cdot \nabla)\mathbf{u}^* = -\nabla p^* + \frac{1}{Re}\nabla^2 \mathbf{u}^*. \tag{2.128}$$

We drop the stars and write

$$\nabla \cdot \mathbf{u} = 0 \tag{2.129}$$

$$\frac{\partial \mathbf{u}}{\partial t} + (\mathbf{u} \cdot \nabla)\mathbf{u} = -\nabla p + \frac{1}{Re}\nabla^2 \mathbf{u}. \tag{2.130}$$

Note p_∞ is a reference pressure in the system to be thought of as the pressure in the fluid at infinity. The dimensionless parameter Re is called the Reynolds number for the flow and is given by

$$Re = \frac{Ul}{\nu}. \tag{2.131}$$

In the case where the Reynolds number is a small parameter, the incompressible Navier-Stokes equations are often replaced with their zero Reynolds number limit, the Stokes equations. This limit is particularly important for microfluidics. A detailed discussion of the relevance of the zero Reynolds number limit to microsystems appears in Chapter 3. We rescale the pressure with Re, that is, let $\hat{p} = p/Re$, and take the limit as $Re \to 0$ in equations (2.129) and (2.130) to obtain

$$\nabla \cdot \mathbf{u} = 0 \tag{2.132}$$

$$\nabla \hat{p} = \nabla^2 \mathbf{u}. \tag{2.133}$$

Equations (2.132) and (2.133) are known as the Stokes equations or the equations of Stokes flow.

2.6.5 Boundary and Initial Conditions

As with other continuum theories, the Navier-Stokes equations require boundary and initial conditions. Generally speaking initial conditions are easily dealt with. If it is of interest to study the evolution of a flow from some quiescent state, we simply specify the velocity and pressure field in the fluid at time $t = 0$. More common is the study of developed or so-called steady flows. Confusingly enough this does not mean the fluid is not moving, but rather that the velocity field is not changing in time. Poiseuille flow, studied below, is an example of a steady flow.

Boundary conditions must be formulated with more care. We must examine boundary conditions at two types of interfaces, a solid-fluid interface, and a fluid-fluid interface. Both of these situations occur frequently in the study of micro- or nanofluidics. At the boundary between a fluid and a solid, such as occurs in a glass of water between the water and the bottom of the glass, the *no-penetration* and *no-slip* boundary conditions are generally used. No penetration is easy to grasp; it simply says that fluid is not flowing into the wall. Hence flow in directions normal to the wall must be zero at the wall. If we consider the general fluid-solid interface shown in Figure 2.10, the boundary conditions would be formulated in terms of the normal, $\mathbf{n}$, and the velocity vector, $\mathbf{u}$, as

$$\mathbf{u} \cdot \mathbf{n} = 0. \tag{2.134}$$

The no-slip boundary condition is rooted in observation. It has been experimentally observed that at the interface between a fluid and solid, the fluid is not moving *tangentially* to the solid surface. This is the idea of no-slip at the surface. Again in terms of a unit normal to a surface as shown in Figure 2.10, the no-slip boundary condition is stated as

$$\mathbf{u} \times \mathbf{n} = 0. \tag{2.135}$$

Notice that the two conditions taken together, no-slip and no-penetration, may be simply stated as

$$\mathbf{u} = 0 \tag{2.136}$$

at the boundary between a solid and fluid. Equation (2.136) is often referred to as simply the no-slip boundary condition. It should also be noted that for inviscid fluid flow, where the order of the spatial derivatives on $\mathbf{u}$ has decreased

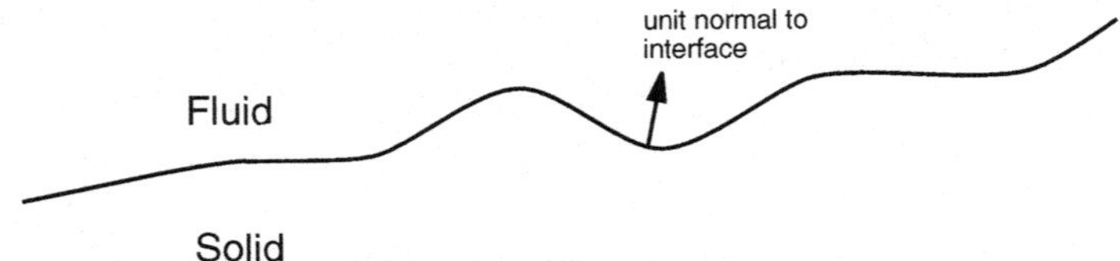

FIGURE 2.10: A fluid-solid interface.

by one from the viscous Navier-Stokes equations, it is only necessary to specify no-penetration into a solid and not the full no-slip condition.

At a fluid-fluid interface it is also necessary that the no-slip boundary condition hold. In order to state conditions at such an interface, we will refer to Figure 2.11 and assume the interface is described by the equation $f(x, y, z, t) = 0$. Usually, determining the function f is part of the problem. That is, the location of the interface between the two fluids is not usually known *a priori*. A problem where the interface is unknown is often called a *free surface* or *moving boundary* problem. The first condition imposed at a free surface says that the change in momentum across the interface is balanced by the tensile force or *surface tension* of the interface. If superscripts denote the fluid region

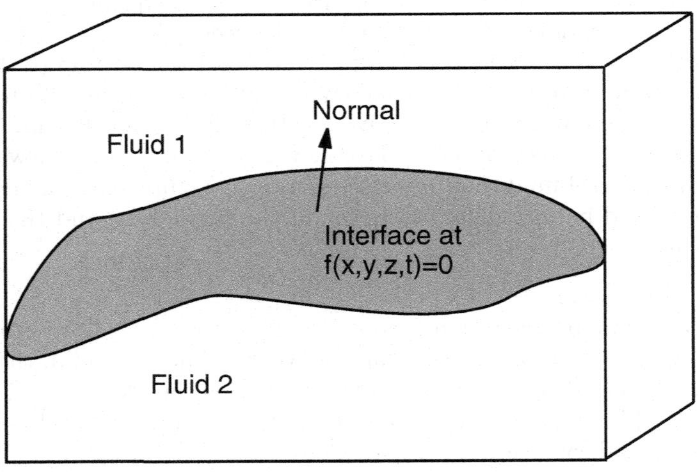

FIGURE 2.11: A fluid-fluid interface.

in Figure 2.11 and the normal points as shown, then this conditions says

$$(\sigma^1 - \sigma^2) \cdot \mathbf{n} = -\gamma \left(\frac{1}{R_1} + \frac{1}{R_2} \right) \mathbf{n}, \tag{2.137}$$

where γ is the surface tension at the interface. The R_i denote the radii of curvature of the interface and the σ^i are the stress tensors in each fluid. If the fluid is quiescent, the stress in each fluid becomes purely normal or hydrostatic and this expression reduces to the familiar Laplace-Young law,

$$p_1 - p_2 = \gamma \left(\frac{1}{R_1} + \frac{1}{R_2} \right), \tag{2.138}$$

where the p_i are the pressures in each fluid.

Since in general, the location of the interface is unknown, we need an additional condition to determine its position. This condition, called the *kinematic condition*, says that fluid that starts on the boundary remains on the boundary. In terms of the interface f as shown in Figure 2.11, this condition may be stated as

$$\frac{\partial f}{\partial t} + \mathbf{u}^i \cdot \nabla f = 0. \tag{2.139}$$

Here $\mathbf{u}^i$ is the velocity vector of the ith fluid.

As mentioned several times in this text, the study of MEMS and NEMS forces the researcher to confront unfamiliar parameter regimes and the limits of continuum theory. The study of microfluidics and in particular the behavior at fluid-solid interfaces in microscale devices presents one such challenge. Above we presented the no-slip boundary condition imposed on the Navier-Stokes equations at a fluid-solid boundary. We mentioned that this boundary condition was based on experimental observation. However, as is well known in the study of rarefied gas dynamics, fluids do sometimes slip along a solid surface. The Knudsen number, a dimensionless parameter denoted Kn, provides a measure of how close a particular system is to the slip regime. The Knudsen number is defined as

$$Kn = \frac{\lambda}{l}, \tag{2.140}$$

where λ is the molecular mean free path and l is a characteristic length of the system under consideration. Recall that the mean free path is the average distance traveled by a molecule between collisions. In rarefied gas dynamics the mean free path becomes large, comparable to system size, and the Knudsen number approaches one. In microfluidics, the Knudsen number becomes large—not because the mean free path is large—but because the system size becomes small. As a rule of thumb, if Kn is less than 10^{-4}, the no-slip boundary condition is applicable. If Kn becomes larger than 10^{-4}, fluid will slip along an interface. In this situation, a modified slip boundary condition is often used. For example, for a wall coinciding with the x axis and moving with velocity V_w in the direction of the x axis, the condition

$$u_1 - V_w = \left(\frac{2-\sigma}{\sigma}\right)\left(\frac{Kn}{1-bKn}\right)\frac{\partial u_1}{\partial y} \tag{2.141}$$

may be imposed.

The parameter σ is called the *accommodation coefficient* while b is called the *slip coefficient.* These coefficients must be determined experimentally or from molecular level simulations. This is an active area of microfluidics research. The reader is urged to consult the literature for an up-to-date look at this topic. References [11, 47, 151, 170] provide a starting point for further reading on the Knudsen number and slip boundary condition.

2.6.6 Poiseuille Flow

A simple but important exact solution of the incompressible Navier-Stokes equations concerns flow in a pipe. Consider a cylinder with uniform circular cross-section of radius R and a flow driven by a constant pressure gradient. We assume the pressure gradient acts in the z direction, i.e., along the axis of the pipe so that we may prescribe

$$p_z = -A, \tag{2.142}$$

where A is a constant. See Figure 2.12. Now, we assume the velocity vector

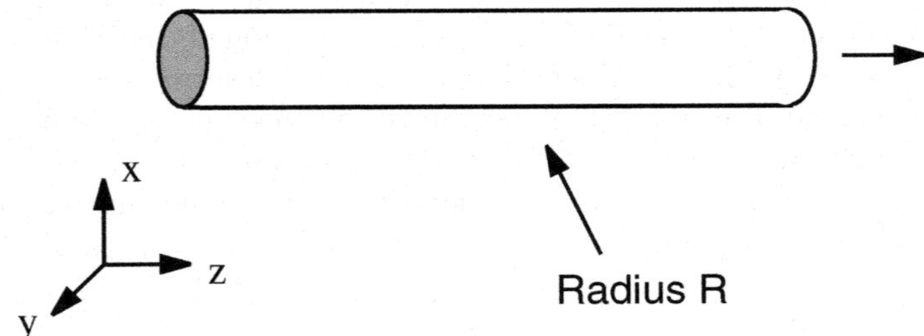

FIGURE 2.12: Poiseuille flow in a pipe.

has the form $\mathbf{u} = (0, 0, u_3(r))$, that is there is flow only in the z direction, and the velocity varies only with r. Then, the incompressible Navier-Stokes equations are satisfied provided

$$-\frac{1}{\rho} = p_z = \frac{\nu}{r}\frac{\partial}{\partial r}\left(r\frac{\partial u_3}{\partial r}\right), \tag{2.143}$$

and provided p is a function of z alone. We can use equation (2.142) in equation (2.143) and integrate twice with respect to r to obtain

$$u_3(r) = -\frac{Ar^2}{4\mu} + c_0 \log(r) + c_1, \tag{2.144}$$

where c_0 and c_1 are constants of integration. We must choose c_0 to be zero in order to keep the velocity bounded at $r = 0$ and c_1 can be found by applying the no-slip boundary condition at $r = R$. This yields

$$c_1 = \frac{AR^2}{4\mu}, \tag{2.145}$$

so that

$$u_3(r) = \frac{A}{4\mu}(R^2 - r^2). \tag{2.146}$$

That is, the velocity profile in Poiseuille flow is parabolic.

2.6.7 Poiseuille Flow with Slip

The same pressure-driven flow studied in the previous subsection may be analyzed for the slip boundary condition, equation (2.141), rather than the condition of no-slip on the walls of the pipe. We simplify the discussion by assuming that the slip coefficient, b, is identically zero. Then, the boundary condition to be imposed on equation (2.144) is

$$u_3(R) = Kn\left(\frac{2-\sigma}{\sigma}\right)\left.\frac{\partial u_3}{\partial r}\right|_{r=R}. \tag{2.147}$$

We still must require that $c_0 = 0$ in order to keep the velocity bounded at the origin. However, the slip boundary condition implies that

$$c_1 = \frac{AR^2}{4\mu} - \frac{AR^2}{2\mu}Kn\left(\frac{2-\sigma}{\sigma}\right). \tag{2.148}$$

Hence the solution for the velocity, $u_3(r)$, is

$$u_3(r) = \frac{A}{4\mu}(R^2 - r^2) - \frac{AR^2}{2\mu}Kn\left(\frac{2-\sigma}{\sigma}\right). \tag{2.149}$$

Notice that this reduces to the no-slip solution when Kn tends to zero. Further, the profile, while modified, retains the basic parabolic structure.

2.7 Electromagnetism

Electromagnetism plays a central role in MEMS and NEMS. Electrostatic forces are the most commonly used forces for actuation of MEMS devices.

Magnetic systems, while generally more difficult to fabricate, are nonetheless used in systems such as micropumps and magnetorheological fluid mixers. Of course, the interface between any MEMS or NEMS device and the macroscopic world is electronic. Electromagnetic phenomena are described by a set of partial differential equations known as the Maxwell equations.

2.7.1 Units for Electromagnetism

In this book we will use the International System (S.I.) of units for electromagnetism. In this system, a unit of electrical current, the *ampere*, joins the meter, kilogram, and second as a fundamental unit. The ampere is defined as the current necessary to produce a force of 2×10^{-7} Newtons per meter between two parallel and very long wires a distance of one meter apart. Experimentally, the force between two such wires is found to obey

$$F = k_1 \frac{I^2}{d}, \tag{2.150}$$

where I is the current carried by each wire and d the distance between them. The ampere is defined such that k_1 is exactly 2×10^{-7}, but, for historical reasons, $\mu_0 = 2\pi k_1 = 4\pi \times 10^{-7}$ is used instead of k_1 itself, where μ_0 is called the *magnetic permeability of free space* or simply the *permeability constant.*

The S.I. unit of electrical charge, the *Coulomb*, is considered a derived unit and is defined as that amount of charge that passes through a wire carrying one ampere of current in one second. For two point charges of magnitude Q a distance d, apart it is then also experimentally determined that the force between them obeys

$$F = k_2 \frac{Q^2}{d^2}. \tag{2.151}$$

Since the Coulomb has already been defined, k_2 must be determined experimentally, its value being approximately $k_2 = 1.129409067 \times 10^{11}$ Nm^{2}C^{-2}. Again for historical reasons, $\epsilon_0 = 1/(4\pi k_2)$ is used instead of k_2 itself; hence,

$$\epsilon_0 = 8.854187818 \times 10^{-12} \ \ \mathrm{C^2N^{-1}m^{-2}}. \tag{2.152}$$

ϵ_0 is called the *electrical permittivity of free space* or the *permittivity constant.*

2.7.2 The Maxwell Equations

All classical electromagnetic phenomena, the propagation of light through a glass fiber, the flow of current in a resistive heating element, the Coulomb force in an electrostatic actuator, are described by Maxwell's equations. This system of four coupled partial differential equations describes the spatial and temporal evolution of the four vector fields, **E**, **D**, **B**, and **H**:

$$\nabla \cdot \mathbf{D} = \frac{\rho}{\epsilon_0} \tag{2.153}$$

$$\nabla \times \mathbf{E} = -\frac{1}{c}\frac{\partial \mathbf{B}}{\partial t} \tag{2.154}$$

$$\nabla \cdot \mathbf{B} = 0 \tag{2.155}$$

$$\nabla \times \mathbf{H} = \frac{1}{c}\frac{\partial \mathbf{D}}{\partial t} + \mathbf{J}. \tag{2.156}$$

Here ρ is the electrical charge density, $\mathbf{J}$ is the current density, and c is the speed of light in vacuum. To complete the description of the dynamics of charged particles in the presence of electromagnetic fields we must also supply the *Lorenz force law*,

$$\mathbf{F} = q\left(\mathbf{E} + \mathbf{v} \times \mathbf{B}\right), \tag{2.157}$$

where q is the electric charge of the particle and $\mathbf{v}$ its velocity.[11] Note that the *conservation of charge*,

$$\frac{\partial \rho}{\partial t} + \nabla \cdot \mathbf{J} = 0 \tag{2.158}$$

is a consequence of Maxwell's equations.

The Maxwell equations are a system of eight equations in the 12 components of **E**, **D**, **B**, and **H**. To close the system the *constitutive relations* of the materials involved are necessary. For *linear* materials these are usually written:[12]

$$D_i = \epsilon_{ij} E_j \tag{2.159}$$

$$B_i = \mu_{ij} H_j. \tag{2.160}$$

The tensors ϵ_{ij} (the *permittivity tensor*) and μ_{ij} (the *magnetic permeability tensor*) are properties of materials which contain information about how they respond to the presence of external electric and magnetic fields. Note that this form does not apply to materials with permanent electric or magnetic moments (*ferroelectric* and *ferromagnetic* materials, respectively).

As an example, many materials respond to an electric field by generating a *dipole* moment. In addition, the dipole moment is independent of the direction of the applied field and its position in space. In this case, the material is said to be *homogeneous* and *isotropic* as well as linear and the permittivity tensor takes on the simple form $\epsilon_{ij} = \epsilon\delta_{ij}$, hence

$$\mathbf{D} = \epsilon \mathbf{E}. \tag{2.161}$$

In this case, ϵ is the called the *dielectric constant* of the material. Note that while a large number of materials (called, naturally, *dielectrics*) fall into this category, most magnetic materials used in current technology are ferromagnetic as well as having a nonlinear response to external magnetic fields. *Paramagnetic* and *diamagnetic* materials, which do not have permanent magnetic moments and are linear in their response, are more the exception than the rule.

2.7.3 Electrostatics

The equations of electrostatics are obtained from the Maxwell equations by setting the time derivatives of both $\mathbf{B}$ and $\mathbf{D}$ to zero and assuming that $\mathbf{J} = 0$. This results in

$$\nabla \times \mathbf{E} = 0 \tag{2.162}$$

$$\nabla \cdot \mathbf{D} = \rho. \tag{2.163}$$

Equation (2.162) immediately implies that $\mathbf{E}$ may be written as the gradient of a potential, conventionally written $\mathbf{E} = -\nabla\phi$. The curl of a gradient is identically zero and hence equation (2.162) is satisfied. For linear, homogeneous, and isotropic materials (e.g., the vacuum) we may use (2.161) to obtain

$$\nabla^2\phi = \frac{\rho}{\epsilon_0}, \tag{2.164}$$

which is Poisson's equation for the electrostatic potential ϕ. Note that $\mathbf{E}$ is unchanged if we add a constant function to ϕ. This type of change is called a *gauge transformation* and the choice of the constant is called the *gauge.*

2.7.4 Magnetostatics

The equations of magnetostatics are obtained similarly, except that now ρ vanishes instead of $\mathbf{J}$. Hence,

$$\nabla \times \mathbf{H} = \mathbf{J} \tag{2.165}$$

$$\nabla \cdot \mathbf{B} = 0. \tag{2.166}$$

In constrast to the electrostatic case, these equations tend to be difficult to solve in realistic cases since for many important materials the relationship between $\mathbf{H}$ and $\mathbf{B}$ will be complicated. However, like the electrostatic case, linear, isotropic, and homogeneous materials for which

$$\mathbf{B} = \mu\mathbf{H} \tag{2.167}$$

can be treated generally. Since the divergence of a curl is identically zero, we introduce the *vector potential* $\mathbf{A}$ by

$$\mathbf{B} = \nabla \times \mathbf{A}, \tag{2.168}$$

whereby (2.165) becomes

$$\nabla(\nabla \cdot \mathbf{A}) - \nabla^2\mathbf{A} = \mu\mathbf{J}. \tag{2.169}$$

As in the electrostatic case, note that $\mathbf{B}$ is unchanged if we add the gradient of a scalar function ψ to $\mathbf{A}$. This is the gauge transformation of the magnetic potential. For magnetostatic problems the most convenient choice is known as the *Coulomb gauge* in which

$$\nabla \cdot \mathbf{A} = 0. \tag{2.170}$$

This is possible since given a vector potential $\mathbf{A}$, we can always choose ψ so that it satisfies $\nabla^2\psi = -\nabla \cdot \mathbf{A}$.

2.7.5 Boundary Conditions

In electromagnetism the boundary conditions on the $\mathbf{E}$, $\mathbf{D}$, $\mathbf{B}$, and $\mathbf{H}$ fields may be computed by integrating the Maxwell equations over a suitably chosen volume or line element at the boundary between two media. The reader may consult virtually any text on electromagnetism for the details of this derivation. Here, we simply state the boundary conditions for each field in terms of conditions at the surface between two materials denoted by the subscripts 1 and 2:

$$(\mathbf{D}_2 - \mathbf{D}_1) \cdot \mathbf{n} = \sigma \tag{2.171}$$

$$(\mathbf{B}_2 - \mathbf{B}_1) \cdot \mathbf{n} = 0 \tag{2.172}$$

$$(\mathbf{E}_2 - \mathbf{E}_1) \times \mathbf{n} = 0 \tag{2.173}$$

$$(\mathbf{H}_2 - \mathbf{H}_1) \times \mathbf{n} = -\mathbf{J}_s. \tag{2.174}$$

In equations (2.171)–(2.174) $\mathbf{n}$ is a unit normal pointing from medium 1 to medium 2, σ is the surface charge density, and $\mathbf{J}_s$ is the surface current density.

These conditions result in boundary conditions on the electric and magnetic potentials which vary according to the particular problem at hand. For example, in electrostatics a metal-dielectric interface would require that $\mathbf{E}$ be normal to the surface (since an electrostatic field field vanishes inside a metal), which is just (2.171) in disguise. Hence ϕ must be constant along the interface. This is known as a *Dirichlet* boundary condition. Other boundary conditions will be derived as needed in subsequent chapters.

2.7.6 Integral Formulation

For electro- and magnetostatics there are integral formulations of Gauss's and Ampere's laws that are often more useful in practice than their differential counterparts (2.153) and (2.156). For the electrostatic case, consider a continuous charge distribution $\rho(\mathbf{x})$ and a closed surface of spherical topology S with outward pointing unit normal vector $\mathbf{n}$. Integrating (2.153) over the volume enclosed by S and using the divergence theorem (A.1) gives the integral formulation of Gauss's law,

$$\int_S \mathbf{E} \cdot \mathbf{n} = \frac{q}{\epsilon_0}, \tag{2.175}$$

where q is the total charge enclosed by S (and we have assumed that $\epsilon_{ij} = \delta_{ij}$).

One standard example of how this may used is the parallel plate capacitor. Recall that a capacitor is a device that will store a certain charge Q when a voltage V is placed across it. For many capacitors the relationship between Q and V is linear and the constant of proportionality is called the *capacitance* of the device and is denoted C;

$$Q = CV. \tag{2.176}$$

Now consider two parallel square metal plates of side length L, separation d with $d \ll L$. Let the top plate be held at potential $\phi = V > 0$ and the bottom plate held at $\phi = 0$. Intuitively we know the electric field must be relatively constant between the plates and mostly pointing from the top plate to the bottom. Let the surface S be the rectangular solid of the same shape as the top plate and with thickness such that one end is contained inside the top plate while the other lies between the two plates. There are six contributions to the integral in (2.175) that are accounted for by the following. The electric field on the top side of S is zero since it is completely contained inside a conductor. The integrals over the vertical faces of S will be neglected since the product $\mathbf{E} \cdot \mathbf{n}$ will be more or less zero there. On the bottom face $\mathbf{E}$ will be roughly constant and downward pointing. Hence the integral over this face is EA, where $E = |\mathbf{E}|$ and A is the area of each plate (and hence the area of the bottom surface of S). On the left-hand side, the total charge enclosed by S is simply the charge on the top face, call it q. Hence

$$E = \frac{q}{\epsilon_0 A}. \tag{2.177}$$

Note that the differential version of Gauss's law tells us that

$$V = \frac{E}{d}; \tag{2.178}$$

hence, the capacitance is

$$C = \frac{Q}{V} = \frac{\epsilon_0 A}{d}. \tag{2.179}$$

The useful integral formulation for magnetostatics is called *Ampere's law* and is often derived from (2.156) although historically the reverse is true: (2.156) is a modification of the integral form made by Maxwell in order to form a consistent set of equations. In a vacuum magnetostatic system (2.156) reduces to

$$\nabla \times \mathbf{B} = \mu_0 \mathbf{J}. \tag{2.180}$$

Consider again the general surface S with an additional path (i.e., curve) C defined on it. Apply Stokes's theorem (A.2) to (2.180) on S and along C to obtain

$$\oint_C \mathbf{B} \cdot d\mathbf{l} = \mu_0 I, \tag{2.181}$$

where I is the total current that passes through the loop C and the line element $d\mathbf{l}$ is defined such that the unit normal to S is outward pointing. The archetypal application of (2.180) is the calculation of the magnetic field around a long straight wire carrying current I. Let S be a cylindrical surface of radius r with the wire at its center and let C be a circular path on S perpendicular to the wire. By symmetry the magnetic field is everywhere perpendicular to the wire and cannot vary in the direction of the wire. Since C is uniformly distant from the wire, we have from (2.180)

$$B = \frac{\mu_0 I}{2\pi r}, \tag{2.182}$$

which gives us the magnitude of $\mathbf{B}$. From (2.155) and the rotational symmetry of the problem, we also see that $\mathbf{B}$ cannot have a component in the radial direction (i.e., directly away from the wire). Hence $\mathbf{B}$ is circulating about the wire in the direction given by the current I and the right-hand rule.

Another useful integral formulation for the magnetic field, which will come up in Chapter 8, is the *Biot-Savart law*,

$$d\mathbf{B} = \frac{\mu_0 I}{4\pi} \frac{d\mathbf{l} \times \mathbf{r}}{r^3}. \tag{2.183}$$

Here $d\mathbf{l}$ is parallel to the current I and the vector $\mathbf{r}$ points from $d\mathbf{l}$ to the point at which $d\mathbf{B}$ is to be evaluated. As an example, we recalculate the field around the wire. We use Cartesian coordinates and let the wire be located along the z axis. If the point in space at which $d\mathbf{B}$ is to be evaluated is (x, y, z), then we have

$$d\mathbf{l} = (0, 0, dz), \quad \mathbf{r} = (x, y, z - z'), \tag{2.184}$$

where z' is the location of the element $d\mathbf{l}$. Now

$$d\mathbf{l} \times \mathbf{r} = dz'(-y, x, 0), \tag{2.185}$$

so that

$$\mathbf{B} = \frac{\mu_0 I}{4\pi} \int_{-\infty}^{\infty} \frac{(-y, x, 0)dz'}{(x^2 + y^2 + (z - z')^2)^{3/2}}. \tag{2.186}$$

Using

$$\int (u^2 + a^2)^{-3/2} du = \frac{u}{a^2\sqrt{u^2 + a^2}} \tag{2.187}$$

gives

$$\mathbf{B} = \frac{\mu_0 I}{2\pi} \frac{(-y, x, 0)}{x^2 + y^2}, \tag{2.188}$$

which the reader can verify is indeed the Cartesian expression of (2.182).

2.8 Numerical Methods for Continuum Mechanics

The various fundamental equations of continuum physics are differential equations governing fields. That is, the independent variables are all continuous functions of time and space. Very often these equations will be difficult or impossible to solve analytically with current mathematical techniques. This is especially true for microsystem models, which often reduce to insoluble nonlinear equations. In cases like this scientists often turn to computational methods to obtain approximate solutions to the equations generated by such models. This book is primarily concerned with how to make realistic models of microsystems amenable to mathematical analysis. However, it is a general rule that interesting models are never completely amenable, so computation will occasionally be necessary. The purpose of this section is to describe several of the more common methods of obtaining numerical solutions to differential equations, some of which are used in the examples in this book. Since each of these methods has a long history and has generated a significant body literature, we will not go into very much detail but rather leave it to the numerically inclined reader to explore the literature indicated in the references.

Fundamentally, all numerical methods for solving field equations rely on two approximations: i) the degrees of freedom of the field are reduced from an infinite number to a finite number through a process known as *discretization*, and ii) the remaining degrees of freedom are computed to a finite precision. These two approximations give rise to two types of error, truncation error and round-off error, respectively. With the advent of 64 bit word length, truncation error is normally the larger of the two, although this is not guaranteed to be the case (for instance, when inverting ill-conditioned matrices). In addition, since digital computers are capable only of algebraic numerical operations, all numerical methods for approximating solutions of differential equations rely on some method of transforming the differential operator into an algebraic one.

By far, the three most commonly used approximation methods are the finite difference method, the finite element method, and the spectral method. We will discuss each method in the context of solving

$$L(u) = S(u, \mathbf{x}) \tag{2.189}$$

on a domain D with boundary ∂D, where u is the field, L is a differential operator, and S is a source term. Normally, the numerical method chosen to solve a given problem is highly dependent on the nature of L (elliptic, hyperbolic, nonlinear, etc.), the topology of D (can it be covered by a non-singular coordinate system?), the underlying physics of the system (can we formulate a method that preserves a constant of the system to within round-off error?), and many other matters.

In the *finite difference method*, the domain of computation is discretized into a finite set of points $\mathbf{x}_i$. The points are normally restricted to lie on the curves of constant coordinates of the domain. The degrees of freedom of the field are then restricted to the values of $u(\mathbf{x})$ at these points: $u_i = u(\mathbf{x}_i)$. One then examines various Taylor series expansions of the field at each point of the domain and replaces the operator L with terms from combinations of these expansions. The result is a set of algebraic equations for the values u_i, which are normally coupled. Clearly, many such schemes may be formed depending on how one combines various Taylor series of differing size about the different points of the domain and varies the placement of the points. Each scheme will have different truncation error terms and different numerical properties, but in essence all finite difference schemes are built using this method. Normally, one creates a scheme tailored to solve the particular problem at hand in the most accurate way possible; hence the finite difference literature is mostly concerned with how to construct specific schemes for solving particular problems and examining their performance on those problems. The advantage of finite difference methods is that they have a long history with applications spanning most of engineering and physics and are equally well suited to solving time-dependent as well as time-independent problems. The main disadvantage is that the mathematical theory behind the method is not very well developed compared to other methods so that properties of a particular scheme often cannot be proved ahead of time and one must resort to experimentation in order to discover a method that works well for a particular problem.

The *finite element method* takes a different point of view. Here, it is not the points that are important, but the spaces between the points. These spaces, called *elements*, can take on many shapes in two and three dimensions, but are normally restricted to various types of polyhedra with a relatively small number of triangular and quadrilateral faces, e.g. tetrahedra, pyramids, rectangular solids (called "bricks"), etc. Inside each element one specifies a basis function that can be completely determined by a small number of parameters. These functions are defined to be nonzero inside each element and to vanish everywhere else. The finite number of elements combined with the finite set of basis function parameters per element is what reduces the problem to a finite number of degrees of freedom. The differential equation is then replaced by an integral equation (called the *weak form* of the problem) and reduced to a set of algebraic equations by performing an inner product of the weak form with each basis function.[13] The solution is then expressed as an expansion in terms of the basis functions for each element. The advantage of finite elements over other methods is that the mathematical theory behind the method (which lies in the field of functional analysis) has been very well developed and the domain D can be of arbitrary topology and geometry. The disadvantages are that for some nonlinear problems a weak form of the operator L may be difficult or impossible to find and that the elements must be created by a separate algorithm (this is known as "mesh generation" and is an entire field in itself).

Spectral methods are similar to the finite element method in that a set of basis functions is chosen and the solution is expanded in terms of these functions. An appropriately defined inner product is then taken, which again results in a set of algebraic equations for the parameters of the basis functions. The difference is that there are no elements, i.e., the basis functions are globally defined over the entire domain D. The advantage of spectral methods is the rate of convergence can be very high; hence very accurate solutions to some problems can be obtained with a relatively small number of basis functions. The disadvantages are that the method is not good at solving problems whose solutions have discontinuities or other very small features and that the basis functions are often eigenfunctions of the domain that may be difficult or impossible to compute exactly unless the domain has a very regular or standard shape.

While these three methods are by far the most commonly used and the most likely to be found in commercial form, many other methods have been developed over the years. Among these are finite volume methods, the boundary element method, various gridless and meshless methods, the fast multipole method, methods based on wavelets, cellular automata, and so on.

2.9 Chapter Highlights

- In the MEMS and NEMS modeling communities continuum models enjoy widespread use. The study of MEMS and NEMS brings us to the limits of the validity of continuum models.

- The basic continuum theories are heat conduction, elasticity, fluid dynamics, and electromagnetism.

- The equations of continuum mechanics may be derived in many ways. The calculus of variations is one important tool used for deriving continuum theories.

- Models of MEMS and NEMS often result in coupled domain problems that must be solved numerically. The finite difference, finite element, and spectral methods are the most commonly used numerical techniques.

2.10 Exercises

Section 2.3

1. Apply the principle of conservation of mass to a control volume within a moving fluid to derive the equation of continuity,

$$\frac{\partial \rho}{\partial t} + \nabla \cdot (\rho \mathbf{v}) = 0,$$

where ρ is fluid density and $\mathbf{v}$ is the fluid velocity field. You may proceed as in the derivation of the heat equation by considering an integral statement of the conservation law for an arbitrary volume.

2. When you burn your finger your natural impulse is to blow on the burned area. Why does this cool your finger? Discuss your answer in terms of the Newton cooling boundary condition.

3. Consider equation (2.26). How does the temperature at the center of the sphere behave? That is, what is the limit of equation (2.26) as $r \to 0$? Define a suitable average temperature for the sphere and compute it from equation (2.26).

4. Again consider equation (2.26). Assume the sphere is made of silicon, initially the temperature of the sphere is 1000 K, and that the ambient environment is at 300 K. If the sphere has radius 10μm, how long does it take for the center of the sphere to cool to 99 percent of its steady-state value? Repeat the calculation if the sphere is 100μm, 1cm, and 1m in radius.

5. Consider two-dimensional heat flow through the annular region shown in Figure 2.13. Assume the inner cylinder is maintained at a constant temperature T_0 and the outside environment is maintained at a constant temperature T_A. Find the steady-state temperature distribution through the annular region. Use your answer to compute the heat flux through the outer surface. Plot this heat flux as a function of $b - a$, the thickness of the outer cylinder. What do you observe?[14]

6. Solve for the temperature field in a one-dimensional rod with constant heat source. That is, solve

$$\rho c_p \frac{\partial T}{\partial t} = k \frac{\partial^2 T}{\partial x^2} + q$$

$$T(0,t) = T_0, \quad T(L,t) = T_0, \quad T(x,0) = T_0.$$

Note that this is the temperature field needed in the discussion of the one-dimensional thermoelastic rod.

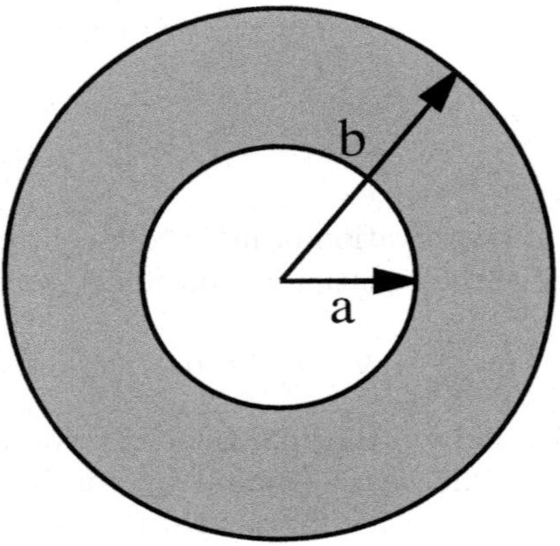

FIGURE 2.13: The annular region for Problem 5.

Section 2.4

7. Consider a vibrating elastic string that is released from rest with an initial displacement of

$$u_2(x,0) = \sin\left(\frac{\pi x}{L}\right).$$

Assuming fixed endpoints find the displacement of the string at all times $t > 0$.

8. Consider a beam with tensile forces acting on its ends. Assuming no body forces and assuming that no forces act on the sides of the beam, set up the equations of elasticity governing the static behavior of this system. Solve for the displacement field.

9. Consider static deflections of a loaded circular elastic membrane. Modify the membrane approximation of Section 2.4.4 by assuming a constant uniform load. Solve for the displacement field assuming fixed boundary conditions.

10. Find the steady-state displacement of the gravitationally loaded elastic string, which is fixed in position at $x = 0$ and $x = L$.

11. In the plane, the shortest curve connecting two points is a straight line. Show that the variational formulation of this minimization problem is to minimize the integral

$$I = \int_{x_1}^{x_2} \sqrt{1 + y'^2}\, dx,$$

where $y(x)$ is the curve connecting the points (x_1, y_1), (x_2, y_2). Apply the Euler-Lagrange equation to this problem to prove that the minimizing curve is a straight line.

12. Consider a curve connecting the points (x_1, y_1) and (x_2, y_2). Revolve this curve about the x-axis. What is the curve which minimizes the surface area of the resulting body of revolution? You may observe this surface by forming a soap film between two concentric circular rings. What happens if you move the rings further and further apart? What happens to your minimizer?

13. Consider a curve of fixed length L with starting point and ending point on the x-axis. What curve maximizes the enclosed area? This is an example of an *isoperimetric* problem with a constraint. Here the constraint is that the curve have fixed length. In order to solve this problem, you will need to recall the technique of Lagrange multipliers.

14. The hanging rope is another classic problem of the variational calculus. Find the shape of a rope of fixed length L, which is perfectly elastic and hangs under the influence of gravity. This is again a problem with a constraint and the method of Lagrange multipliers will be needed. You can observe the answer to this problem by looking out your window at hanging telephone or electric wires.

15. Derive the Euler-Lagrange equation for integrals of the form

$$I = \int_{x_1}^{x_2} F(x, y, y', y'')\, dx.$$

16. Redo the analysis of Section 2.4.7 for the static beam, this time assuming that a constant gravitational force acts on the beam. Compute the shape of a gravitationally loaded beam of length L with pinned end conditions.

17. Redo the analysis of Section 2.4.7 for the vibrating beam this time assuming that a constant gravitational force acts on the beam.

18. Consider an elastic beam in the absence of tension that is pinned at the endpoints $x = 0, L$. Assume the beam is given an initial displacement of $f(x)$ and is released from rest. Using separation of variables solve for the motion of the beam.

Section 2.5

19. In 1950 V.I. Danilovskaya carried out one of the earliest investigations of dynamic thermoelasticity investigating thermal stress waves in an elastic half-space. Repeat her analysis by considering an elastic half-space subject to sudden heating on the boundary. Assume the uncoupled linear theory is valid, but do not make the quasistatic approximation. Solve for the elastic displacement wave propagating into the half-space.

20. Consider a one-dimensional thermoelastic rod that is anchored at its left end and free at its right end. Assume the right end of the rod is immersed in a temperature bath of temperature T_R and that the temperature of the left end of the rod is varied in time according to

$$T(0,t) = T_R + T_p \sin(\omega t).$$

Find the displacement at the right end of the rod.

Section 2.6

21. Consider the steady flow of an incompressible viscous fluid between a pair of infinite parallel plates a distance d apart. Assume the top plate is moving with velocity U and that the bottom plate is stationary. This type of flow is known as plane Couette flow. See Figure 2.14. By assuming that the velocity field has the form $\mathbf{u} = (u(y), 0, 0)$, that no body forces are acting, and that the no-slip boundary condition holds at the walls, reduce the incompressible Navier-Stokes equations to an exactly solvable system. Solve. Sketch the velocity profile $u(y)$.

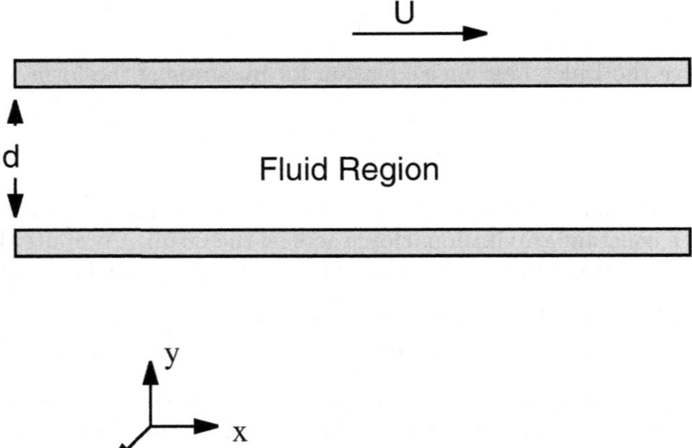

FIGURE 2.14: Plane Couette flow.

22. Again consider plane Coutte flow as in the previous problem, but this time impose the slip boundary condition on both plates. Assuming the velocity field has the form $\mathbf{u} = (u(y), 0, 0)$ and that no body forces are acting, reduce the incompressible Navier-Stokes equations to an exactly

solvable system. Solve. Compare the velocity profile obtained to the case of no-slip at the walls.

23. Consider steady, pressure-driven flow of an incompressible viscous fluid between a pair of infinite parallel plates a distance d apart. Assume the plates are stationary and that there is a constant pressure gradient in the fluid. That is, assume $p_x = A$, where A is a constant. As in the previous problem, assume $\mathbf{u} = (u(y), 0, 0)$, that no body forces are acting, and that the no-slip boundary condition holds at the walls, and reduce the incompressible Navier-Stokes equations to an exactly solvable system. Solve. Sketch the velocity profile $u(y)$. This type of flow is called plane Poiseuille flow. Contrast this flow with plane Couette flow.

24. Repeat the analysis of the previous problem, this time applying the slip boundary condition rather than the condition of no-slip on the plates. Compare your solution with the no-slip plane Poiseuille flow solution.

25. Using the solution for Poiseulle flow in a circular cylinder with a constant pressure gradient compute the mass flow rate through a pipe of radius R. If the fluid is water at room temperature and the pipe is a microchannel of radius 10μm, what is the pressure gradient needed to obtain a Reynolds number of 1? 10? 100?

26. Using the solution for Poiseulle flow in a circular cylinder with a constant pressure gradient and allowing for slip at the walls, compute the mass flow rate through a pipe of radius R. If the fluid is water at room temperature and the pipe is microchannel of radius 10μm, what is the pressure gradient needed to obtain a Reynolds number of 1? 10? 100? Compare with the solutions of the previous problem.

27. Consider a semi-infinite region of fluid above an infinite horizontal flat plate. Assume that initially the fluid is quiescent and the plate is motionless. Assume that at time $t = 0$, the plate is suddenly set into uniform motion with constant velocity U. This type of flow is called Rayleigh impulsive flow. Assume the velocity field in the fluid has the form $\mathbf{u} = (u(y,t), 0, 0)$, assume no-slip boundary conditions, and reduce the incompressible Navier-Stokes equations to an exactly solvable system. Solve. Assuming the fluid is water, how long does it take for fluid 10μm away from the wall to reach 99 percent of its limiting velocity? How does this change with viscosity?

Section 2.7

28. Consider a square wire loop of side length L, confined to the $z = 0$ plane, and having sides parallel to the x and y axes. Assume the loop is carrying a current I in the counterclockwise direction when viewed from $z > 0$. If the edges of the loop are located at $x = 0, L$ and $y = 0, L$, use

the Biot-Savart law to show that the magnetic field around the edge at $y = 0$ is

$$\mathbf{B}_{\text{bot}} = \frac{\mu_0 I}{4\pi}\frac{y\hat{\mathbf{z}} - z\hat{\mathbf{y}}}{y^2+z^2}\left(\frac{x-L}{\sqrt{(x-L)^2+y^2+z^2}} - \frac{x}{\sqrt{x^2+y^2+z^2}}\right).$$

Use this to compute the magnetic field created by the entire loop.

2.11 Related Reading

An excellent introduction to continuum mechanics and related mathematical methods may be found in the series of books by Lin and Segel.

C.C. Lin and L.A. Segel, *Mathematics Applied to Deterministic Problems in the Natural Sciences*, Philadelphia: SIAM, 1988.

L.A. Segel, *Mathematics Applied to Continuum Mechanics*, New York: Dover, 1987.

Carslaw and Jaeger is a useful compendium of information on heat transfer. They include a careful discussion of boundary conditions and the solution to the heat equation in many situations.

H.S. Carslaw and J.C. Jaeger, *Conduction of Heat in Solids*, Oxford: Oxford, 1959.

A somewhat more modern compendium of information on diffusion is provided by Crank.

J. Crank, *The Mathematics of Diffusion*, Oxford: Clarendon Press, 1975.

A very accessible introduction to the calculus of variations can be found in Lemmons.

D.S. Lemmons, *Perfect Form*, Princeton: Princeton University Press, 1997.

The accessible but slightly more advanced classic by Weinstock is now available in a Dover addition. Weinstock gives a full treatment of engineering elasticity via the variational approach.

R. Weinstock, *Calculus of Variations, With Applications to Physics and Engineering*, New York: Dover, 1974.

The text on thermoelasticity is now available in a Dover addition:

B.A. Boley and J.H. Weiner, *Theory of Thermal Stresses*, New York: Dover, 1988.

Jackson is considered the definitive work on electromagnetism.

J.D. Jackson, *Classical Electrodynamics*, Second Edition, New York: Wiley, 1975.

Stratton is also a classic.

J.A. Stratton, *Electromagnetic Theory*, New York: McGraw-Hill, 1941.

There are many useful texts on fluid mechanics. The classics by Lamb and Batchelor are among the best.

Sir H. Lamb, *Hydrodynamics*, Cambridge: Cambridge, 1993.

G.K. Batchelor, *An Introduction to Fluid Dynamics*, Cambridge: Cambridge, 1967.

Landau and Lifshitz is also a useful reference on fluid mechanics.

L.D. Landau and E.M. Lifshitz, *Fluid Mechanics*, Butterworth Heinemann, Oxford, 1959.

Several texts on microfluidics have been recently published.

G.E. Karniadakis and A. Beskok, *Microflows: Fundamentals and Simulations*, Springer Verlag, 2001.

M. Koch, A. Evans, A. Brunnschweiler, and A. Brunschweiler, *Microfluidic Technology and Applications*, Research Studies Press, 2000.

2.12 Notes

1. There are however engineering systems for which one may not ignore the generation of heat by elastic waves or the inertial terms in the elastic equations. The study of laser drilling and electron beam welding provide examples where these terms are important.

2. The divergence theorem appears in Appendix A.

3. This relies on the Dubois-Reymond lemma that is stated in Appendix A.

4. The fact that we need only specify that the function be bounded at a point, rather than specify its actual value, is a feature of singular boundary value problems.

5. The method of separation of variables is briefly reviewed in Chapter 5.

6. This result is called the fundamental lemma of the calculus of variations. It is stated precisely in Appendix A.

7. We note that this introduction to the calculus of variations is very incomplete. We have derived a necessary condition that an extremum of our energy must satisfy. However, we have not addressed the question of existence or whether the extremum is a minimizer or maximizer.

8. We use the notation ∇^4 to denote the biharmonic operator.

9. One of the most intriguing examples of thermoelastic behavior concerns the Colossi of Memnon, a pair of gigantic statues in Thebes representing the Greek hero Memnon. Memnon is said to have fought in defense of Troy and was slain by Achilles. Prior to about 199 A.D., the Colossi of Memnon would produce a musical sound when struck by the morning sun's rays. In 199 A.D. the Roman emperor Septimus Severus ordered repairs made to the statue and the music stopped.

10. An excellent discussion of the limits of the uncoupled quasistatic theory is available in the text by Boley mentioned in the related reading section in this chapter.

11. To be more precise, we must determine whether the particle's motion is relativistic or not and then use the appropriate equation of motion.

12. The asymmetry in these formulas is a historical artifact. Here we write them in their traditional form, which is the one most often encountered in the literature.

13. The inner product referred to here is defined for functions rather than vectors. But once the problem has been discretized the functional inner product is related to the usual (N-dimensional) vector inner product in a very direct manner.

14. The nonmonotonic dependence of heat flux on the thickness of the outer cylinder is one example of what is known as the critical radius of insulation effect in the heat transfer literature. It is a nice example of how crucial behaviors, such as heat flux, can depend nonlinearly on the size of a system.

Chapter 3

Small is Different

The initial mystery that attends any journey is: How did the traveler reach his starting point in the first place?

Louise Bogan

3.1 The Backyard

The range of human sensory experience is in some ways surprisingly large. In terms of length, we have experience of objects from 0.1 millimeters (a hair) to 10 meters (a tree), or about five orders of magnitude. In terms of time, from 0.3 seconds (your reaction time) to about a year, about eight orders of magnitude. In terms of mass, from 0.001 grams (a very small insect), to 1000 kg (a large animal or boulder), nine orders of magnitude. Other quantities have similar ranges, for instance, the viscosity of liquids ranges from 0.02 cP (air) to 10,000 cP (honey)[1], the kinetic energy of a moving object ranges from 10^{-2} Joules (a bee) to 10^{5} Joules (a buffalo), and the intensity and frequency of sound waves ranges from 10^{-12} to 10^{1} Wm^{-2} and 20 to 20,000 Hz.[2]

There are other areas in which we are more limited. Our experience of temperature essentially spans less than one order of magnitude, roughly 230 K (or −41°F) to perhaps 1200 K (the middle part of a candle flame). Similarly, our visual perception of electromagnetic radiation covers only a factor of two in wavelength, from 0.4 to 0.7 microns.[3] In addition, we rarely experience optical diffractive effects, that is, we experience light not as waves but as rays. Our direct sensory experience of static electric and magnetic fields is so weak and infrequent that few people can claim to have any intuition about them at all.[4]

This chapter is concerned with how the qualitative characteristics of some physical phenomena change as length scales decrease from meter (human) sizes to micron (cell) sizes,[5] that is, how they change in magnitude and proportion when things go from large to small. This is commonly called *scaling*, and the exact manner in which a particular quantity changes with respect to a characteristic length, or in proportion to another quantity, is often called a *scaling law*. While human beings do not have direct sensory experience of the mesoscale world that single-celled organisms inhabit,[6] we do, thanks to the relatively large ranges above, have an intuitive feeling for how many physical phenomena scale with respect to length. This indirectly gives us intuition about the microworld that we would not otherwise have. An example of this is the cooling of small objects. When you take the turkey out of the oven, the very thin aluminum foil covering cools from 400°F to room temperature within a minute or so while the turkey itself stays hot for perhaps an hour. Another everyday example is the strength of beams: the ratio of the height of a tree to the radius of its trunk is much smaller when the tree is fully grown than when it is a sappling. The rather dramatic differences in our sensory ranges and capabilities will fully reveal themselves when we try to apply our intuition about scaling to the microworld. Some things that we have a large range of experience with, heat conduction for instance, will remain more or less familiar, while other things that we have very little experience with (electrostatics) will yield a seemingly endless series of surprises.

In the end we will learn two important lessons. The first is that scaling is about ratios: it is only important to look at how one quantity, say a force, changes with respect to another. Hence we will examine scaling in terms of *systems* of forces, rather than the individual forces themselves. The second is that the only hard-and-fast scaling rule to remember is geometrical: for a device with fixed proportions, the surface area to volume ratio *always* increases as the length scale decreases. Hence, as things become smaller surface effects become relatively more important. While the transition length at which surface effects take over from volume (body) effects depends on the system under consideration, a good rule of thumb is that millimeter-scale devices are small enough for surface effects to be important while in the micron regime these effects will be dominant.

We begin with Section 3.2 by looking at a variety of very simple systems, fluidic, electrostatic, elastic, etc, and examining the scaling of important characteristic effects in each of them. We revisit the concept of nondimensionalization of variables introduced in Section 1.3 and discuss how the particular choice of the scaling of the variables affects the model under consideration. In Section 3.3 we take a closer look at the issue of scaling of systems. We show how constraints arising from side-effects within a system often limit the scaling of a particular effect.

3.2 Scaling

In this section we revisit the various fields of continuum mechanics reviewed in the previous chapter and look at the scaling of a few characteristic phenomena from each of them. In the course of this we will use two key ideas:

the use of dimensionless quantities and equations, as in Section 1.3, and the reduction of an expression to one involving constants and functions of length only.

3.2.1 Viscosity of Fluids

Consider a sphere of radius a and uniform density ρ_0 floating in a liquid of density ρ and *dynamic viscosity* ν.[7] The *Reynolds number* of this system is

$$Re = \frac{\rho v a}{\nu}, \tag{3.1}$$

where v is the velocity of the sphere. Re is a dimensionless number that characterizes the flow of the fluid: systems for which Re is small generally exhibit laminar flow while large Reynold's number flows are turbulent.[8] Suppose the velocity is such that the sphere always moves a fixed number, α, times its radius in a given time T. Then

$$v = \frac{\alpha a}{T} \tag{3.2}$$

and

$$Re = \frac{\rho \alpha a^2}{\nu T} = \beta a^2. \tag{3.3}$$

Note that here ρ, α, ν, and T have all been fixed, so that β is a constant, independent of a. Thus Re is proportional to a constant times a^2. For this system, the Reynolds number decreases quadratically with a: for a sphere half the size of another, given sphere, and moving at half its velocity, the Reynold's number is four times smaller. Thus for a small enough sphere, the flow will always be laminar and in microsystems one will often encounter flows for which $Re \ll 1$. This type of very low Reynolds number flow is called *creeping flow.*

What have we done to reach this conclusion? We used the definition of the Reynolds number (3.1) along with (3.2) to arrive at (3.3). The key step was the second one, where we assumed a *scaling law* for the velocity of the sphere. Note that we could have assumed a different law, say $v \propto a^{-2}$, which would have led to the opposite conclusion ($Re \propto a^{-1}$) but we chose a scaling for v that seemed reasonable. Such scaling laws typically come from one of three sources: assumptions (like this one), separate calculations, or experimental data.

That being said, the conclusion that the flow is laminar is true for the sphere even if we assume that the velocity is independent of the length scale, in this case, the radius of the sphere. The velocity of the sphere through the gas must actually *increase* with decreasing a in order for a very small sphere to have a turbulent wake.

Now consider the same sphere in a constant gravitational field, falling through the fluid. We assume that the sphere has reached terminal velocity and compute the time it takes to fall a given distance d. Since we are going to look at the limit of very small spheres we may assume, by the above, that the flow is laminar. In this case the viscous drag force on the sphere is given by *Stoke's law*,

$$F = 6\pi \nu a v. \tag{3.4}$$

Balancing this with the gravitational and buoyant forces on the sphere gives

$$6\pi a v \nu = \frac{4}{3}\pi a^3 g(\rho_0 - \rho), \tag{3.5}$$

where we have assumed that $\rho_0 > \rho$. Hence the terminal velocity is given by

$$v_\infty = \frac{2ga^2(\rho_0 - \rho)}{9\nu}, \tag{3.6}$$

and the time it takes for the sphere to travel the given distance d is

$$T = \frac{9d\nu}{2ga^2(\rho_0 - \rho)}. \tag{3.7}$$

We take a as the characteristic length and let $d = \alpha a$. Since everything else is already constant we have

$$T = \frac{9\nu\alpha}{2g(\rho_0 - \rho)}\frac{1}{a}. \tag{3.8}$$

Again, T is expressed as a constant multiplied by a function of the length scale only. The conclusion is that T scales as a^{-1}, in other words, very small particles take a very long time to precipitate out of solution. In this calculation we had to be given two scaling laws: the one for d, which is an assumption that the system retains its proportions as the length scale is decreased, and one for the viscous drag force (Stoke's law), which can either be considered experimental data or the result of solving the Stoke's equations for a sphere in an infinite fluid.

Now suppose the sphere contains a small engine and is capable of moving itself through the fluid. We assume the engine generates a force of the form

$$F = \epsilon a^n, \tag{3.9}$$

where $n = 3$ for an engine that produces a force per unit volume (e.g., an internal combustion engine) or $n = 2$ for a force per unit area (a biological muscle). As per the above, we will look at the limit in which the sphere becomes very small and assume laminar flow and ignore gravity. Assuming that the sphere starts from rest and the engine produces a constant thrust, how far does the sphere travel before it reaches a given fraction, say 0.95, of its terminal velocity? Assuming the sphere accelerates straight along the x' axis, the equation of motion is

$$m\frac{d^2x'}{dt'^2} = \epsilon a^n - 6\pi a\nu\frac{dx'}{dt'} \tag{3.10}$$

with initial conditions

$$x(0) = \frac{dx'}{dt'}(0) = 0. \tag{3.11}$$

At this point we make use of another useful tool: the nondimensionalization of independent variables first discussed in Section 1.3. We make the substitution,

$$x = \frac{x'}{x_0}, \quad t = \frac{t'}{t_0}, \tag{3.12}$$

where x_0 and t_0 are our chosen scales of length and time. Here, we have used the convention that primed variables have dimensions while unprimed variables are dimensionless. After this transformation the variables x and t are dimensionless. One way to think about this is the following: so far we have not discussed the units of length that x' is to be measured in for the good reason that these units are irrelevant to the discussion. However, just as we choose to measure the height of human beings in feet and not nanometers or light years, some units are more irrelevant than others. It would be nice to choose units of length and time, x_0 and t_0, which are "natural" and express our answers in terms of them and not others, which are perhaps less suited to the system. Another way to think about this is that by nondimensionalizing the independent variables, we shift all discussion of units to the quantities x_0 and t_0 and hence all matters involving units and measurements can be deferred until we actually need them.

The tradeoff is that at this point, we must choose x_0 and t_0. For now we'll make the suspiciously serendipitous choice,

$$x_0 = a, \quad t_0 = \frac{2a^2\rho_0}{9\nu}. \tag{3.13}$$

We assume the sphere is made of a constant-density material so that the scaling law for m is $m = 4\pi a^3\rho_0/3$. After using (3.12) we obtain

$$\frac{d^2x}{dt^2} = k - \frac{dx}{dt}, \quad k = \frac{\epsilon\rho_0 a^n}{27\pi\nu^2}. \tag{3.14}$$

Note that all the terms in this equation are dimensionless, including k, which neatly encapsulates ϵ, ρ_0, a, n, and ν, all in one parameter

We can solve the ordinary differential equation (3.14) with initial conditions (3.11) for $x(t)$,

$$x(t) = k(t - 1 + e^{-t}). \tag{3.15}$$

We thus see that k is simply the terminal velocity, and the distance the sphere moves before reaching a fraction, f, of the terminal velocity is

$$x_f = -k\,(f + \log(1 - f)) \tag{3.16}$$

with $0 < f < 1$. Now x_f is already in units of the sphere radius a, so we can immediately conclude that for $n > 0$ and a small enough the sphere will reach velocity $v = fk$ before travelling even a small fraction of its own diameter. Clearly the same conclusion will be reached if we consider the distance it takes for the sphere to stop from a given velocity. Since any realistic engine will satisfy $n > 0$, we see that on a small enough length scale, viscosity is the dominant force in this system.

Before leaving this section we note that the choice (3.13) is certainly not unique. We chose x_0 because it is intuitively reasonable to pick the sphere's radius as a length scale and we specifically chose a "viscosity-limited" time scale in anticipation of the final result. We could have, for instance, chosen $x_0 = a$ and the "inertial" time scale, $t_0 = \sqrt{4\pi\rho_0 a^{4-n}/3\epsilon}$ instead. In this case the solution for $x(t)$ would look like

$$x(t) = \frac{1}{c^2}\left(ct - 1 + e^{-ct}\right), \quad c = \sqrt{\frac{27\pi\nu^2}{\rho_0 \epsilon a^n}}. \tag{3.17}$$

Note that the constant c scales as $a^{-n/2}$ so that the graph of $x(t)$ would be incoveniently compressed for a very small. In fact, instead of plotting x vs. t one would be sorely tempted to plot it against ct instead, which of course, would simply be the graph of (3.15).

3.2.2 Heating and Cooling

Any engine will generate waste heat along with thrust. Suppose the sphere's engine generates a constant amount, Q, of waste heat per unit volume per unit time. If the sphere has heat capacity c, and thermal conductivity k, then, following Section 2.3, the temperature is governed by[9]

$$\kappa\nabla^2 T' + \frac{Q}{\rho_0 c} = \frac{\partial T'}{\partial t'}, \tag{3.18}$$

where $\kappa = k/(\rho_0 c)$ is called the *thermal diffusivity*. Let's assume that the sphere is at rest and in contact with a heat reservoir at temperature T_b' and loses heat to the reservoir by conduction. Since Q is constant and the sphere is at rest, T' will depend only on the radial coordinate r'. Hence the equilibrium state will obey

$$\frac{\partial^2 T'}{\partial r'^2} + \frac{2}{r'}\frac{\partial T'}{\partial r'} + \frac{Q}{\kappa\rho_0 c} = 0. \tag{3.19}$$

We choose to scale r' and T' by

$$r = \frac{r'}{a}, \quad T = T'\frac{\kappa\rho c}{Qa^2}, \tag{3.20}$$

which gives

$$\frac{\partial^2 T}{\partial r^2} + \frac{2}{r}\frac{\partial T}{\partial r} + 1 = 0. \tag{3.21}$$

We assume the reservoir is capable of absorbing heat at a very high rate, so that the boundary conditions on the temperature are

$$T(1) = \tau = T_b' \frac{\kappa \rho_0 c}{Q a^2}, \quad \frac{\partial T}{\partial r}(r = 0) = 0. \tag{3.22}$$

As discussed in Section 5.4, the second of these is actually a consequence of the spherical symmetry of the problem and the requirement that the temperature be bounded. As before, T, r, and τ are all dimensionless. The solution is

$$T(r) = \frac{1 - r^2}{6} + \tau. \tag{3.23}$$

Again, the scaling has been chosen so that the solution is as simple as possible. Observe that the central temperature is always a constant above the temperature of the bath. Reversing the scaling gives

$$\Delta T' = T'(0) - T_b' = \frac{Q a^2}{6 \kappa \rho_0 c}. \tag{3.24}$$

Since Q is a volume heat generation, rate a reasonable scaling is $Q \propto a^0$; that is, the *total* heat generation rate of the engine scales as a^3. Thus $\Delta T'$ scales as a^2; it vanishes in the limit as a vanishes: no matter how large Q is the temperature difference between the center and surface of the sphere can be made arbitrarily small by making a small enough.

Now suppose we turn the engine off and let the sphere cool down to the bath temperature. How long does this take? We have introduced a new variable, t', which needs to be scaled. We choose

$$t = t' \frac{\kappa}{a^2} \tag{3.25}$$

so that after the engine has been turned off ($Q = 0$) the temperature obeys

$$\frac{\partial^2 T}{\partial r^2} + \frac{2}{r} \frac{\partial T}{\partial r} = \frac{\partial T}{\partial t} \tag{3.26}$$

with boundary conditions given in (3.22) and initial condition (3.23). Using the method of separation of variables, it can be shown that the time required for the central temperature difference to drop to within a given fraction f of its initial value of $1/6$ is approximately

$$t_f' = -\frac{a^2 \log f}{\pi^2 \kappa}, \tag{3.27}$$

where we have assumed that $f \ll 1$. Hence t_f' also scales as a^2 so that not only does the central temperature not rise very much above the bath, but it drops back to the bath temperature very rapidly once the heat source is removed.

The heat equation (3.18) governs not only the conduction of heat but diffusive phenomena in general. We normally think of diffusion as a relatively weak and slow phenomenon, but the conclusion we reach here is that on small enough length scales diffusion can be very fast. It is only our macroscopic perspective that sees diffusion as slow.

Suppose we change the boundary condition so that the sphere loses heat by convection to the surrounding fluid bath. Since convection is normally a much more effective way of transferring heat, we might expect the time scale t'_f to be even smaller. However, we need to be a little careful about this. First we must distinguish between *free convection*, in which the cooling is achieved by heating of the fluid around the sphere and the subsequent motion of that fluid away from the sphere by its buoyant force, and *forced convection* in which the fluid is forced past the sphere by an external means (say a pump). Based on what we have discovered so far, it is clear the free convection will be negligible on the microscale since temperature differences in the surrounding fluid will be small (diffusion is fast) and buoyant forces, which are gravitational, are weak compared to viscosity.

So suppose we are cooling the sphere by forced convection and the fluid free stream velocity (its velocity far from the sphere) is v. If we assume the no-slip boundary condition discussed in Section 2.6, then the fluid has zero tangential velocity at the surface of the sphere. This creates a region near the sphere known as a *fluid boundary layer* in which the fluid velocity is significantly less than the free stream velocity.

The existence of the fluid boundary layer means that convection cannot take place exactly at the surface of the sphere. Instead, heat is conducted away from the sphere through the fluid until the velocity of the fluid becomes significant, at which point convection can be considered to begin. In the region in which conduction is dominant, the temperature of the fluid will drop and the length scale over which it drops significantly is called the *thermal boundary layer*.

In order to model convection, we replace the boundary condition (3.22) at the surface of the sphere by the Newton cooling condition,

$$k\frac{\partial T'}{\partial r'}(a) = h(T'_b - T'(a)), \tag{3.28}$$

where h is called the *boundary conductance* or *surface heat transfer coefficient.* If we use the scaling above then this becomes

$$\frac{\partial T}{\partial r}(1) = Bi(\tau - T(a)), \tag{3.29}$$

where

$$Bi = \frac{ha}{k} \tag{3.30}$$

is another dimensionless number called the *Biot number.* Now according to Fourier's law (2.8), the temperature of any system must be a continuous function of space, that is, there can be no discontinuous jumps in temperature. But this is exactly what (3.28) requires. In this case, the jump in temperature refers to the change in temperature across the thermal boundary layer, so that for (3.28) to be useful the thermal boundary layer must be much smaller than the dimensions of the system, in this case the radius of the sphere.

The relationship between the fluid boundary layer and the thermal boundary layer is governed by the *Prandtl number*,

$$Pr = \frac{\mu}{\kappa_f}, \tag{3.31}$$

that is, the ratio between the kinematic viscosity and the thermal diffusitivity of the fluid. For laminar boundary layers the relationship is of the form

$$\frac{\delta'_F}{\delta'_T} = Pr^n, \tag{3.32}$$

where δ'_F and δ'_T are the fluid and thermal boundary layer thicknesses, respectively [92]. The exponent n varies according to type of fluid but generally speaking, it lies in the range $0 < n < 1$. For instance, liquid metals like mercury typically have very low Prandtl numbers, on the order of 10^{-2}, since they have low viscosity and large thermal conductance. In this case $n = 1/2$. On the other extreme, high viscosity, thermally insulating liquids like motor oils can have $Pr \approx 5000$ and for this case $n = 1/3$. Water (at 300 K) lies in between with $Pr = 5.8$ (also $n = 1/3$).

Hence the range of Pr^n for many fluids is roughly $0.1 < Pr^n < 10$ so that the thermal boundary layer is more or less of the same order of magnitude as the fluid boundary layer. We shall see in Chapter 9 that for microsystems the fluid boundary layer can be very large compared to the dimensions of the system, which implies that the thermal boundary layer is large as well. This in turn means that true forced convection will be rarely encountered.[10]

Returning to conduction, we claimed that the time scale $t' \approx a^2$ for cooling was fast. To quantify this, consider the equation of motion (3.14) from the previous section. Suppose the sphere is moving along at its terminal velocity and suddenly turns its engine off, so that it begins to slow down and cool at the same time. The time required for it to slow down to a fraction f_v of the terminal velocity is

$$t'_v = -\frac{2a^2 \rho_0 \log f_v}{9\nu}, \tag{3.33}$$

while the time require to cool by conduction to a fraction f_c of the central temperature is

$$t'_c = -\frac{a^2 \log f_c}{\pi^2 \kappa}. \tag{3.34}$$

We see that these scale at the same rate and the ratio of cooling to slowing times is

$$\tau = \frac{9\kappa_f \rho Pr}{2\pi^2 \kappa \rho_0} \frac{\log f_c}{\log f_v}. \tag{3.35}$$

Assuming that $\kappa_f \approx \kappa$ and $\rho \approx \rho_0$, the coefficient here is of the order of unity for many liquids, so that if we set $f_c \approx f_v \ll 1$, then $\tau \approx 1$. In other words by the time the sphere reaches rest from terminal velocity (which happens after traveling a distance small compared to its own radius), it has already cooled off.

3.2.3 Rigidity of Structures

Consider a long thin beam of length L and square cross-section of side length a with $a = \epsilon L$ and $\epsilon \ll 1$. Let the beam have uniform Young's modulus E, density ρ, and let one face of the beam support an evenly distributed load (i.e., a force per unit length) of magnitude w'. From Section 2.4, the equation governing the deflection, u', of the beam is

$$\frac{d^4u'}{dx'^4} = -\frac{w'}{EI}, \tag{3.36}$$

where $I = a^4/12$ and x' is a coordinate along the length of the beam. If the beam is fixed at one end (i.e., attached to a wall) and free at the other (a cantilever beam), then the boundary conditions on u' are

$$u'(0) = \frac{du'}{dx'}(0) = 0, \quad \frac{d^2u'}{dx'^2}(L) = \frac{d^3u'}{dx'^3}(L) = 0 \tag{3.37}$$

with $x' = 0$ at the fixed end and $x' = L$ at the free end. If we scale according to

$$x = \frac{x'}{L}, \quad u = \frac{u'}{a}, \tag{3.38}$$

we obtain

$$\frac{d^4u}{dx^4} = -w, \quad w = \frac{12w'}{\epsilon^5 LE} \tag{3.39}$$

with the boundary conditions.

$$u(0) = \frac{du}{dx}(0) = 0, \quad \frac{d^2u}{dx^2}(1) = \frac{d^3u}{dx^3}(1) = 0. \tag{3.40}$$

The solution of (3.39) and (3.40) is

$$u(x) = -\frac{w}{24}x^2(6 - 4x + x^2). \tag{3.41}$$

In particular the maximum deflection occurs at the tip of the beam and is

$$u_{\max} = u(1) = -\frac{w}{8} = -\frac{3w'}{2\epsilon^5 LE}. \tag{3.42}$$

In order to get the scaling of $u_{\max}$ we have to assume a scaling law for w'. Let this be of the form

$$w' = \kappa L^n \tag{3.43}$$

which gives

$$u_{\max} = -\frac{3\kappa L^{n-1}}{2\epsilon^5 E}. \tag{3.44}$$

For a beam deflecting under its own weight the force per unit length is $w' = \rho g a^2 = \rho g \epsilon^2 L^2$ so that $n = 2$ and

$$u_{\max} = -\frac{3\rho g L}{2\epsilon^3 E}. \tag{3.45}$$

Hence $u_{\max}$ scales as L. That is, the maximum deflection, as a fraction of the side length of the beam, goes to zero as $L \to 0$ when the beam is deflecting under its own weight in a constant gravitational field. Small beams are relatively stiff compared with their own weight.[11] Now suppose that the load w' is caused by a constant pressure P' on one of the long faces of the beam. In this case $w = P'a = \epsilon P' L$, and we see that $n = 1$ and $u_{\max}$ is independent of L. Thus pressure, a surface force, will still be capable of bending small beams. Hence the conclusion we reach here is that the strength of beams scales as the square of the length scale so that on small length scales, body forces will be relatively ineffective at deforming structures compared to surface forces.

3.2.4 Electrostatics

Suppose we wish to levitate a small sphere of radius a and mass density ρ in a constant, upward pointing electric field of magnitude E. In order for levitation to occur, the charge on the sphere charge must satisfy

$$Q = \frac{4\pi a^3 \rho g}{3E}. \tag{3.46}$$

If the sphere is made of a dielectric material and has an embedded volume charge density μ, then $Q = 4\pi a^3 \mu/3$ and the condition for levitation is

$$\mu = \frac{\rho g}{E}. \tag{3.47}$$

In other words with E constant, μ is independent of the length scale a. On the other hand, if the levitating force is generated at the surface of the sphere, that is, if it has a net surface charge density σ, then we get

$$\sigma = \frac{\rho g a}{3E}. \tag{3.48}$$

Again, if E is independent of the length scale, then σ scales as a: for very small spheres we need a very low surface charge density to create levitation.

Now suppose we have a conducting wire in the presence of a constant magnetic field. Let the wire have length L and radius a with $a = \epsilon L$ and $\epsilon \ll 1$ and let the field have constant magnitude B. Let the wire be carrying current

I and suppose the field is oriented parallel to the surface of the Earth and the wire is perpendicular to both the magnetic and gravitational fields. Then the force exerted by the magnetic field on the wire is ILB and the wire will levitate if

$$I = \frac{\pi\epsilon^2 L^2 \rho g}{B}. \tag{3.49}$$

If the wire carries the current throughout its volume, then the current density is $J = I/(\pi a^2)$ and the current density that causes levitation is

$$J = \frac{\rho g}{B}, \tag{3.50}$$

which, like the volume electrostatic charge, is independent of L if B is constant.

Suppose, as in the electrostatic case, that we are able to generate a surface current in the wire, that is, a current which is confined to a constant thickness region near the surface of the wire.[12] Let the surface current be $K = I/(2\pi a)$. Levitation occurs at

$$K = \frac{\epsilon \rho g L}{2B} \tag{3.51}$$

and again, levitation becomes possible with a very small surface current. The situation here follows the same general rule as in the electrostatic case: levitation only becomes practical when the opposing force is generated on surfaces rather than in volumes.

The commonly stated rule of thumb is that electrostatic actuation is practical on small length scales because of the relative ease in creating surface charge densities using conductors, while magnetic actuation is impractical because of the relative difficulty in creating surface currents. In other words, the situation for magnetostatics is the reverse of that for electrostatics. However, in the above we have been looking at the onset of levitation as a function of charge and current density only, and the reader may well wonder whether this is reasonable. This question is taken up in Section 3.3.

3.2.5 Fluid Interfaces

The *surface tension* of a fluid interface is defined as the energy required to increase the area of the interface by a unit amount. As pointed out by Myers [139], surface tension is an intrinsic property of a fluid interface, rather than something applied externally (like the tension in a drumhead or a rubber membrane). Since surface tension is caused by an imbalance of Van der Waals forces on the molecules of the fluid at the interface, its value depends on the properties of both of the fluids. Table 3.1 gives surface tension values for some common fluid–air and fluid–water interfaces.

Since surface tension is a material property, in essentially the same way as the Young's modulus of a bulk material is, it does not scale with length.[13] From the equations describing interface dynamics given in Section 2.6, one

TABLE 3.1: Surface tension values for a variety of fluids. The units are the standard milliNewtons per meter, which are equivalent to the cgs unit Dynes per square centimeter. All values measured at 15°C.

Liquid	Formula	Surface Tension	Value vs. Water
Ammonia	NH_3	21	—
Ethanol	C_2H_6O	23	—
Acetone	C_3H_6O	24	—
Kerosene	$C_{13}H_{26}$	28	—
Benzene	C_6H_6	29	35
Glycerine	$C_3H_8O_3$	63	—
Water	H_2O	73	—
Mercury	Hg	485	390

can derive certain properties of interfaces. For instance, the fact that surface tension is an effect due to the imbalance of forces normal to the surface means that there must be a pressure drop across a fluid interface in order for the interface to be in equilibrium. The change in pressure is given by the *Young-Laplace* equation,

$$\Delta P = \sigma\left(\frac{1}{R_1} + \frac{1}{R_2}\right), \tag{3.52}$$

where σ is the surface tension and R_1 and R_2 are the principal radii of curvature of the interface.[14] In the case of a spherical surface $R_1 = R_2 = R$ and (3.52) reduces to

$$\Delta P = \frac{2\sigma}{R}. \tag{3.53}$$

One consequence of the Young-Laplace equation is the capillary effect. Consider a long, slender tube, open at both ends and with one end immersed in a fluid bath, which itself has a surface open to the atomsphere. If the fluid wets the tube, i.e., adheres to the surface of the tube, then the fluid–air interface inside the tube will consist of a concave spherical cap of radius a, which is of the order of the tube radius. Then the height h of the column of fluid is such that its weight is balanced by the pressure lifting the spherical cap,

$$h = \frac{4\sigma}{a\rho g}, \tag{3.54}$$

where ρ is the density of the fluid. Hence h scales as a^{-1}. The constant pressure drop across the surface of a fluid interface can be exploited in some applications. In particular, microfluidic pumps are conveniently self-priming due to the capillary effect.

Bubbles and droplets provide more examples.[15] Using (3.53) we see that very small bubbles must have large internal pressures in order to support themselves against the surface tension of the surrounding fluid. Consider an air bubble one micron in diameter in water. Using Table 3.1 the surface tension is 73 millinewtons per meter and hence the pressure drop across the bubble surface is 1.5×10^5 Pa or about 1.5 times standard atmospheric pressure. Thus if the water itself is at atmospheric pressure, the air inside the bubble is at 2.5 times atmospheric pressure.

Another consequence of (3.53) is that the force holding a bubble against a flat surface to which it is clinging is

$$F = \pi d\sigma, \tag{3.55}$$

where d is the diameter of the (assumed circular) region of contact between the bubble and the surface. We see that detaching a bubble from such a surface becomes a very difficult task since the detaching force must scale as length. For instance, the bouyant force on the bubble is approximately $4\pi a^3 \rho g/3$ where ρ is the density of the liquid. If $d = \alpha a$ then there is a critical radius at which bubbles will no longer detach from the surface and float freely in the liquid,

$$a_{\text{crit}} = \sqrt{\frac{3\alpha\sigma}{4g\rho}}. \tag{3.56}$$

For an air bubble in water this is about 3mm (using $\alpha = 1$), which is a very large bubble by microscale standards.

3.3 Systems

Real devices are always systems that involve the interaction of different physical domains.[16] Sometimes, as in an internal combustion engine, the operation of the device relies on this interaction, while in other cases, as when a computer chip heats up during operation, the interaction is a side effect. Even though they are highly idealized, the examples above are also all systems. For instance, the cooling sphere is a thermal-fluidic system since it relies on heat loss by conduction to a surrounding fluid for the cooling. The nature of the engine in the sphere adds an additional component (mechanical, biological) to the system that causes the heating. The levitating sphere is a gravitational-electrostatic system, etc.

Normally scaling is considered when the main concern is about optimizing a particular aspect of the performance of a device. For actuators, that aspect might be the total force or torque the device can supply, for a heater it might be a temperature difference or power dissipation. Because of the differences in scaling of different phenomena, an actuator that works efficiently on centimeter-length scales may be very inefficient or perhaps not work at all when scaled down to microns. The reason for this is that effects in different domains scale differently so that the operation of the device as a whole changes. Hence scaling is really about how effects in different physical domains interact in a particular system and how this interaction changes with the length scale.

Now suppose we have a particular system in mind, say an actuator, and we want to investigate the scaling of the force it can exert. In general there are two classes of effects that we need to consider: those which *generate* the force and those which *constrain* the force. For instance, if the actuator is electrostatic, the force may be generated by surface charges, by volume charges, or by some other mechanism. Each of these will have a different scaling with respect to length. An example of a constraint is that the electrostatic pressure generated by surface charges not exceed a certain limit beyond which the actuator material fractures. If the actuator is immersed in a gas, another constraint is that the electric field outside the device not exceed that which would cause the gas to ionize (assuming that is not pertinent to the operation of the actuator). Hence the scaling of the force is not determined solely by the mechanism by which it is generated but also by how this mechanism interacts with the other components of the system. In many systems it is the constraints that ultimately limit the scaling of the effect under investigation, independently of how it is generated.

3.3.1 An Electrostatic Actuator

Suppose we have a device that depends for its operation on a pair of parallel, electrostatically charged rods that repel each other. Let's isolate this part of the system and examine the scaling of the force on each rod. As per the above, in order to compute the scaling we need to include as many effects as we can within the limitations of a simplified model. Here we'll assume the rods are made of a real material with finite strength, and similarly the fluid between the rods is made of a material that will break down (ionize) if the external electric field becomes too large. Hence the system under consideration is an electrostatic-mechanical-fluidic system where the properties of all three physical domains must be taken into account.

Let the rods have length L, radius a, and be distance d apart. We let the length scale be L and define $a = \alpha L$, $d = \beta L$ with $\alpha \ll \beta \ll 1$. Suppose each rod has a linear charge density $\sigma = Q/L$, where Q is the total charge per rod. The force felt by each rod is

$$F = \frac{\sigma^2}{2\pi\beta\epsilon_0} = \frac{Q^2}{2\pi\beta\epsilon_0 L^2}. \tag{3.57}$$

Hence the scaling of F is determined by the scaling of Q. In an idealized system in which only electrostatics is considered, we could let Q be anything, that is, there are no physical laws that limit it and it would be up to the engineer to simply choose a scaling for it as he or she wished. However, we will see that the scaling of F is in fact constrained by the other components of the system, and this will set an upper limit on how slowly F can scale with length, and hence how strong the actuator can be.

For instance, the electric field will be strongest at the surface of each rod,

$$E_s = \frac{\sigma}{2\pi\epsilon_0 a} = \frac{Q}{2\pi\epsilon_0 \alpha L^2}, \tag{3.58}$$

and the system is constrained by the fact that E_s cannot be too large; otherwise the molecules that make up the fluid will become ionized. This will cause an electrical current to flow between the rods and whatever the nearest ground is, which will short out the actuator. Hence E_s can at best be constant as the system decreases in size. If we take this for the scaling law for E_s, that is, $E_s = E_c$ is a constant, then

$$F = \frac{2\pi\epsilon_0 \alpha^2 E_c^2 L^2}{\beta}. \tag{3.59}$$

Thus $F \propto L^2$ is the best that can be done for this system, independently of how the charges are generated or distributed in the rods.[17] Note that this in turn gives us a maximum scaling for Q,

$$Q = 2\pi\epsilon_0 \alpha E_c L^2. \tag{3.60}$$

Now we have some information about how Q can be scaled: in the limit of small L the function $Q(L)$ must decrease at least as fast as L^2.

Now suppose the rods have fixed ends. Another constraint on this device is that the rods are made of a real material with a finite strength. The charge on each rod induces a stress that puts it under compression. Since the rods are thin circular beams, we can apply beam theory and conclude that the rods will buckle if the compressive load exceeds the critical load,

$$F_b = \pi^3 E \alpha^4 L^2, \tag{3.61}$$

where E is the Young's modulus of the beam material. The compressive load caused by the charge on each rod is

$$T = \frac{\sigma^2 \pi}{24\epsilon_0} = \frac{Q^2 \pi}{24\epsilon_0 L^2}. \tag{3.62}$$

If we want to prevent the beams from buckling, then we need to scale the system so that the tension never exceeds the buckling threshold. In other words,

$$\tau = \frac{T}{F_b} = \frac{Q^2}{24\epsilon_0 \pi^2 \alpha^4 E L^4} \tag{3.63}$$

must always be less than unity. Let the scaling law for τ be such that it is constant. Then we see that the force is given by

$$F = \frac{12\pi\alpha^4 E\tau L^2}{\beta}; \tag{3.64}$$

and we see that again, F is limited to $F \propto L^2$.

Since it has been determined from a constraint, expression (3.59) does not tell us the actual scaling of the force with respect to length; it merely sets an upper limit to how good the scaling can be. To get the actual scaling we need to consider how the charge is really generated. The two cases we will consider are when Q is distributed over the surface (as in a charged conductor) and when Q is distributed over the volume (as in a dielectric with embedded charges). In these cases we have

$$Q = \kappa_n L^n, \tag{3.65}$$

where $\kappa_2 = 2\pi\alpha\sigma$ and $\kappa_3 = \pi\alpha^2\mu$, where σ is the surface charge density and μ the volume charge density. In the case of a constant surface charge density, we actually achieve the maximum scaling whereas for a constant volume charge density the scaling of the force is $F \propto L^4$. Note that in the former case, the constants α, β, E, etc., need to be chosen carefully in order for the actuator to function properly since both E_s and τ will be independent of L, while in the latter case, the constants need not be so carefully chosen since both τ and E_s vanish in the limit $L \rightarrow 0$.

3.3.2 A Similar Magnetostatic Actuator

Now replace the rods with wires carrying antiparallel currents. Retaining the geometry and notation of the previous section, the wires repel each other with force,

$$F = \frac{\mu_0 L I^2}{2\pi d} = \frac{\mu_0 I^2}{2\pi\beta}, \tag{3.66}$$

where I is the current carried by each wire. Again, we need to look at the constraints in order to begin to get a handle of the scaling of F. One constraint is that the total current carried by each wire cannot be too large; otherwise Joule heating will cause the wires to heat up beyond their melting point. If we assume that, for a given volume heat generation rate Q, the central temperature of the wire obeys a law of the same form as the sphere discussed above (Section 3.2.2), then

$$\Delta T = \gamma Q L^2, \tag{3.67}$$

where γ is a constant. Let the resisitivity of the wire be ρ. The total resistance of the wire is

$$R = \frac{\rho L}{\pi a^2} \tag{3.68}$$

and Q is given by

$$Q = \frac{I^2 R}{\pi a^2 L} = \frac{I^2 \rho}{\pi^2 \alpha^4 L^4}. \tag{3.69}$$

Hence

$$\Delta T = \frac{\gamma \rho I^2}{\pi^2 \alpha^4 L^2} \tag{3.70}$$

where we assume for the moment that ρ is not dependent on temperature.[18] If we let ΔT be constant, then we see that

$$F = \frac{\mu_0 \pi \alpha^4 \Delta T L^2}{2\beta\gamma\rho} \tag{3.71}$$

so that $F \propto L^2$. Note that in this case the current scales as

$$I = \pi \alpha^2 L \sqrt{\frac{\Delta T}{\gamma\rho}} \tag{3.72}$$

and the current density scales as

$$J = \frac{1}{L}\sqrt{\frac{\Delta T}{\gamma\rho}}, \tag{3.73}$$

which is unbounded as $L \to 0$. The reason that small wires can support these extremely large current densities is only because they are cooled so quickly. Note that the total power consumption by each wire in this case is

$$P = I^2 R = \frac{\pi \alpha^2 \Delta T L}{\gamma} \tag{3.74}$$

and the operational voltage is

$$V = IR = \sqrt{\frac{\rho \Delta T}{\gamma}}. \tag{3.75}$$

Now suppose that we design the actuator so that instead of straight wires we use parallel circular loops. As long as the aspect ratio of the loop, d/L, remains small, all of the above results hold. However, in this case there will be tension in the wires caused by the action of the magnetic field of each loop on itself,

$$T = \chi \mu_0 I^2, \tag{3.76}$$

where χ is a geometrical constant. If the wires are made of real materials, then they will only be able to withstand a certain tension per unit area before breaking. Hence if this critical tension density is kept constant,

$$\tau_c = \frac{\chi \mu_0 I^2}{\pi a^2} = \frac{\chi \mu_0 I^2}{\pi \alpha^2 L^2}, \tag{3.77}$$

then the force is

$$F = \frac{\alpha^2 \tau_c L^2}{2\chi\beta}. \tag{3.78}$$

Hence this constraint does not change the scaling of the force.

3.4 Chapter Highlights

- The physics of the microworld is dominated by surface effects because of the fundamental geometrical rule that the surface area to volume ratio increases as objects are scaled down proportionally. Intuitively, for microscopic objects "most of the space" consists of surfaces, while for macroscopic objects, volumes are much more important than surfaces.

- Microfluidics is dominated by viscosity. Reynolds numbers are typically very low, often much smaller than unity.

- Diffusion is relatively fast at small length scales. This results in the rapid cooling of small objects. As a result, the dissipation of waste heat is not problematic in many cases.

- Conduction is the dominant form of heat transfer. Because of the large size of thermal boundary layers, the Newton cooling condition (3.28) should be used with care when describing heat transfer by convection to liquids. Note that the Newton condition is often used to model other heat transfer mechanisms and each of these will have an associated boundary layer, which may or may not be small compared to the system under consideration.

- Because the strength of structures varies according to their cross-section, small structures are relatively rigid and difficult to deform.

- Fluid interfaces, including bubbles and droplets, play a prominent role. The pressure drops across such interfaces can be very large.

- Electrostatic surface charges, whether intentional or not, can exert large forces on structures and create large electric fields.

- Scaling is all about ratios, so that it only makes sense to talk about the scaling of effects within a given system. The magnitude of constraints (often side effects) often limit the scaling of an effect, independently of how it is produced.

3.5 Exercises

Section 3.1

1. The temperature range of sensitivity is in Kelvin rather than Farenheit or Celsius. Why does it make a difference in this case and not in the

others? Can you think of units for other quantities in which something similar happens?

Section 3.2.1

2. Given the properties of air at standard temperature and pressure, calculate approximately the size of the sphere moving at a reasonable velocity when the flow turns from turbulent to laminar. If we assume that the dynamic viscosity of air varies with pressure as

$$\nu = \nu_0 \frac{P}{P_0},$$

then what would the pressure have to be in order to create turbulent flow around a sphere of radius one micron moving at 100 microns per second? Estimate the spacing between air molecules at this pressure. What is wrong with this picture?

3. Can you construct length and time scales for model (3.10) that are *independent* of the sphere radius a? If so, what do these quantities tell you?

4. It is often said that inertia is unimportant on the microscale because mass scales as the cube of length,

$$m = \kappa l^3,$$

which is a volume effect. Consider a device which consists in part of a small ball connected by a string to a motor which twirls it around. Compute the scaling of the rotational frequency of the ball considering the fact that the string can withstand a certain tension per unit cross-sectional area before it will break. In your opinion, how important is inertia in this system?

5. Not all objects in the universe have a mass vs. size relationship of the form given in Problem 4. For instance, black holes have the highly counterintutive relationship,

$$m = \frac{rc^2}{2G},$$

where r is the radius of the (spherical) hole. Come up with an example of how a relationship of this form could be exploited in a microscale device.

6. Because of the difficulty in manufacturing and handling black holes, they are not yet considered a suitable material for use in commerical engineering. On the other hand, single-walled carbon balls and tubes (Buckeyballs and Buckeytubes) are fast becoming useable materials. These

objects have a mass vs. size relationship of the form,

$$m = \kappa r^2,$$

where r is the radius of the ball or tube (i.e., cylinder). Approximate the constant κ for both balls and tubes using known properties of carbon (not included!). Can this relationship be exploited in the same device that you came up with for exercise 5? If not, come up with another example. What are the limits on r in the expression above? Does this restrict the usefulness of your device in any way?

7. For a very wide variety of animals, the strength of a muscle scales as its cross-section; in other words, the maximum force that a given muscle can exert is given by

$$F = \kappa r^2,$$

where r is a characteristic length of the muscle. Suppose a five-millimeter-long ant weighing in at 0.003 grams can lift the weight of 25 ants over its head. Given the scaling above, how much could the same ant, scaled up to 1.6 meters in length, lift above its head? Now suppose we take a human being 1.6 meters in height and 50 kilograms in weight who can lift 25 kilograms over her head. After scaling her down to five millimeters in height, how much weight can she now lift proportional to her body mass? Using this, estimate the ratio of κ for humans to that of ants. Are ants really that strong?

Section 3.2.2

8. In the conduction boundary condition (3.21), we assumed that the fluid had infinite thermal diffusivity. Let the thermal diffusivity of the fluid be κ_f and suppose that the fluid itself is in contact with a thermal reservoir at temperature T_R and located at radius $R \gg a$. Compute the time-independent, spherically symmetric solution to the heat equation with heat source,

$$Q(r) = \begin{cases} Q \,, r \leq a \\ 0 \,\,, r > a \end{cases},$$

where Q is a constant. The boundary conditions on the temperature are

$$\frac{\partial T}{\partial r}(0) = 0, \quad T(R) = T_R$$

and note that the temperature and its first derivative must match at the surface of the sphere. From your solution define a "thermal boundary layer" and examine its scaling for two cases: constant Q and constant ΔT. Now suppose at time $t = 0$ both the sphere and the fluid are in equilibrium with the reservoir. At this time the engine is turned on. Compute the approximate time required for the boundary layer to form and examine its scaling.

9. In the text it was stated that free convection is not important in microsystems because the bouyant force is gravitational in origin. Construct a very simple model in which bouyant forces are centrifugal in origin and examine the scaling in this case. Could this be an effective method for cooling microsystems?

Section 3.2.3

10. Consider the same beam but with a point force F acting on the free end. The deflection of the tip of the beam is

$$u = \frac{4F}{E\alpha^4 L}.$$

If we think of the tip of the beam as a spring being acted on by the force F, what is the effective spring constant? Why does this not contradict the conclusion of this section that small structures are stiff?

11. The equation that governs the static deflection of a one-dimensional string under tension T is

$$T\frac{d^2u}{dx^2} = F,$$

where F is the distributed load (force per unit length) placed on the string. Investigate the scaling of the deflection of the string under various assumptions on the scaling of F and T. (Hint: make sure the string doesn't break or deform plastically.)

Section 3.2.4

12. Levitation can also be achieved by directing an upward stream of air against an object. Examine the scaling of this mechanism using a simple model. Airplanes essentially use this effect by redirecting a stream of air downward. You are charged with building a highly efficient glider. In terms of optimizing its glide ratio, what is your main objective? (The glide ratio is how far the glider flies forward for every meter it descends in still air.) How does this mechanism fair in comparison to balloons and zeppelins when it comes to scaling?

13. Suppose we try to detach a bubble stuck to a wall by placing an electrostatic charge on it and applying an external electric field. Examine the scaling of this mechanism. What assumptions do you have to make about the scaling of the field and the charge density on the bubble?

14. Now suppose we try to detach the bubble by shaking the device to which it is attached. Using a simple model, compute the scaling of an appropriate quantity. What are the constraints of this method?

Section 3.3

15. Let's attempt to compute the scaling of a purely electrostatic system. We can eliminate all material, thermal, and elastic effects by considering two idealized point charges of magnitude q and mass m. The force on one charge due to the presence of the other is

$$F = \frac{1}{4\pi\epsilon_0}\frac{q^2}{a^2},$$

where a is the distance between the particles. In order to get a scaling law for F we need one for q. Propose one and justify it. Suppose that q is constant. What is the difference between scaling the system and simply moving the particles closer together? Can we really say that F scales as a^{-2} in this case?

16. Suppose we have two square pads of Velcro, one each of hooks and rings. Let the pads have side length L and suppose that each hook and ring pair occupy an area l^2 on each pad. If the force required to separate a single hook and ring is given by

$$F_0 = \kappa l^n,$$

then show that the force required to separate the pads is

$$F = \epsilon^{n-2}\kappa L^n,$$

where $l = \epsilon L$. Using a simple model of the hook and ring part of the connector, make an argument that the exponent n above is equal to two. Based on this, what is your prediction of the effectiveness of micro-Velcro?

Section 3.3.1

17. We can consider two additional forces for this system. The charged rods will experience a pressure, which, if too large, will cause them to explode. Calculate an appropriate quantity in each case (for surface and volume charge densities) and determine if either of them change the scaling in any way.

Section 3.3.2

18. As in exercise in magnetostatics, compute the constant χ in (3.76).

3.6 Related Reading

The most widely referenced work on scaling laws is Trimmer [199].

Chapters on scaling that build on Trimmer's work can be found in Kovacs [108], Madou [128], and Elwenspoek and Wiegerink [46].

The concept of scaling of forces within a given system is, of course, not a recent idea. Thompson [192] and Galileo [63] are two of the many predecessors of Trimmer.

3.7 Notes

1. Both measured at 20 degrees centigrade. The unit here is the more common cgs unit the *centi-Poise* (cP) rather than the SI unit, the *Poiseuille* (PI). The conversion is 1 cP = 0.01 Poise = 1 dyne second per cm^2 = 0.1 PI. The viscosity of normal liquids is usually very temperature dependent; for instance, that of water decreases by a factor of five from 0°C to 100°C.

2. The range of intensity is measured at an average frequency of 1000 Hz, where the human ear is most sensitive.

3. The range in *intensity*, however, is very large, an amazing 16 orders of magnitude. Our thermal sensitivity to infrared wavelengths covers only the wavelengths from about 1 to about 10 microns and has a much smaller intensity range.

4. This can be roughly illustrated by the following two examples. The electric field created by a typical static charge excess on your body is sufficient to cause a spark to form across a small gap (e.g., the one between your fingertip and the doorknob). The field is on the order of 10^5 V/m, which is not even noticeable before the spark occurs. Contrast this with the ability of some sharks to detect fields as small as 10^{-6} V/m. On the magnetostatic side, consider the fact that a human being can be placed in an MRI scanner, which typically uses a field on the order of 1 Tesla, and not feel a thing. On the other hand, there are many animals capable of sensing and responding to the Earth's magnetic field, which is on the order of 10^{-4} Tesla in magnitude at the planet surface.

5. Cells, of course, come in a wide range of sizes, from about 0.3μ for the smallest bacterial cells to 90μ for amoebas, to several feet in length for human nerve cells (or a kilogram in weight for an ostrich egg cell). When we refer to cells, we are talking about an average or standard cell, which is roughly round and about 30 microns in diameter.

6. For example, the human eye has a resolving power of about one minute of arc or about 100 microns at 30 centimeters. The tactile acuity of an average fingertip is about 0.1 millimeters and the smallest wavelength of sound we can hear is approximately one centimeter.

7. This is the viscosity mentioned in the introduction. It is an intrinsic, i.e., material, property of a fluid. The *kinematic viscosity* is the dynamic viscosity divided by the fluid density and is not an intrinsic property. The SI unit of kinematic viscosity is the *Stokes*.

8. A Reynolds number less than about 2000 indicates a laminar flow while one greater than about 4000 indicates turbulence. In between is a "transition" region, the large width of which is due to the approximate nature of the Reynolds number itself.

9. As in subsequent chapters, we use notation whereby the del operator, ∇, in this equation remains unprimed even though it is with respect to primed (i.e., dimensional) variables. The context of each particular equation will make it clear which ∇ is intended.

10. Note that this conclusion only applies to forced convection by a liquid. Gases, which do not obey the no-slip condition, have different boundary layer structures, which may make convective cooling relevant. Gases typically have Prandtl numbers of about 1.

11. This result goes back to Galileo, see [63].

12. What we are doing here is assuming the same sort of effect for magnetostatics as happens in electrostatics. For charged conducting bodies, the surface charge density actually occupies a volume, which extends a few angstroms into the body, independent of the size of the body. This is sometimes the case in magnetostatics as well, for instance high-frequency alternating currents are essentially restricted to the surfaces of conducting wires.

13. In much of the MEMS literature one can find statements to the effect that surface tension scales linearly with length. Taken literally, this is incorrect. What is intended is that certain interface effects, such as the force on an interface due to the pressure drop across it, scale as length.

14. The student of differential geometry will recognize the mean curvature of the interface.

15. In what follows, we define a bubble as a gas-liquid interface in which the gas is contained inside the interface, while a droplet is the version containing a liquid. Things like soap bubbles fall into a different category because there are two interfaces involved.

16. This term is commonly used to refer to what we have been calling effects, e.g., the thermal domain, the mechanical or structural domain, the electromagnetic domain, etc.

17. For the sake of simplification, we have assumed the breakdown field is independent of d, which is untrue for some materials. For a gas exposed to an electrical gap of size d with voltage drop V, the breakdown voltage is given by the *Paschen curve.* For most gases at atmospheric pressure, the curve has a minimum when d is on the order of a few tens of microns and increases steeply for smaller d.

18. For ordinary metals ρ has a temperature dependence of the form

$$\rho(T) = \rho_0(1 + k(T - T_0)), \tag{3.79}$$

where T is in degrees centigrade and ρ_0 is normally measured at 20°C. For instance, for copper $\rho_0 = 1.67 \times 10^{-8}$ ohm m and $k = 0.0039°\text{C}^{-1}$.

Chapter 4

Thermally Driven Systems

Cold morning on Aztec Peak Fire Lookout. First, build fire in old stove. Second, start coffee. Then, heat up last night's pork chops and spinach for breakfast. Why not? And why the hell not?

Edward Abbey

4.1 Introduction

There is an abundance of devices that utilize thermal effects. From measuring mass-flow rates with a thermal anemometer [115, 144, 145, 161, 164, 185, 197], to writing data onto a polycarbonate disk using heated cantilevers [33, 104], to providing a motive force for six-legged, insect-like microrobots [105, 106], researchers are continually exploring new ways of using thermal and thermally driven effects.

In this chapter, we explore the mathematical modeling of thermal effects in MEMS and NEMS. We begin in Section 4.2 with an overview of eight devices that rely upon thermal effects for their operation. We briefly describe the operation of each device, how thermal effects are used, and how mathematical modeling can contribute to the design process. Next, in Section 4.3, we ease our way into modeling by considering the problem of cooling of an arbitrarily shaped body. The fact that the body is arbitrarily shaped precludes an exact solution to the equation of heat conduction. We introduce the notion of a perturbation expansion and use the perturbation method to reduce the PDE based model to a "lumped" or reduced-order, ODE-based model. Such reduced-order models are pervasive in the engineering literature. We see that perturbation methods provide a powerful tool for deriving lumped models and for uncovering their regions of validity. Then, in Section 4.4, we examine Joule heating. This leads to a coupled domain problem where the thermal problem is coupled to the solution of a problem in electrostatics. In general, due to the temperature dependence of the electrical conductivity, this is a nonlinear problem. We utilize the perturbation techniques introduced in Section 4.3 to simplify the model. The result is a nonlinear lumped model that is then studied analytically and numerically. During the study of this lumped model, we introduce the idea of a *bifurcation*. Roughly, a bifurcation is the change in the behavior of a system as a parameter is varied. This idea will reoccur in numerous places throughout the text. We also introduce the idea of *stability* of solutions to a model. Basic notions and methods of linear stability analysis are introduced. The lumped model is also used as a toy system to discuss iterative coupled-domain numerical solvers often used by MEMS/NEMS researchers. The contraction mapping theorem, a key tool in the analysis of such schemes, is presented. Finally, in Section 4.5, we put our toy model to work in predicting write speeds for a thermomechanical data storage system.

4.2 Thermally Driven Devices

In this section we take a brief look at eight devices that use thermal effects for their operation. Our selections are representative rather than comprehensive. Researchers are continuously using thermal effects in novel ways in MEMS and NEMS devices. Many examples appear in the bibliography; the reader is directed to the literature for other examples of thermally driven systems.

4.2.1 The Thermoelastic V-Beam Actuator

Perhaps the simplest use of thermal effects in MEMS is in actuation based directly on thermal expansion. A typical system is sketched in Figure 4.1.

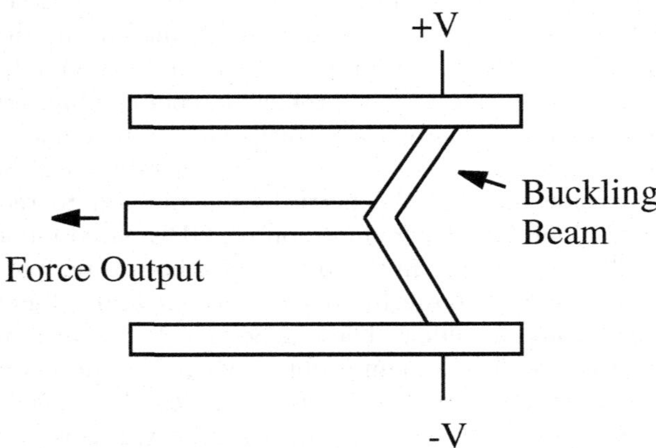

FIGURE 4.1: Sketch of a simple thermoelastic actuator.

Here an elastic beam is held fixed between two rigid supports. A current is passed through the beam resulting in a temperature rise. The temperature increase causes thermal expansion in the beam. Since the beam cannot expand in the direction along its axis, it buckles. If the beam is notched on one side or fabricated in a "V" shape, it will preferentially buckle in one direction. A rod or other mechanical structure attached to the beam allows the buckling force to be redirected and used to perform work.

We make two observations concerning this system. First, the use of an electric current to provide a heat source is typical in thermally driven MEMS and NEMS. In fact, all eight of the devices discussed in this chapter utilize Joule heating. With this in mind, we make developing a model of Joule heating a goal for this chapter. Ultimately, we would like the model to relate the temperature rise in the system to geometric and material properties of the design. Second, we note that the thermal actuator shown in Figure 4.1 is little used in practice. The ratio of output force to system size is unfavorable for this design. We include this device because of its simplicity. The thermal bimorph discussed in the following subsection is more commonly used in practice.

4.2.2 The Thermal Bimorph Actuator

To overcome the low force output of the thermal actuator discussed in the previous subsection, a thermal bimorph[1] design is often used. A typical system is sketched in Figure 4.2. In this system the hot and cool arms are anchored at one end and free to move elsewhere. A current is passed through the entire system and travels from one anchor to the other. Since the hot arm is very thin compared to the cool arm, its resistance to current flow is much greater. This means that Joule heating will cause a large temperature increase in the hot arm while the temperature of the cool arm will remain relatively unchanged. In turn large thermal stresses will develop in the hot arm as it tries to expand due to the temperature increase, and the differential stress between the arms will cause the entire system to deflect as indicated in Figure 4.2. This type of actuator has been fabricated and studied by several research groups. The reader is directed to [23, 38, 39, 88, 89, 121, 129] and other such references at the end of this text for further reading on thermal bimorph actuators. A particularly novel use of thermal actuators in the design of MEMS "insects" is discussed in [105, 106]. Another novel use of thermal actuators in the design of MEMS "cilia" for a microsatellite docking system appears in [191].[2]

Notice that once again an electric current is used to provide a heat source for this system. Here, in addition to understanding Joule heating, we need to understand how the geometry of the arms affects the temperature increase. A successful model will allow the researcher to optimize the ratio of arm widths and other geometric properties of the design.

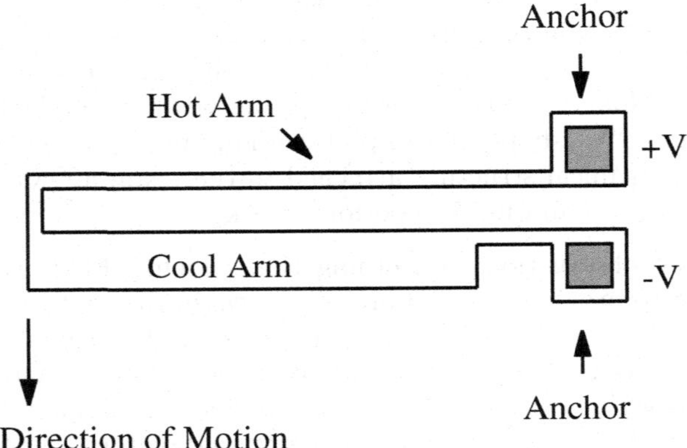

FIGURE 4.2: Sketch of a thermoelastic bimorph actuator.

4.2.3 The Thermopneumatic Valve

A third thermally actuated system is shown in Figure 4.3. In contrast to the previous two systems, the thermopneumatic valve relies upon the change in volume of a heated fluid to provide a force. In this design, a resistive heating element is embedded in a rigid substrate. A fluid resides in a cavity above the heating element. The top of the cavity is covered by an elastic membrane.

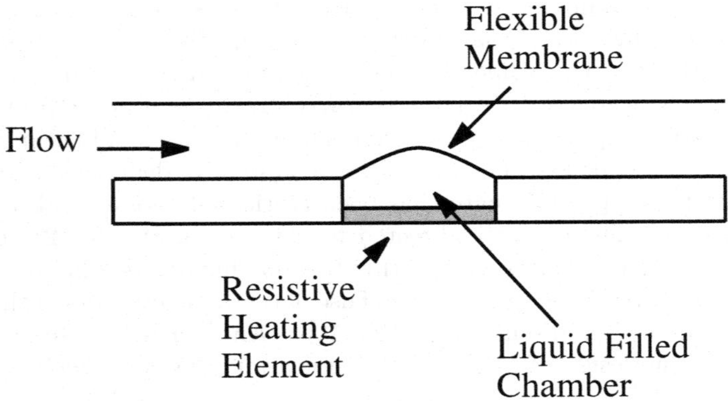

FIGURE 4.3: Sketch of a typical thermopneumatic valve.

As a current passes through the heating element Joule heating occurs and the fluid's temperature increases. This increases the pressure in the fluid and the elastic membrane is pushed upward. The membrane then acts as a valve regulating the flow in a microchannel.

Once again, notice that Joule heating is central to the design of this device. Here, a model relating the temperature increase in the heating element to the temperature rise in the fluid cavity and ultimately the deflection of the elastic membrane would be desirable.

4.2.4 The Thermal Anemometer

In addition to providing a motive force, thermal effects are useful in sensor applications. A cartoon of one such device, the thermal anemometer or mass flow sensor, is shown in Figure 4.4. In a thermal anemometer, a heat source

FIGURE 4.4: Cartoon illustrating the idea behind the thermal anemometer.

is placed in a flow between a pair of temperature sensors. By measuring the difference in temperature upwind and downwind of the heat source, the speed of the flow can be inferred. As with the actuators discussed above, in MEMS, Joule heating provides the heat source. Here, modeling provides the relationship between the mass flow rate and the measured difference in temperature. Several research groups have attacked the fabrication and study of MEMS anemometers of various design. References [115, 144, 145, 161, 164, 185, 197] provide a good introduction to the literature concerning this device.

4.2.5 The Metal Oxide Gas Sensor

A second sensor utilizing thermal effects is the metal oxide gas sensor. The basis of such sensors is the fact that the electrical resistance of metal oxides (SnO_2, TiO_2, etc.) is a function of gases adsorbed at their surface. Different metals are useful for detecting different types of gases. Typically the sensor surface is maintained at a constant high temperature on the order of 100°C. This helps reduce the deleterious effects of humidity on the sensor surface. This constant high temperature is created and maintained through Joule heating of a polysilicon layer, placing this type of sensor in the class of thermally driven devices. The reader is directed to [128, 130, 175] for more information on metal oxide gas sensors.

4.2.6 The Thermal Data Storage Device

A recent exciting application of thermal effects in MEMS/NEMS makes use of the resistive heating of an atomic force microscope (AFM) tip to create a thermal data storage system. In this device a resistively heated AFM tip is

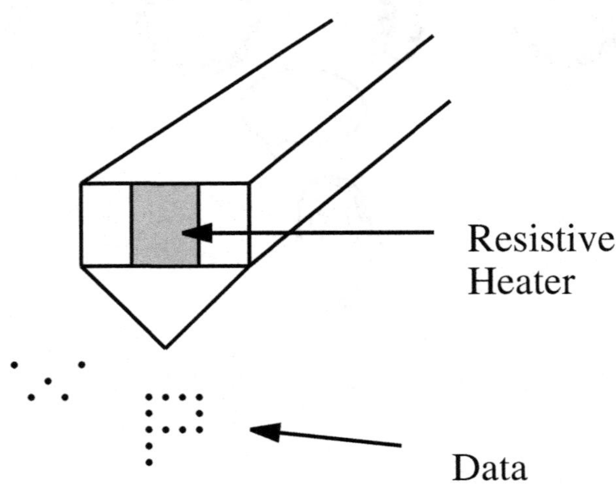

FIGURE 4.5: Sketch of a typical thermal data storage device.

used to make indentations in a polycarbonate disk. The presence of an indentation corresponds to a 1, while the absence of an indentation corresponds to a 0, making it possible to record binary data. The data can be read using the same system as the presence of an indentation changes the local heat transfer characteristics of the system. That is, if the temperature of the tip is measured, the presence or absence of an indentation can be inferred. Data density near 100 Gb/in^2 has been demonstrated using this system. A typical CD-ROM records data with a density of about 30 Gb/in^2. This thermal data storage system has been pushed to even higher densities by utilizing arrays of read-write heads. More details on this device are available in [33, 104].

Again we note that Joule heating is central to the operation of this device. Additionally, the optimization of the design of the AFM tip presents an interesting challenge in the coupled electrostatic/heat transfer domain. In Section 4.5 we use our first model of Joule heating to address the question of write speeds in thermal data storage devices.

4.2.7 The Shape Memory Actuator

The shape memory effect has been employed in the design of several MEMS actuators. In this effect, a material undergoes a phase change in response to a temperature change. The basic cycle is sketched in Figure 4.6. In the

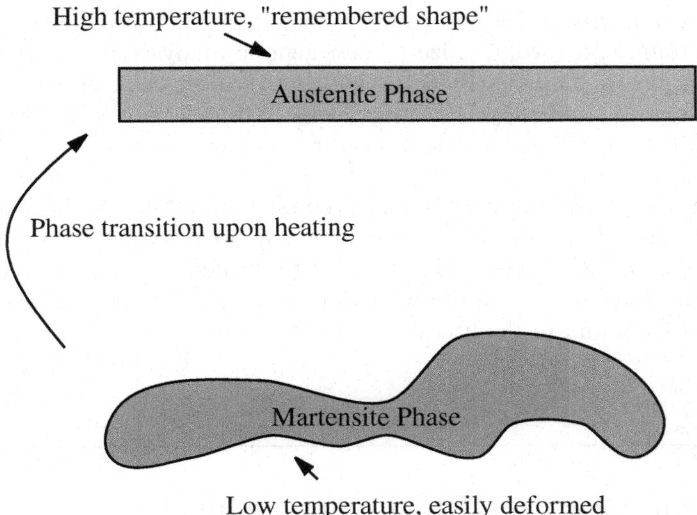

FIGURE 4.6: The basic cycle of the shape memory effect.

austenite, or "remembered" phase, which occurs at high temperature, the material is stiff and not easily deformed. On the other hand, in the martensite phase, which occurs at low temperature, the material deforms plastically. Upon heating a deformed material in the martensite phase, it transforms to the austenite phase and assumes its previous high-temperature shape. This process creates large forces that may be utilized in an actuator.

Typically, in an actuator that uses a shape memory alloy, the temperature change is achieved by applying a current. That is, resistive heating is once again utilized.

4.2.8 The Microscale PCR Chamber

The wildly popular movie, *Jurassic Park* (Universal Studios, 1993), features a group of renegade scientists cloning dinosaurs on a remote island somewhere off the coast of South America. In the movie, dinosaur DNA is obtained from prehistoric dinosaur-loving mosquitoes trapped in amber and preserved with minute quantities of dinosaur DNA in their system. This movie arguably marks the first time that the so-called polymerase chain reaction (PCR) entered the public consciousness. The PCR method allows molecular biologists, and their fictional counterparts, to produce any quantity of DNA with a desired sequence starting from only a few molecules. From a very small sample such as might be contained in a mosquito's belly, enough DNA can be produced to unravel the DNA sequence and in theory create a clone of a related organism. In the real world the polymerase chain reaction has become an indispensable part of the molecular biologist's toolbox. Coupled with the miniaturization advantages of MEMS technology, PCR based "lab-on-a-chip" systems promise to revolutionize on-site genetic analysis and testing. The underlying idea behind the PCR method is to utilize *thermal cycling* to repeatedly separate and copy DNA sequences. A typical protocol calls for the sample to be heated from room temperature to approximately 95°C, cooled to 65°C, reheated to 72°C, and the process repeated. Microscale approaches to this protocol vary from direct thermal cycling of a fluid in a fixed chamber to flow of the sample through a sequence of chambers held at different temperatures. As with other devices discussed in this chapter, Joule heating is the source of heat for microscale PCR systems. Details of various lab-on-a-chip designs can be found in the literature [40, 107, 132].

4.3 From PDE to ODE: Lumped Models

The engineering literature is littered with "lumped" models. But where do these models come from? How are they derived? When are they valid?

To begin to answer these questions, we examine the cooling of an arbitrarily shaped body. On the surface this problem is simple. In fact, cooling of bodies is a subject often tackled in introductory calculus courses. However, when the mathematical model of the cooling of a body is formulated in terms of the continuum theory developed in Chapter 2, the resulting equations are generally intractable. How do we reconcile this difficulty with the apparent simplicity of the problem? We shall show that the simple "calculus" model of cooling is in fact a reduced model obtained from the continuum theory via a perturbation expansion. Along the way we shall develop a method for reducing intractable PDEs to tractable, reduced-order models and uncover the regimes of validity of such "lumped" systems.

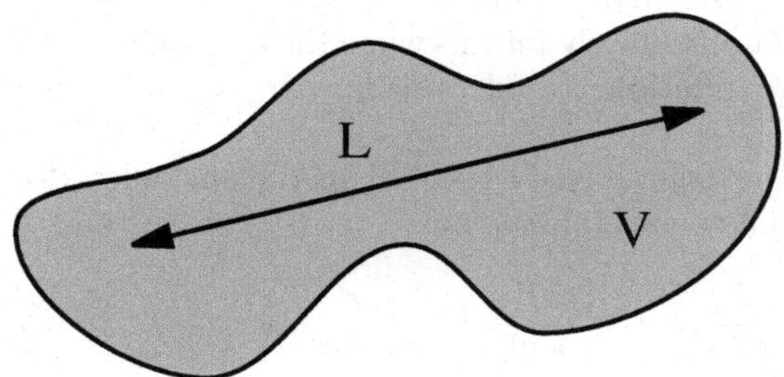

FIGURE 4.7: Cooling of an arbitrary body.

To begin, suppose the body shown in Figure 4.7 is initially at temperature T_0 and exchanges heat via convection with the outside world. Assume the outside world is maintained at temperature T_A. From our study of heat conduction in Chapter 2, we can immediately write down a model governing this system:

$$\rho c_p \frac{\partial T}{\partial t'} = k\nabla^2 T \quad \text{in } V \tag{4.1}$$

$$k\nabla T \cdot \mathbf{n} = -h(T - T_A) \quad \text{on } \partial V \tag{4.2}$$

$$T(x', y', z', 0) = T_0. \tag{4.3}$$

Here we are modeling the exchange with the outside world by a Newton cooling-type boundary condition.

Also, we've put primes on all the independent variables. We'll remove them shortly by introducing dimensionless variables. Before we proceed, let's take a hard look at our model, equations (4.1)–(4.3). For the body shown in Figure 4.7, do we have any hope of constructing the exact solution to equations (4.1)–(4.3)? No! The geometry, unless it happens to be very simple and symmetric, precludes an exact solution. So, what do we do? One answer is to turn to numerical simulation. A second option is to argue as follows: If the body in Figure 4.7 is a good thermal conductor and the convective transfer to the outside world relatively slow, we can approximate the temperature in the body as being constant in space. Then, we can derive a simple model from an energy balance:

$$\Delta\text{Body's Energy} = mc_p\frac{dT}{dt'} = -hS_A(T-T_A) = \text{Energy lost at surface.} \quad (4.4)$$

Here m is the body's total mass and S_A is the body's total surface area. Equation (4.4) is certainly easier to handle than equations (4.1)–(4.3). But, how good is our approximation? And, most importantly, when is it valid? To answer these questions we return to our PDE-based model, equations (4.1)–(4.3), and introduce a method for deriving the simplified lumped model, equation (4.4), from the more complicated system. As we shall see, this *perturbation method* immediately tells us when the approximate model is valid, how good the approximation is and even how to derive corrections to the simplified model.

To begin, we recast equations (4.1)–(4.3) in dimensionless form by introducing the dimensionless variables,

$$u = \frac{T-T_A}{T_0-T_A}, \quad t = \frac{ht'}{\rho c_p L}, \quad x = \frac{x'}{L}, \quad y = \frac{y'}{L}, \quad z = \frac{z'}{L} \quad (4.5)$$

into equations (4.1)-(4.3). This yields

$$B\frac{\partial u}{\partial t} = \nabla^2 u \quad \text{in } V, \quad (4.6)$$

$$\nabla u \cdot \mathbf{n} = -Bu \quad \text{on } \partial V, \quad (4.7)$$

$$u(x,y,z,0) = 1 \quad (4.8)$$

where $B = hL/k$ is a dimensionless number known as the *Biot number*[3]. Now, let us consider our choice of dimensionless variables, equation (4.5). The scaling of the spatial variables, x', y', and z', is natural; we have scaled with a characteristic length of the body. Also, the scaling of the temperature T is natural, we have used the initial and ambient temperatures to define u. Note that this scaling implies that u lives on the interval $[0,1]$. The choice of timescale is a bit harder. Notice that we have defined a dimensionless time by using h, the heat transfer coefficient. This means that we are looking at the problem on the convective time scale.[4] Since we are interested in how the body goes from its initially hot temperature to the ambient temperature, this is a reasonable choice. This choice of scale has the effect of putting the

Biot number in two places. First, it appears in front of the time derivative term in the heat equation. Second, it appears in the boundary condition. The Biot number is a dimensionless ratio expressing the relative importance of convection and conduction. So, the Biot number measures how well heat is conducted through the body in comparison with how well heat leaves the body via convection at the surface. In our hand-waving arguments above we were really claiming that the Biot number is small! Of course, this won't always be the case; the approximate model we derived above isn't always valid. But, if we work in the limit where B is small, we should obtain the approximate model from the full PDE model. Also notice, without doing any work other than scaling, we've learned something about the validity of the approximate model.

To proceed, we assume that the solution to equations (4.1)–(4.3) can be expanded in an infinite series[5] in powers of the Biot number. That is, we write

$$u(x,y,z,t) \sim u_0(x,y,z,t) + Bu_1(x,y,z,t) + B^2u_2(x,y,z,y) + \cdots; \qquad (4.9)$$

substitute this expansion into equations (4.1)-(4.3), collect together terms multiplying like powers of B, and equate these coefficients of powers of B to zero. This yields an infinite set of equations that sequentially determine the u_n. The first two are

$$\nabla^2 u_0 = 0 \qquad (4.10)$$

$$\nabla u_0 \cdot \mathbf{n} = 0 \qquad (4.11)$$

$$u_0(x,y,z,0) = 1 \qquad (4.12)$$

and

$$\nabla^2 u_1 = \frac{\partial u_0}{\partial t} \qquad (4.13)$$

$$\nabla u_1 \cdot \mathbf{n} = -u_0 \qquad (4.14)$$

$$u_1(x,y,z,0) = 0. \qquad (4.15)$$

Now, how is this progress? We've turned one partial differential equation into an infinite sequence of partial differential equations! Luckily, each equation in our sequence is easier to deal with than the original problem. Consider equations (4.10)–(4.12). These are just the equations governing steady-state heat flow in a perfectly insulated body with initial temperature equal to one. The solution must be constant in space and is simply a function of time, that is $u_0 = u_0(t)$. This immediately satisfies equations (4.10)–(4.11). Equation (4.12) is satisfied by setting $u_0(0) = 1$. Notice, this conclusion is the same as our hand-waving argument that the temperature across the body should be nearly independent of position within the body. Also, notice that we now know how big the correction to this approximation is; it's the same size as the second term in our expansion, which is measured by the Biot number.

At this point we do not know what equation $u_0(t)$ satisfies. In order to determine this equation, we turn our attention to the $O(B)$ system[6], equations (4.13)–(4.15). Integrating equation (4.13) over the entire volume V yields

$$\int_V \nabla^2 u_1 \, dV = \int_V \frac{du_0}{dt} \, dV. \tag{4.16}$$

Keep in mind that V is a dimensionless volume. In fact, $V = V_A/L^3$, where V_A is the actual volume of the body. The integral on the right is easily evaluated since u_0 is independent of space, while the integral on the left may be rewritten as a surface integral by using the divergence theorem. That is,

$$\int_S \nabla u_1 \cdot \mathbf{n} \, dA = V \frac{du_0}{dt}. \tag{4.17}$$

But, equation (4.14) relates the normal derivative of u_1 to the function u_0. Hence, the integral on the left may be evaluated to yield

$$-Su_0 = V \frac{du_0}{dt}, \tag{4.18}$$

where S is the dimensionless surface area of the body. Again, S may be related to the actual surface area through $S = S_A/L^2$. Rewriting equation (4.18) in terms of dimensional variables yields

$$mc_p \frac{dT}{dt'} = -hS_A(T - T_A), \tag{4.19}$$

which is the same as equation (4.4). That is, we have derived the simple energy balance model from the full PDE-based model using perturbation methods. We reiterate our reasons for deriving equation (4.4) using this perturbation approach. We have learned precisely when the reduced model is valid; the Biot number must be small. We have learned the size of corrections to the reduced model; corrections are the size of the Biot number. We have learned how to compute correction terms to the reduced model; compute the next term in the perturbation expansion. Most importantly, we have developed a method allowing us to simplify complicated systems when the reduced order or lumped model is not as obvious as it was here.

4.4 Joule Heating of a Cylinder

Now that we have seen how perturbation methods can reduce a complicated problem to a simple one and uncover parameter ranges of validity, we begin an investigation of Joule heating[7]. In this section we will model and analyze the Joule heating of a cylinder. This simple geometry is chosen because it lends itself to analysis. However, our model will be sufficiently sophisticated to capture many of the real features and phenomena of the devices discussed at the start of the chapter. The coupling of the thermal and electrostatic fields will be included as will the temperature dependence of certain material properties. We note that while here we have in mind applications to MEMS and NEMS, the analysis that follows has appeared many times in many contexts in the applied mathematics and engineering literature. After all, Joule heating is a fundamental problem in devices as common as the light bulb! Nonlocal problems arising in Joule heating have been studied by Lacey [118, 119]. The analysis of microwave heating is similar to Joule heating and has been examined exhaustively by Kriegsmann and his coworkers [109, 110, 111, 112, 113]. Also, the analysis of combustion and chemical reactor theory closely resembles that of Joule heating [8, 34, 35, 36, 37]. Much of the material in this section follows the references above.

4.4.1 The Electromagnetic Problem

We begin by formulating the electromagnetic problem for the cylinder shown in Figure 4.8. Recall from Chapter 2 that the **E**, **D**, **B**, and **H** fields

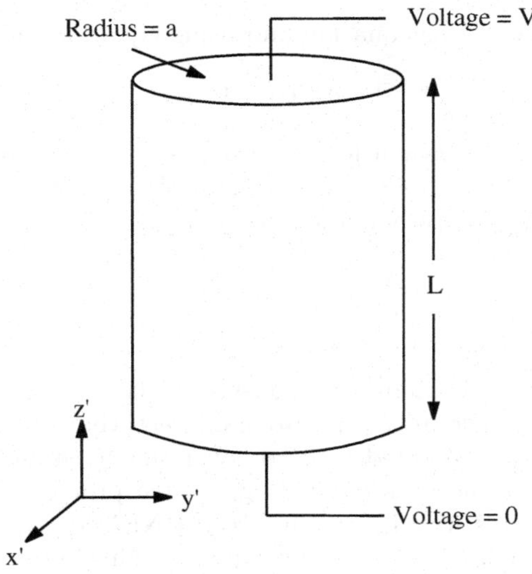

FIGURE 4.8: The cylinder for Joule heating study.

satisfy the four Maxwell equations:

$$\nabla \cdot \mathbf{D} = \frac{\rho}{\epsilon_0} \tag{4.20}$$

$$\nabla \times \mathbf{H} - \frac{1}{c}\frac{\partial \mathbf{D}}{\partial t'} = \mathbf{J} \tag{4.21}$$

$$\nabla \times \mathbf{E} + \frac{1}{c}\frac{\partial \mathbf{B}}{\partial t'} = 0 \tag{4.22}$$

$$\nabla \cdot \mathbf{B} = 0. \tag{4.23}$$

In a conductor with conductivity σ, we may relate the current and the $\mathbf{E}$ field using *Ohm's law*, i.e.,

$$\mathbf{J} = \sigma \mathbf{E}. \tag{4.24}$$

Now, assuming electrostatics is valid and dropping the time derivatives, equation (4.22) becomes

$$\nabla \times \mathbf{E} = 0, \tag{4.25}$$

which implies that $\mathbf{E}$ may be written as the gradient of a potential. That is,

$$\mathbf{E} = -\nabla \psi. \tag{4.26}$$

Next, dropping time derivatives and using Ohm's law, equation (4.21) becomes

$$\nabla \times \mathbf{H} = \sigma \mathbf{E}. \tag{4.27}$$

Taking the divergence of this equation and using equation (4.26) yields

$$\nabla \cdot (\sigma \nabla \psi) = 0, \tag{4.28}$$

which is a single scalar equation for the electrostatic potential. A potential difference is applied across the cylinder in Figure 4.8. This implies that the potential, ψ, must satisfy the boundary conditions,

$$\psi(x', y', 0) = 0 \tag{4.29}$$

$$\psi(x', y', L) = V. \tag{4.30}$$

Here we have chosen to examine the constant voltage source case of Joule heating. The constant current case is also of interest; the reader is encouraged to modify the analysis presented here to cover the constant current source system. We note that the constant voltage source problem is typically the appropriate model for resistive heating in MEMS/NEMS systems. We assume no current flows through the sides of the cylinder. This implies that $\mathbf{J} \cdot \mathbf{n} = 0$ on the sides where $\mathbf{n}$ is a normal vector. Relating $\mathbf{J}$ to $\mathbf{E}$ through Ohm's law and expressing $\mathbf{E}$ as the gradient of the potential, we obtain

$$\frac{\partial \psi}{\partial r'} = 0 \Big|_{r'=a}. \tag{4.31}$$

Equations (4.28)–(4.31) comprise the complete electrostatics problem that must be solved to understand the Joule heating of the cylinder shown in Figure 4.8. Note that, in general, this problem is coupled to the thermal problem through the electric conductivity, σ. For most materials σ is a function of temperature, i.e., $\sigma = \sigma(T)$. As we shall see in Section 4.5, this dependence leads to a nonlinear response of heating time on applied voltage for thermal data storage systems.

4.4.2 The Thermal Problem

Now we formulate the thermal problem for the Joule heating of the cylinder in Figure 4.8. Our heat equation governing the evolution of the temperature field within the cylinder must include a source term that captures the electrical energy converted to thermal energy. To compute this source term we note that in an arbitrary volume, the work done by the electric field in moving electrons through a conductor is given by

$$-\int_V \mathbf{J} \cdot \mathbf{E}. \tag{4.32}$$

Relating $\mathbf{J}$ and $\mathbf{E}$ to the potential ψ as in the previous subsection allows us to rewrite as

$$\int_V \sigma \left|\psi\right|^2. \tag{4.33}$$

Since the volume is arbitrary, this also yields a local law. That is, at a point the energy of the electric field that is transformed to heat is given by

$$\sigma \left|\psi\right|^2. \tag{4.34}$$

This expression serves as the source term in the heat equation. So, the temperature field, T, in the cylinder satisfies

$$\rho c_p \frac{\partial T}{\partial t'} = k\nabla^2 T + \sigma \left|\psi\right|^2. \tag{4.35}$$

We assume that initially the cylinder is at the ambient temperature of the environment and impose

$$T(x', y', z', 0) = T_A. \tag{4.36}$$

For simplicity we assume the ends of the cylinder are insulated[8] and impose

$$k\frac{\partial T}{\partial z'} = 0\Big|_{z'=0,L}, \tag{4.37}$$

while on the sides of the cylinder we allow for heat loss and model the loss using the Newton cooling condition

$$k\frac{\partial T}{\partial r'} = -h(T - T_A)\Big|_{r'=a}. \tag{4.38}$$

4.4.3 Scaling

Next, we take our model, equations (4.28)–(4.31) and equations (4.35)–(4.38), and recast them in nondimensional form by introducing the scaled variables,

$$x = \frac{x'}{a}, \quad y = \frac{y'}{a}, \quad z = \frac{z'}{L}, \quad u = \frac{T - T_A}{T_A}, \quad \phi = \frac{\psi}{V}, \quad t = \frac{ht'}{\rho c_p a}, \tag{4.39}$$

into equations (4.28)–(4.31) and (4.35)–(4.38). Notice we have scaled the axial coordinate with the length of the cylinder, the perpendicular coordinates with cylinder radius, the temperature with the ambient temperature, the potential with the applied voltage, and have chosen to work on the time scale defined by cooling at the surface. The choice of time scale reflects our expectation that the temperature in the cylinder will equilibrate as a balance between Joule heating and cooling is attained. Since we expect cooling to occur on a slower time scale than Joule heating, we must look to the cooling time scale to see the balance achieved. It is also convenient to rewrite the electrical conductivity as

$$\sigma(T) = \sigma_0 f(u), \tag{4.40}$$

where σ_0 is the conductivity at ambient temperature, implying that $f(0) = 1$ and f is a dimensionless function of the dimensionless temperature u.

Our rescaling yields

$$B\frac{\partial u}{\partial t} = \nabla_\perp^2 u + \frac{a^2}{L^2}\frac{\partial^2 u}{\partial z^2} + Bpf(u)\left(|\nabla_\perp \phi|^2 + \frac{a^2}{L^2}\frac{\partial^2 \phi}{\partial z^2}\right) \tag{4.41}$$

$$u(x, y, z, 0) = 0 \tag{4.42}$$

$$\frac{\partial u}{\partial z}(x, y, 0, t) = 0 \tag{4.43}$$

$$\frac{\partial u}{\partial z}(x, y, 1, t) = 0 \tag{4.44}$$

$$\left.\frac{\partial u}{\partial r} + Bu\right|_{r=1} = 0 \tag{4.45}$$

$$\nabla_\perp \cdot (f(u)\nabla_\perp \phi) + \frac{a^2}{L^2}\frac{\partial}{\partial z}\left(f(u)\frac{\partial \phi}{\partial z}\right) = 0 \tag{4.46}$$

$$\phi(x, y, 0) = 0 \tag{4.47}$$

$$\phi(x, y, 1) = 1 \tag{4.48}$$

$$\left.\frac{\partial \phi}{\partial r}\right|_{r=1} = 0. \tag{4.49}$$

Here, we have defined the dimensionless groupings,

$$B = \frac{ha}{k}, \quad p = \frac{\sigma_0 V^2}{haT_A}. \tag{4.50}$$

The parameter B is the Biot number, a parameter we've already encountered in this chapter. The parameter p is new, but is easily interpreted. We see that

$$p = \frac{\sigma_0 V^2}{haT_A} = \frac{\text{energy in}}{\text{energy out}}. \tag{4.51}$$

That is, p, is a dimensionless version of the electrical power being converted to heat. Keep in mind that p is the dimensionless parameter that contains the applied voltage. To think experimentally, we should think of varying p. Also, we have introduced the notation $\nabla_\perp$. This is shorthand for

$$\nabla_\perp = \left(\frac{\partial}{\partial x}, \frac{\partial}{\partial y}\right), \tag{4.52}$$

i.e., the gradient operator in the x and y directions only.

4.4.4 An Interesting Limit

Our model of Joule heating of a cylinder, equations (4.41)–(4.49), is a system of nonlinear coupled partial differential equations. Even for our simple cylindrical geometry, we have no hope of finding an exact solution. However, we can employ the methods developed in Section 4.3 and investigate a restricted parameter regime. As in the previous section, we will assume that the Biot number, B, is small. Next, we'll assume that our cylinder is long and thin and introduce

$$\frac{a^2}{L^2} = \delta B, \tag{4.53}$$

where δ is assumed $O(1)$. That is, the aspect ratio is assumed to be the same order as the Biot number. Finally, we'll assume that power into the system is significant and take

$$p = \frac{P}{B}, \tag{4.54}$$

where $P = O(1)$. With these assumptions, we seek an asymptotic expansion of u and ϕ as in Section 4.3. We write

$$u(x, y, z, t) \sim u_0(x, y, z, t) + Bu_1(x, y, z, t) + \cdots \tag{4.55}$$

and

$$\phi \sim \phi_0(x, y, z) + B\phi_1(x, y, z) + \cdots; \tag{4.56}$$

substitute these expansions into equations (4.41)–(4.49), collect together terms multiplying like powers of B, and equate these coefficients of powers of B to zero. This yields an infinite set of equations that sequentially determines the u_n and ϕ_n. The first two are:

$$\nabla_\perp^2 u_0 + Pf(u_0)\left|\nabla_\perp \phi_0\right|^2 = 0 \tag{4.57}$$

$$u_0(x, y, z, 0) = 0 \tag{4.58}$$

$$\frac{\partial u_0}{\partial z}(x, y, 0, t) = 0 \tag{4.59}$$

$$\frac{\partial u_0}{\partial z}(x, y, 1, t) = 0 \tag{4.60}$$

$$\left.\frac{\partial u_0}{\partial r}\right|_{r=1} = 0 \tag{4.61}$$

$$\nabla_\perp \cdot (f(u_0)\nabla_\perp \phi_0) = 0 \tag{4.62}$$

$$\phi_0(x, y, 0) = 0 \tag{4.63}$$

$$\phi_0(x, y, 1) = 1 \tag{4.64}$$

$$\left.\frac{\partial \phi_0}{\partial r}\right|_{r=1} = 0 \tag{4.65}$$

and

$$\nabla_\perp^2 u_1 = -\frac{\partial u_0}{\partial t} + \delta\frac{\partial^2 u_0}{\partial z^2} + Pf(u_0)\left(\frac{\partial \phi_0}{\partial z}\right)^2 \tag{4.66}$$

$$u_1(x, y, z, 0) = 0 \tag{4.67}$$

$$\frac{\partial u_1}{\partial z}(x, y, 0, t) = 0 \tag{4.68}$$

$$\frac{\partial u_1}{\partial z}(x, y, 1, t) = 0 \tag{4.69}$$

$$\left.\frac{\partial u_1}{\partial r} + u_0\right|_{r=1} = 0 \tag{4.70}$$

$$\nabla_\perp \cdot (f(u_0)\nabla_\perp \phi_1) + \delta\frac{\partial}{\partial z}\left(f(u_0)\frac{\partial \phi_0}{\partial z}\right) = 0 \tag{4.71}$$

$$\phi_1(x, y, 0) = 0 \tag{4.72}$$

$$\phi_1(x, y, 1) = 0 \tag{4.73}$$

$$\left.\frac{\partial \phi_1}{\partial r}\right|_{r=1} = 0. \tag{4.74}$$

Once again we have turned a single set of equations into an infinite set of coupled equations. However at each stage the system is tractable.

Consider equation (4.62). If we multiply this equation by ϕ_0, we can rewrite as

$$\nabla_\perp \cdot (f(u_0)\phi_0\nabla_\perp \phi_0) - f(u_0)\left|\nabla_\perp \phi_0\right|^2 = 0. \tag{4.75}$$

Then integrating across a cross-section, C, of the cylinder we obtain

$$\int_C \nabla_\perp \cdot (f(u_0)\phi_0\nabla_\perp \phi_0) - \int_C f(u_0)\left|\nabla_\perp \phi_0\right|^2 = 0. \tag{4.76}$$

The first integral can be rewritten as a surface integral by using the divergence theorem,

$$\int_{\partial C} (f(u_0)\phi_0 \nabla_\perp \phi_0) - \int_C f(u_0) \left|\nabla_\perp \phi_0\right|^2 = 0, \tag{4.77}$$

and by the boundary condition (4.74) this integral vanishes. We are left with

$$-\int_C f(u_0) \left|\nabla_\perp \phi_0\right|^2 = 0. \tag{4.78}$$

But this immediately implies that $\phi_0 = \phi_0(z)$. That is, to leading order the potential is a function of z only. Using this result in equation (4.57), we see that $u_0 = u_0(z,t)$ as well. That is, to leading order, the temperature is independent of the perpendicular coordinates, x and y. Next, consider equation (4.66). If we integrate this across a cross-section, use the divergence theorem and the boundary condition, equation (4.70), we obtain an equation for u_0, namely,

$$\frac{\partial u_0}{\partial t} = \delta \frac{\partial^2 u_0}{\partial z^2} - 2u_0 + P f(u_0) \left(\frac{\partial \phi_0}{\partial z}\right)^2. \tag{4.79}$$

Additionally, u_0 must satisfy the leading order initial and boundary conditions,

$$u_0(x,y,z,0) = 0 \tag{4.80}$$

$$\frac{\partial u_0}{\partial z}(x,y,0,t) = 0 \tag{4.81}$$

$$\frac{\partial u_0}{\partial z}(x,y,1,t) = 0. \tag{4.82}$$

At this point, the leading order potential, ϕ_0, is still unknown. If we return to equation (4.71) and again play the game of integrating across a cross-section, using the divergence theorem and boundary conditions, we see that ϕ_0 must satisfy

$$\frac{\partial}{\partial z}\left(f(u_0)\frac{\partial \phi_0}{\partial z}\right) = 0. \tag{4.83}$$

This may be integrated with the aid of the boundary conditions to obtain

$$\phi_0(z) = \frac{\displaystyle\int_0^z \frac{dz}{f(u_0)}}{\displaystyle\int_0^1 \frac{dz}{f(u_0)}}. \tag{4.84}$$

Finally, inserting the expression for the potential, equation (4.84), into equation (4.79), we obtain a reduced model for the leading order behavior of the temperature:

$$\frac{\partial u_0}{\partial t} = \delta \frac{\partial^2 u_0}{\partial z^2} - 2u_0 + \frac{P}{f(u_0)} \left(\int_0^1 \frac{dz}{f(u_0)}\right)^{-2} \tag{4.85}$$

$$u_0(x, y, z, 0) = 0 \tag{4.86}$$

$$\frac{\partial u_0}{\partial z}(x, y, 0, t) = 0 \tag{4.87}$$

$$\frac{\partial u_0}{\partial z}(x, y, 1, t) = 0. \tag{4.88}$$

This reduced model is similar in spirit to the reduced model obtained in the previous section. Notice that we have eliminated two of the spatial variables, x and y. In the previous sections we eliminated all three. Here, variations in the z direction cannot be ignored at leading order. Also notice that our reduced model captures the effects of heat loss through the sides of the cylinder. This is embodied in the $-2u_0$ term. The potential problem has completely disappeared and the source term expressed solely in terms of the leading order temperature. Keep in mind that we know that this model is valid under the assumptions of a small Biot number and small aspect ratio. Finally, notice that while our reduced model is far simpler than our original system, it is still a nonlinear partial differential equation. We must now use other mathematical tools to uncover the behavior of solutions to this problem. One mathematical tool we might turn to is numerical simulation. While a numerical method could handle the full nonlinear coupled PDE model of Joule heating, implementing a numerical method for our reduced model is much simpler and faster. Even if one ultimately wishes to rely upon numerical simulation, the construction of a reduced-order model is still beneficial.

4.4.5 Spatially Independent Solutions

In this section we study spatially independent solutions to our reduced-order model of Joule heating of a cylinder, equations (4.85)–(4.88). That is, we look for solutions for which u_0 is independent of z, i.e., $u_0 = u_0(t)$ only. Such solutions satisfy the boundary conditions, equations (4.87)–(4.88), automatically. Further, the spatial derivative in equation (4.85) drops out and the integral can be computed. Hence equation (4.85) becomes the ordinary differential Equation,

$$\frac{du_0}{dt} = -2u_0 + Pf(u_0), \tag{4.89}$$

with initial condition $u_0(0) = 0$. To make the discussion in the remainder of this chapter concrete and to illustrate key ideas in bifurcation theory, we choose a particular form for $f(u_0)$. Borrowing from microwave heating theory, the theory of combustion, or the theory of chemical reactions, we assume an exponential form for f:

$$f(u_0) = e^{u_0}. \tag{4.90}$$

In the study of combustion or in the theory of chemical reactions, this form of f models the case when the reaction is exothermic. In an exothermic reaction heat is released. The hotter the system the more rapidly the reaction proceeds and the more heat released. For such systems the source term in the heat equation should increase in magnitude with increasing temperature, hence the assumed form for f. The reader will recognize this as a system with *positive feedback.* In the theory of microwave heating, the positive feedback occurs because the electrical conductivity of many materials is an increasing function of temperature. The hotter the system the more microwave energy absorbed and the hotter the system grows. In our study of Joule heating, we are also assuming that the cylinder's electrical conductivity is an increasing function of temperature. While this is certainly not the case for all materials used in MEMS, it is the case for many materials used to fabricate Joule heaters. For example, silicon nitride, SiN, is such a material. On the other hand, un-doped silicon has an electrical conductivity that decreases with temperature.

Many of the devices discussed earlier in the chapter are operated in a *steady-state* regime. A switch is thrown, a current applied, Joule heating occurs, and the device comes to rest at some new higher temperature. Mathematically, to capture this equilibrium configuration, we seek steady-state solutions for equation (4.89) by setting the time derivative to zero. We find that spatially independent, steady-state solutions satisfy

$$Pf(u_0) = 2u_0. \tag{4.91}$$

This equation should come as no surprise. The left-hand side is the energy input from the electric field while the right-hand side is the energy output through cooling at the surface. This equation simply says that in the steady state, the input and output of energy must be in balance. Defining

$$F(u_0) = Pf(u_0) - 2u_0 \tag{4.92}$$

allows us to characterize steady-state solutions for our system as real roots of $F(u_0)$. In Figure 4.9 we plot $F(u_0)$, assuming an exponential form for f. Plots for various values of P are shown. We see that F either has two real roots, precisely one real root, or none at all. The transition occurs as P increases. For small values of P we have two real roots. Why? Recall that P is a dimensionless measure of the power input into the system. It is proportional to the square of the applied voltage. When P is small, the energy input into the system is sufficiently small so that the linear Newton cooling on the surface may remove enough heat to establish an equilibrium. As P is increased the power absorbed increases, eventually overwhelming the linear Newton cooling and no equilibrium can be established. In microwave heating or combustion, this phenomenon is called *thermal runaway.* Whether this phenomenon occurs in a given MEMS device depends on the range of P in which the device is operated and the actual material's dependence of electrical conductivity on temperature. In this chapter, the value of this phenomenon is that it serves to illustrate the idea of a *bifurcation.* That is, the system changes behavior as a parameter is varied. Here, we transition from the existence of steady-state solutions to no steady-state solution as our parameter, P, is varied.

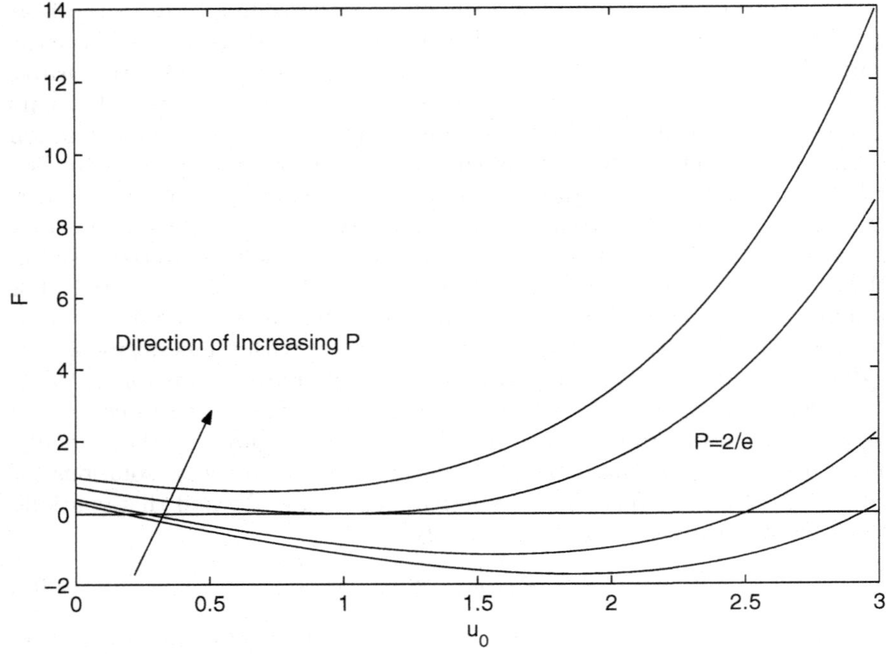

FIGURE 4.9: Plot of $F(u_0)$. Real roots define steady states.

We will encounter the bifurcation phenomenon repeatedly throughout this text. Buckling of elastic beams and the "pull-in" phenomenon in electrostatic actuation are other examples of bifurcations relevant to MEMS and NEMS.

Another characterization of this bifurcation is possible. If we solve equation (4.91) for P, we obtain

$$P = \frac{2u_0}{f(u_0)}. \tag{4.93}$$

For the moment let's think of P as being a function of u_0, defined by equation (4.93) and plot $P(u_0)$. This is done in Figure 4.10. if we now flip the axes in Figure 4.10 as is done in Figure 4.11, we obtain what is known as the *bifurcation diagram* for our model. Along the x axis, we pick the value of P that corresponds to our experiment and then read off the possible steady-state solutions along the u_0 axis. When P passes the value of P^* in Figure 4.11, no such solutions exist. We can think of P^* as being a critical input power, beyond which lies thermal runaway. Here, we can compute P^* exactly. To do so, we simply compute the location of the maxima of the curve in Figure 4.10. We find that $P^* = 2/e$.

Mathematically, the object picture in Figure 4.11 is called a *fold*. The curve of solutions as visualized in the bifurcation diagram folds back upon itself as

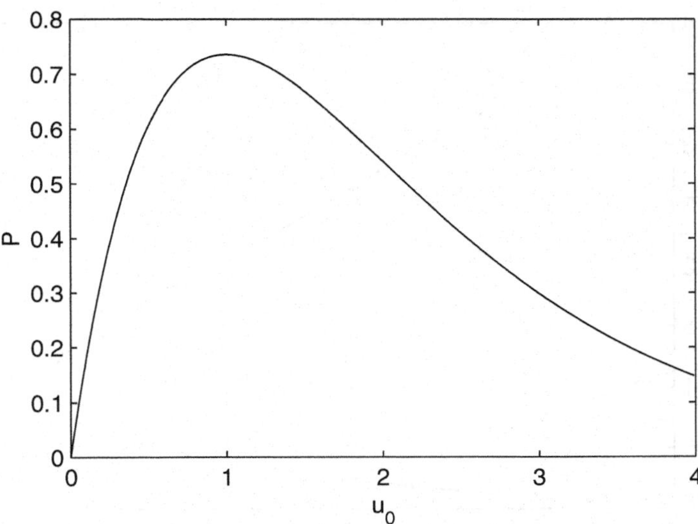

FIGURE 4.10: Plot of P as a function of u_0.

the bifurcation parameter, P, is increased. In later chapters, we will see that other instabilities for other types of devices can be characterized in terms of a fold in the bifurcation diagram. While the u_0 axis in Figure 4.11 may be replaced by some other measure of the solution, the underlying structure will still be a simple fold.

4.4.6 Stability of Steady States

When there is more than one physically relevant steady-state solution to a given model, as is the case with our system, knowing the *stability* of these steady states is crucial. Simply stated stability indicates whether or not a system will stay at an equilibrium point when perturbed from its equilibrium position. Let's work with our spatially independent ODE model and determine the stability of the steady-state solutions in Figure 4.11. It is convenient to rewrite equation (4.89) as

$$\frac{du_0}{dt} = F(u_0). \tag{4.94}$$

Recall that the right-hand side was plotted in Figure 4.9. Stability may be determined by simply examining Figure 4.9. Consider the case where P is bigger than P^*, i.e., P is so large that no roots of F exist. Then, from Figure 4.9, the right-hand side of equation (4.94) is always positive. So, starting from the initial condition $u_0(0) = 0$, the temperature will continuously increase, tending to infinity as time tends to infinity. This should come as no surprise;

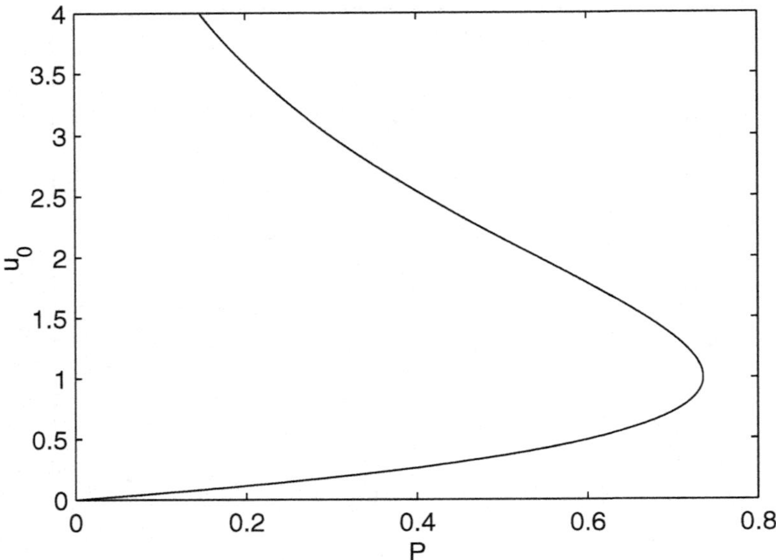

FIGURE 4.11: Bifurcation diagram for our reduced Joule heating model.

we are beyond the thermal runaway point. Now, consider the case where $P < P^*$ so that two roots for F exist in Figure 4.9. Consider the smaller of the two roots. To the left of this root, $F(u_0)$ is positive, while to the right, provided we are less than the second root, $F(u_0)$ is negative. Hence, from equation (4.94), if we start to the left or to the right and smaller than the second root, we approach this solution. This means that this solution is stable. On our bifurcation diagram, Figure 4.11, the lowest branch of solutions, i.e. before the fold, is a stable branch of solutions. Now, consider the larger of the two roots. It is easy to see that this root is unstable. If we start to the left or to the right, the right-hand side of equation (4.94) is either negative or positive, pushing us away from this solution. On the bifurcation diagram, these roots correspond to the upper branch, and hence we conclude that this branch is a branch of unstable solutions.

Notice that for this simplified first-order model, equation (4.94), we have determined the global *dynamics* of solutions. If we pick an initial value for u_0 and a value for P and consider the motion on our bifurcation diagram, we see that for $P < P^*$, solutions monotonically approach the lower stable branch provided they start below the upper branch. If they start above the unstable upper branch, the temperature increases monotonically. If $P > P^*$, the temperature also increases monotonically.

We would also like to know the stability of these spatially independent steady-state solutions for the full reduced-order PDE model. For partial differential equations global stability results are not as easily obtained as they are in the first-order ODE case. Usually, we rely upon a local result called linear stability theory, which tells us how steady-state solutions respond to small perturbations. There are several possible ways to determine the linear stability of steady-state solutions. Here, we introduce an approach that is useful for determining stability of steady-state solutions to models based upon partial differential equations. We work with equations (4.85)–(4.88). Let us denote any of the steady-state solutions, i.e., roots of $F(u_0)$, by u_0^*. Now, we seek a solution to equation (4.85) in the form,

$$u_0(z,t) = u_0^* + \epsilon e^{-\mu t} v(z), \tag{4.95}$$

where $\epsilon \ll 1$ and μ is an eigenvalue parameter to be determined. The idea is that we are examining a small perturbation from the steady state, u_0^*, and attempting to determine whether solutions move away or toward this steady state. Notice that $\mu > 0$ ($\mu < 0$) corresponds to linear stability (instability). Inserting (4.95) into equation (4.85), we obtain

$$-\epsilon\mu e^{-\mu t} v(z) = \epsilon\delta e^{-\mu t} v'' - 2u_0^* - 2\epsilon e^{-\mu t} v(z) + \frac{P}{f}\left(\int_0^1 \frac{dz}{f}\right)^{-2}. \tag{4.96}$$

Now, we expand f in a Taylor series about $\epsilon = 0$, ignore terms of order ϵ^2, and obtain the characteristic equation for μ,

$$v'' + \frac{\mu - 2 - Pf'(u_0^*)}{\delta} v + \frac{2Pf'(u_0^*)}{\delta} \int_0^1 v(z)dz = 0. \tag{4.97}$$

The boundary conditions on $u_0(z,t)$ imply that

$$v'(0) = v'(1) = 0. \tag{4.98}$$

This eigenvalue problem has the nontrivial solution v constant only when

$$\mu - 2 + Pf'(u_0^*) = 0 \tag{4.99}$$

or equivalently when

$$\mu = -F'(u_0^*). \tag{4.100}$$

Hence the real part of μ is positive if and only if $F'(u_0^*) < 0$. This implies that once again we may read off stability from Figure 4.9. The function F passes through the smallest root with negative slope and hence this solution is stable, while the opposite is true for the other steady-state solution.[9] For this example we have shown that the stability result for the ODE- and PDE-based models are identical. The reader is cautioned that this is not always the case. It can and does occur that solutions stable for the ODE system are unstable to spatially varying perturbations to the PDE system. This is the basis of the famous Turing instability in the study of morphogenesis.[10]

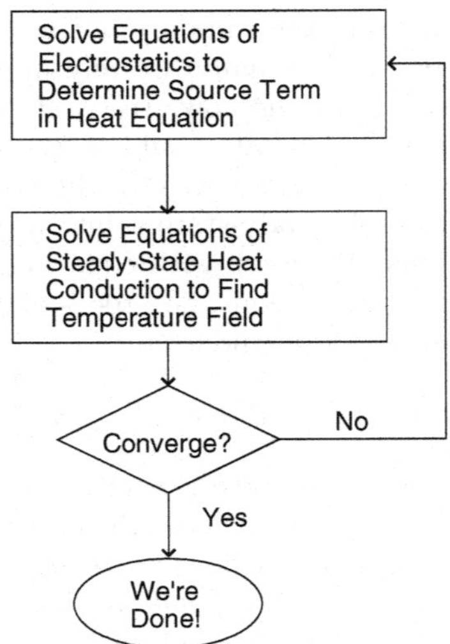

FIGURE 4.12: Flowchart of a basic iterative method for numerical electrostatic-thermal solvers.

4.4.7 Iterative Numerical Schemes

Numerical simulation is a valuable tool for designing and understanding the behavior of microsystems. While our focus in this text is on the construction and analysis of fundamental models rather than upon detailed numerical simulation of realistic devices, the relationship between our analysis and numerical simulation is worth exploring. In addition to developing physical intuition, our analysis develops *numerical* intuition. That is, by applying typical numerical simulation methods to our idealized models, we gain insight into the difficulties with and failures of such methods. Ideally, this will suggest ways to improve numerical techniques.

A flowchart outlining a common iterative algorithm used in MEMS simulation is presented in Figure 4.12. This algorithm is at the core of many methods used to compute the steady-state configuration of microsystems. The idea behind the algorithm is quite natural. We begin with some guess as to the temperature distribution in the device, holding this configuration fixed, we compute the electrostatic field and hence the electrostatic source term, then using the computed source, solve the thermal problem to find the updated state of the device. We iterate until things no longer change, i.e., until convergence. In addition to simplicity, this algorithm has the advantage that it allows the researcher to simply couple together well-developed and well-understood thermal and electrostatic solvers to create a coupled-domain MEMS solver. One can imagine that electrostatic or fluid effects etc., can easily be included in this algorithm. In fact, as we shall see in Chapter 7, this same scheme is the basis for many solvers simulating electrostatic-elastic devices.

We can apply this algorithm to the solution of the steady-state problem for our reduced-order Joule heating model. Here, the configuration of the device is given by the temperature, which we denote v_n. Solving for the electrostatic field simply means computing the value of

$$Pf(v_n) \tag{4.101}$$

and solving for the new temperature, v_{n+1}, simply means balancing the Newton cooling term with the electrostatic source, i.e.,

$$v_{n+1} = \frac{Pf(v_n)}{2}. \tag{4.102}$$

It is useful to introduce the notation, T, for the right-hand side of equation (4.102). That is,

$$Tv_n = \frac{Pf(v_n)}{2}. \tag{4.103}$$

Then, our iterative scheme may be written as

$$v_{n+1} = Tv_n. \tag{4.104}$$

In an abstract sense, the operator T stands for the inversion of the thermal problem. The reader is encouraged to think of T in this way and envision equation (4.104) as shorthand for the algorithm in Figure 4.12. That is, T should be viewed as the operator that allows us to invert the equations of heat conduction and hence solve the thermal problem. The numerically minded reader may think of T as the code, perhaps a finite element code, which solves the thermal problem.

A key tool in the analysis of such iterative schemes is the contraction mapping theorem.

THEOREM 4.1
(Contraction Mapping Theorem) *Let S be a closed set in the Banach space X and let $T : S \to S$ be a contraction. Then, T has one and only one fixed point in S. This fixed point can be obtained from any initial element, v_0, in S as the limit of the iterative sequence, equation (4.104).*

We remind the reader that a contraction satisfies

DEFINITION 4.1 *Let $T : S \to S$. Then T is a contraction on S iff there exists a constant $\rho < 1$ such that $||Tu - Tv|| < \rho||u - v||$ for all u, v in S.*

That is, the operator T is a contraction if it brings elements of S closer together when measured in the metric of the Banach space S. The reader unfamiliar with the notion of a Banach space is referred to Appendix A for a brief discussion or to [184] for an extensive discussion. In this example, it is sufficient to take S to be a closed interval on the real line and the metric, $||u||$, to be the absolute value of a real number. With this in mind, let's consider our iterative scheme, equation (4.104), with T defined by equation (4.103). To investigate the convergence of our numerical scheme, we must investigate the contractivity of T. Let u, v be real numbers, let $P \geq 0$ and consider

$$||Tu - Tv|| = |Tu - Tv| = \frac{P}{2}|f(u) - f(v)| = \frac{P}{2}|e^u - e^v|. \tag{4.105}$$

For our Banach space, X, we take the closed interval $[0, b]$. We take the closed subset S to be all of X. By the mean value theorem, there exists a point $c \in [0, b]$ such that

$$\frac{P}{2}|e^u - e^v| = \frac{P}{2}e^c|u - v|, \tag{4.106}$$

which implies

$$\frac{P}{2}|e^u - e^v| \leq \frac{P}{2}e^b|u - v|, \tag{4.107}$$

since e^x is a monotonically increasing function of x. Hence T is a contraction if and only if

$$\frac{Pe^b}{2} < 1. \tag{4.108}$$

We also need T to map the set S into itself. This is clearly true if and only if

$$\frac{Pe^b}{2} < b. \tag{4.109}$$

Combining these two inequalities we find that the contraction mapping theorem applies to our iterative scheme. However, there is a relationship between P and b. As b increases, the maximum value of P for which existence is guaranteed decreases. The largest P for which existence can be guaranteed is $P = 2/e$ occurring when $b = 1$.

While it is nice to have proven convergence of our numerical method, here the implications of our analysis for numerical difficulties are of most interest. First, we notice that the contractivity of the mapping T depends on P. In particular, as P is increased T becomes "less contractive." This is not surprising, after all, we know that the method must fail once P exceeds P^*. That is, for P beyond P^* there is no solution to the problem in S and hence our method should not converge. What is perhaps surprising is that the rate of convergence of the iterative method will depend on P. As P approaches P^* and the mapping becomes less contractive, the number of iterations to convergence will tend to infinity. This is illustrated in Figure 4.13, where starting

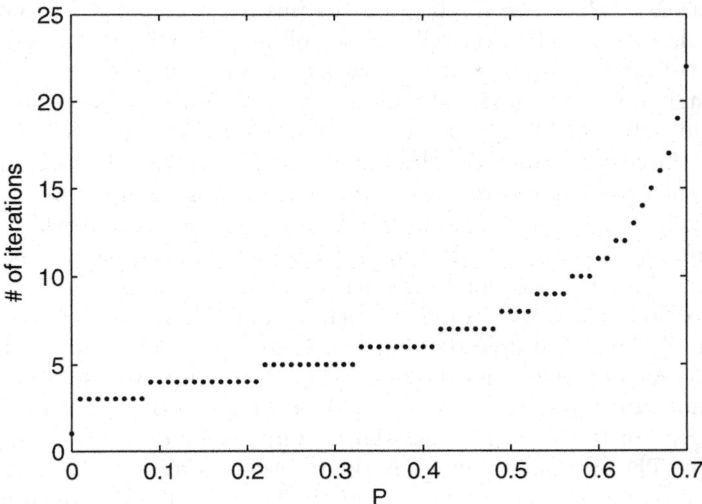

FIGURE 4.13: Iterations to convergence as a function of P.

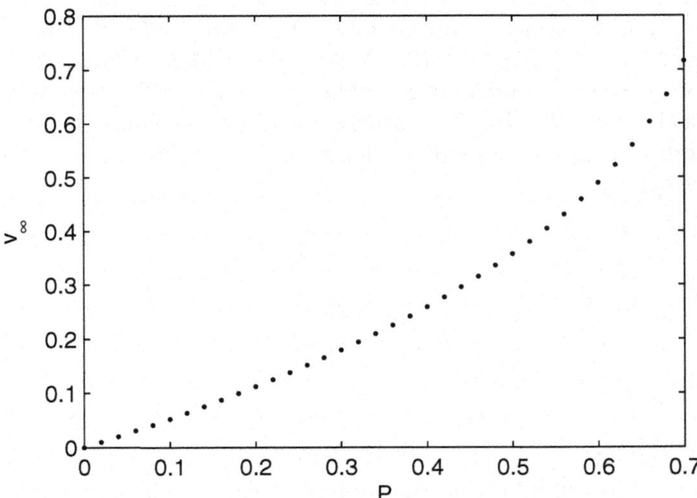

FIGURE 4.14: Output of the iterative method. Note we have captured the lower branch of solutions only.

from the initial guess $v_0 = 0$, we plot the number of iterations to convergence as a function of P. The implications of this are severe. While the simple iterative algorithm will work well at low voltages, it will become slower and slower as voltages are increased. If we wish to numerically attempt to compute the thermal runaway voltage, the iterative algorithm will after considerable work, yield only a crude approximation. Finally, the iterative scheme is only returning the lower branch of solutions. That is, we are only computing the lowest branch of solutions on the bifurcation diagram, Figure 4.11. The results of such a computation using the iterative method are shown in Figure 4.14. Here, one may argue that the upper branch is unstable and hence not of interest. This is a specious argument. *A priori*, one does not know the stability or location of the upper branches. Further stable upper branches can occur. A numerical investigation that ignores this possibility would miss this discovery and hence miss the possibility of utilizing such solutions in the construction or optimization of devices. An example of this scenario occurs in chemical reactor theory, where anS-shaped bifurcation diagram often governs the system. The optimal point to run the reaction is often on an upper branch of the bifurcation diagram. A numerical analysis that fails to capture all but the lower branch would not be able to predict this result.

How do we overcome the difficulties with the iterative algorithm? Our analysis points toward the solution. Consider again the bifurcation diagram, Figure 4.11. When we view u_0 as a function of P, the situation is quite complicated. The function $u_0(P)$ is multivalued and has infinite slope at P^*. The bifurcation diagram contains a fold. On the other hand, when we view P as a function of u_0^*, Figure 4.10, the situation is quite tame. The function $P(u_0)$ is single valued and differentiable everywhere. The key to resolving the difficulties with the iterative scheme lies in this change in perspective. Here, for our reduced-order ODE model, we can ask to find $P(u_0)$ rather than $u_0(P)$. Earlier we wrote

$$P = \frac{2u_0}{f(u_0)} \tag{4.110}$$

and simply plotted $P(u_0)$ rather than solving for $u_0(P)$. The same idea can be used to rescue our numerical method. While for the reduced-order ODE model this is trivial, for more complicated models choosing the correct perspective is not. Much research has focused upon the efficient numerical computation of bifurcation diagrams that contain a fold. Notable is the work of H.B. Keller [100] and the method of pseudo-arc-length continuation.

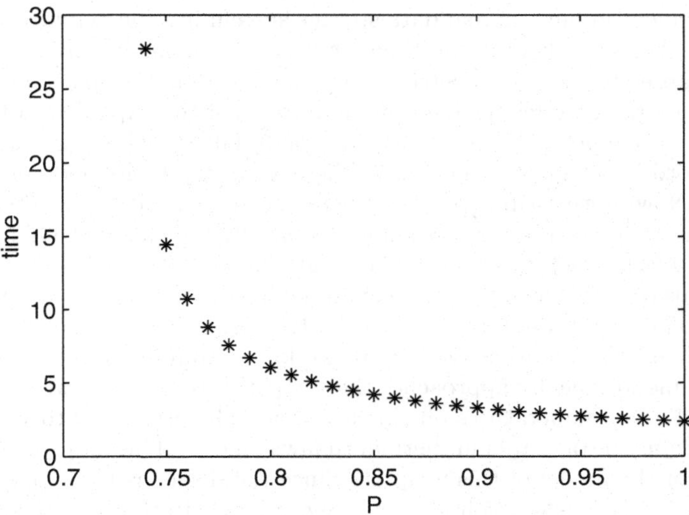

FIGURE 4.15: Dimensionless time to reach data writing temperature of 350°C as a function of P. Recall P is proportional to the applied voltage.

4.5 Analysis of Thermal Data Storage

In this section we outline how our model of Joule heating of a cylinder might be applied to the study of the thermal data storage device discussed in Section 4.2.6. In this system there are many aspects of the design that call for optimization. Of particular interest is optimizing the system to improve data density and read/write speeds. Here, we will focus on data write speed. We will use actual parameters for the system built jointly by Stanford University and IBM Research and available in [33, 104]. We note that our goal is not to obtain detailed quantitative information about the system in [33, 104] but rather to illustrate how models developed in this chapter might be applied in practice and to suggest to the reader avenues for further exploration. A set of problems based on enhancing our basic Joule heating model in directions suggested by the study of thermal data storage systems appears at the end of the chapter.

For our study here we will use the reduced-order, spatially independent ODE model, equation (4.94). First, recall that in the thermal data storage system an AFM tip is heated via Joule heating and melts/vaporizes a small region in a polycarbonate disk. The typical writing temperature for such systems is $T_w = 350°C$ [104]. Hence write speeds are limited by how quickly the AFM tip may be heated to T_w from room temperature. We imagine that the heating element in our thermal data storage system has the cylindrical geometry we've studied. This indicates one obvious extension; extend the analysis of this chapter to other geometries. We imagine that the material used to fabricate the resistive heater has conductivity with the exponential form assumed in this chapter. We note that the system fabricated by groups at IBM and Stanford is designed to be run in the constant voltage mode our model assumed. Now, if we assume room temperature is $T_A = 20°C$, the dimensionless writing temperature for our reduced spatially independent ODE model, equation (4.94), is $u_0^w \approx 16.5$. A glance at the bifurcation diagram, Figure 4.11, reveals that the device must be operated in the "thermal runaway" mode in order to achieve these temperatures.[11] That is, if $P < P^*$ and the system starts with initial condition $u_0(0) = 0$, we know from our previous analysis that $u_0(t)$ monotonically approaches the temperature on the lower branch in Figure 4.11. All temperatures on this lower branch correspond to $u_0 < 16.5$; the device must be operated in thermal runaway mode. This indicates another extension of the model of this chapter. Since the device is to be operated in runaway mode, a constant voltage cannot be applied. Rather a voltage "pulse" must be used. This has the effect of making the parameter P in our model a function of time, $P = P(t)$, perhaps a step function with the width of the desired pulse. The reader is encouraged to explore the implications of such a change for the model of this chapter. Ignoring this complication we can compute heating time for various applied voltages corresponding to various values of P by simply integrating equation (4.94) numerically until $u_0 \approx 16.5$ and then recording the time at which this writing temperature was achieved. The result of such a computation is shown in Figure 4.15. Note the nonlinear dependence of write time on applied voltage. This is a direct consequence of the nonlinear dependence of the electrical conductivity on temperature. The shape of this curve agrees with the shape of the curve obtained from simulation and experiment and presented in [33, 104]. That is, our model, at least in this regard, qualitatively agrees with simulation and experiment. In [104], King et. al., also provide a prediction of the lowest operating voltage the thermal data storage device may use. From our analysis, since we know the device must be operated in runaway mode, this simply means that $P > P^*$. Noting that

$$P = pB = \frac{\sigma_0 V^2}{haT_A}\frac{ha}{k} = \frac{\sigma_0 V^2}{T_A k} \tag{4.111}$$

and using $\sigma_0 = 2120\,\Omega$, $T_A = 293\,\mathrm{K}$, $k = 50\mathrm{W/(mK)}$, our model predicts that the lowest operating voltage is approximately $2.25V$. The prediction of King et. al. [104] is that the lowest operating voltage is $6V$. The discrepancy is undoubtedly due to our assumption that $f(u_0) = e^{u_0}$. A fit of the function $ae^{bu_0} + c$ to the actual electrical conductivity data used in [33, 104] would improve the model. This data is unfortunately unavailable in the literature. The reader is encouraged to explore this and other extensions to the basic model as presented in the exercises.

4.6 Chapter Highlights

- A wide variety of MEMS devices utilize Joule heating for their operation. These include actuators, sensors, and "lab-on-a-chip" systems.
- Reduced-order models are desirable from both an analytical and numerical point of view. Perturbation methods provide a powerful tool for deriving reduced-order models. They reveal the parameter range of validity, the size of the corrections, and how to compute corrections to a reduced model.
- Joule heating is a coupled-domain problem of relevance to MEMS. It couples electromagnetism to heat transfer. In general, a model of Joule heating is a coupled system of nonlinear partial differential equations.
- In devices with a small aspect ratio and small Biot number, the Joule heating model can be reduced using perturbation methods to a tractable form. The reduced-order model is suitable for numerical simulation or further analysis.
- When multiple steady-state solutions to a model are uncovered, it is desirable to know the stability of these steady states. Linear stability theory allows us to determine the stability of steady-state solutions to small perturbations.
- Iterative numerical schemes are commonly used in the solution of coupled-domain problems. The contraction mapping theorem is a useful tool in their analysis.

4.7 Exercises

Section 4.3

1. Consider the simple model describing the cooling of a hot body, equation (4.4). Solve this ordinary differential equation subject to the initial condition, $T(0) = T_h$.
2. The technique of using perturbation methods to derive a "lumped" model is powerful and versatile. Often the reduction is not from PDE to ODE, but from PDE to PDE with fewer independent variables as in Section 4.4. Here, consider heat conduction in a short fat cylinder. That is, consider a cylinder whose radius is large compared to its height. Assume the sidewalls are held at constant temperature, T_A, the tops and

bottoms are insulated, and the initial temperature is T_h. Show that to leading order the temperature is independent of the axial coordinate.

The next five problems are intended to provide practice with perturbation methods. For many more such problems, the reader is encouraged to consult the section on Related Reading in this chapter.[12]

3. Consider the cubic equation,

$$v^3 - 2v^2 + v - \epsilon = 0.$$

Find an asymptotic expansion of all three roots of this equation valid for $\epsilon \ll 1$.

4. Consider

$$\frac{d^2u}{dx^2} - \epsilon u = 1$$

$$u(0) = -1, \quad u(1) = 0$$

for $\epsilon \ll 1$. Solve this problem exactly. Then, compute the first two terms in a perturbation expansion of the solution. Compare your exact and approximate solutions as ϵ is varied.

5. Consider the linear second-order boundary value problem,

$$u'' + \epsilon u' - u = 1$$
$$u(0) = u(1) = 1,$$

where $\epsilon \ll 1$. Solve this problem exactly. Find a two-term asymptotic expansion of the solution for small ϵ. Compare your exact and approximate solutions.

6. Consider the initial value problem,

$$u' + 2xu - \epsilon u^2 = 0, \quad u(0) = 1.$$

Assume $\epsilon \ll 1$ and obtain the first two terms in a perturbation expansion of the solution. Numerically solve this initial value problem and compare the output of your numerical method with the results of your perturbation scheme.

7. In Chapter 1 we discussed the projectile problem of Lin and Segel [124], or Holmes [83]. This problem concerns the motion of a projectile near the surface of the earth. Here $x(t)$ is the height of the projectile and satisfies the equation,

$$\frac{d^2x}{dt^2} = -\frac{gR^2}{(x+R)^2},$$

where g is the gravitational constant and R the radius of the Earth. Assume the object starts from rest with initial velocity $V > 0$. Choose nondimensional variables and recast this problem in nondimensional form following Chapter 1. Show that a single dimensionless parameter, $\epsilon = V^2/RG$, arises. Develop an asymptotic expansion of the solution assuming $\epsilon \ll 1$. Solve the problem numerically and compare your asymptotic and numerical solutions.

Section 4.4

8. Redo the analysis of Section 4.4 for a short, squat cylinder rather than a long, thin cylinder. A clear understanding of problem 2 will be useful here.

9. If $f(u_0)$ is assumed to be equal to one, the reduced model of Section 4.4 may be solved exactly. Solve this system and discuss how solutions differ from the case $f(u_0) = e^{u_0}$.

10. Assume $f(u_0)$ is still nonlinear, but saturates at a constant for high temperatures. In particular, take $f(u_0) = 1 + \tanh(u_0)$. Seek spatially independent solutions to equations (4.85)–(4.88) and examine the resulting ODE. Sketch a bifurcation diagram for the system. Determine the stability of all steady-state solutions.

11. Consider the first order ODE,

$$\frac{dx}{dt} = P - x^2,$$

where P is a parameter. Find all steady-state solutions to this problem. Sketch the bifurcation diagram and show that this system possesses a fold-type bifurcation.

12. Consider the problem of heat conduction in a one-dimensional rod of length one with a nonlocal boundary condition.

$$\frac{\partial\theta}{\partial t} = \frac{\partial^2\theta}{\partial x^2}, \quad \theta(0,t) = 0, \quad \theta(1,t) = \tanh\left(2\delta \int_0^1 \theta(\zeta,t)d\zeta\right).$$

Here δ is a positive parameter. This system arises in the study of thermoelastic contact instabilities [156]. Seek steady-state solutions to this problem. Show that a bifurcation occurs as δ passes through one. Sketch the bifurcation diagram. This type of bifurcation is called a *pitchfork bifurcation.* Perform a linear stability analysis to find the stability of the steady-state solutions.

The remaining exercises lead to the development of simple models of devices discussed in this chapter or enhancement of models discussed in this chapter. Many of these problems are open-ended.

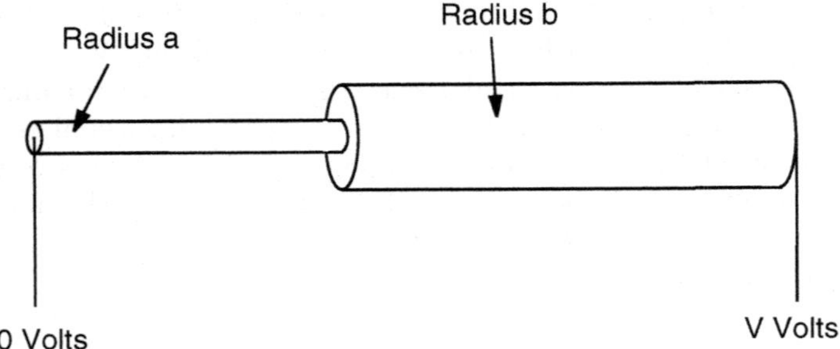

FIGURE 4.16: The two-cylinder setup for problem 13.

Thermal Bimorph Actuator Revisited

13. A key feature of the thermal bimorph actuator is the difference in size of the two arms. Develop a toy model of this system by considering Joule heating of the two cylinders shown in Figure 4.16. Assume that the cylinders behave as in Section 4.4, but that the ratio of their radii, a/b, is small. How does this ratio appear in the model after scaling? How do solutions to this model depend on this ratio? Can you say anything about the design of such devices based on your model?

Thermopneumatic Valve Revisited

14. In the thermopneumatic valve, heat transfer to the fluid above the heating element can be significant. Build a first model of this scenario by considering the two stacked cylinders. Treat the heating element as a short, squat cylinder insulated on the sidewalls and heated via Joule heating. Treat the fluid as a solid cylinder stacked on top of the heating element. Assume that at the interface between the fluid and heating element there is *perfect thermal contact*. (Perfect thermal contact means continuity of temperature and heat flux at the boundary.)

Thermal Data Storage Revisited

15. One shortcoming of the model in this chapter is the limitation to heat conduction in a cylinder. We can easily imagine that the alternate shapes for AFM tips and/or heating elements may lead to improved performance. In this problem you are asked to explore heat conduction in structures that are almost cylindrical but can have a slowly varying shape. Consider steady-state heat conduction in a long, thin rod of variable circular cross-section. You can take the surface of the rod as a function of the axial coordinate z' to be located at $r' = a + bf(z'/L)$, where L is the length of the rod. Assume the sides of the rod are insulated and that the ends are held in temperature baths with different

temperatures. Write down the governing equations for this problem and scale the system. Assume that b/a is a small parameter and develop a perturbation expansion of the solution of this problem. You should find that your leading-order solution is independent of r' and contains the function f in its governing equation.

16. Repeat the analysis of Section 4.4 for a cylinder of varying circular cross-section. See problem 15.

4.8 Related Reading

Many of the devices discussed in this chapter are discussed in greater depth by Maluf.

N. Maluf, *An Introduction to Microelectromechanical Systems Engineering*, Artech House, 2000.

There are several good books on perturbation methods. Lin and Segel introduce the basics.

C.C. Lin and L.A. Segel, *Mathematics Applied to Deterministic Problems in the Natural Sciences*, Philadelphia: SIAM, 1988.

Kevorkian and Cole is comprehensive.

J. Kevorkian and J.D. Cole, *Multiple Scale and Singular Perturbation Methods*, Springer-Verlag, 1996.

Bender and Orszag do a nice job with lots of things, especially boundary layer theory.

C.M. Bender and S.A. Orszag, *Advanced Mathematical Methods for Scientists and Engineers*, Springer-Verlag, 1999.

Hinch is a short, readable introduction to perturbation methods.

E.J. Hinch, *Perturbation Methods*, Cambridge University Press, 1991.

A well-written introduction to nonlinear dynamics and bifurcation theory is the book by Strogatz.

S.H. Strogatz, *Nonlinear Dynamics and Chaos: With Application to Physics, Biology, Chemistry, and Engineering*, Perseus Books, 2001.

A higher-level but still readable text on bifurcation theory is the book by Hale and Kocak.

J.K. Hale and H. Kocak, *Dynamics and Bifurcations*, Springer-Verlag, 1992.

4.9 Notes

1. There is a confusing bit of terminology concerning thermal actuators. Some authors refer to thermal actuators based on material with different CTEs as thermal bimorphs. In this text, we'll call such actuators bimetallic thermal actuators. The term thermal bimorph will always refer to a structure composed of a *single* material but differentially heated.

2. The microcilia mentioned here use bimetallic type actuators.

3. For clarity, in this section we use B rather than the more common Bi to denote the Biot number. In later chapters we return to the Bi notation.

4. If we looked at the problem on the diffusive time scale, we would immediately see our perturbation expansion become nonuniform. However, if the initial temperature distribution varied in space, it would be on the diffusive time scale that this distribution relaxed to a constant. That is, there is a boundary layer in time present in this problem. Here, the boundary layer is trivial.

5. A brief discussion of asymptotic methods including a precise definition of the notation $\sim$ appears in an appendix.

6. The reader unfamiliar with the "big-O" notation will find a discussion in Appendix A.

7. J.A.P. thanks Prof. G.A. Kriegsmann for introducing him to the example in this section.

8. This assumption may be removed without significantly altering the subsequent analysis.

9. There is a second type of solution for this eigenvalue problem, but μ is strictly positive for these solutions and they do not affect stability.

10. The Turing instability is discussed thoroughly by Murray [138].

11. The reader is cautioned to remember that this is a toy calculation; we have assumed a form for the temperature dependence of the electrical conductivity

that may or may not be correct for materials used in fabricating this particular device. In this instance it is a reasonable assumption that the doped silicon used in the thermal data storage system has electrical conductivity increasing with temperature. On the other hand, the exact form of the conductivity is not available in the literature.

12. In this chapter we have really only touched on *regular perturbation expansions*. The world of *singular perturbation expansions* is much larger and richer. Any of the texts mentioned in the Related Reading section will provide the reader with a good introduction to singular perturbation problems.

Chapter 5

Modeling Elastic Structures

My cousin Elroy spent seven years as an IBM taper staring at THINK signs on the walls before he finally got a good idea: He quit.

Edward Abbey

5.1 Introduction

The microelectronics revolution has yielded miracles. In thirty short years personal computers, cellular phones, PDAs, GPS locators, digital cameras, compact discs, and countless other devices went from the drawing board to the marketplace to the commonplace. All because of the microchip. Now imagine the power of microelectronics united with things that *move*. That is the essence of the promise of MEMS and NEMS. In this chapter we turn our attention to the motive side of micro- and nanoscale systems and study the elastic behavior of MEMS and NEMS.

We begin in Section 5.2 by briefly discussing seven micro- and nanoscale systems that illustrate the range of elastic structures and behaviors present in MEMS and NEMS. We examine micromirrors that exhibit the simple behavior of a mass on a spring, we encounter a carbon nanotube that behaves like a nanoscale spring, we meet a pressure sensor with an elastic membrane as a key component, and run into other devices composed of elastic beams and elastic plates.

In Section 5.3 we spend some time with the mass-spring system and review the basics of simple harmonic motion and the phenomenon of resonance. While the mass-spring structure may seem a bit elementary for a chapter that purports to discuss elasticity, a quick glance at the literature reveals that mass-spring-based analyses are very much alive and well in the study of microsystems. In fact, in Chapters 6, 7, and 8 our first pass at modeling thermopneumatic actuators, electrostatic actuators, and magnetic actuators will all involve a mass-spring system. In Section 5.3 we also discuss the phase plane and elementary phase plane theory.

In Section 5.4 we study the next simplest elastic structure, namely, elastic membranes. An elastic membrane is a body that resists stretching but has no resistance to bending. A soap film and a balloon are familiar objects whose elastic behavior is close to that of a membrane. In MEMS and NEMS many devices make use of elastic components that are membrane-like. Our study here will prepare us for later discussions of models of thermopneumatic actuators, electrostatic actuators, etc. In this chapter we investigate both the steady-state deflection of membranes and the vibration of membranes. We review the notion of resonance and provide a quick refresher on the method of separation of variables.

In Section 5.5 we study the deformation and vibration of elastic beams. An elastic beam is an idealized structure that resists both bending and stretching. In MEMS and NEMS many devices and structures resemble elastic beams. Nanotweezers constructed from carbon nanotubes, resonant sensors, and MEMS switches are but a few of the microsystems that use beam-like components. In this chapter we derive the shape of a cantilevered beam subjected to a constant load and a cantilevered beam subjected to a concentrated load. We consider the free vibrations of a beam and derive the natural frequencies of a beam with two pinned ends. This calculation is particularly important for the study of resonant sensors.

In Section 5.6 we study the deformation and vibration of elastic plates. As with beams, many structures in MEMS and NEMS resemble elastic plates. Computing plate deflections is an order of magnitude more difficult than computing beam deflections. Except in simple geometries where the problem can be reduced to a one-dimensional problem, one is forced to solve a partial differential equation even in the static case. We consider the simple geometry of a circular plate and determine the deflection due to a constant load. We consider free vibrations of the circular elastic plate and determine the natural frequencies of vibration.

While models making use of the elastic theory developed in this chapter appear throughout the remainder of the text, in Section 5.7 we consider a simple but illustrative example of the use of elastic models and study the capacitive pressure sensor. In this device, pressure in a given environment causes a deflection of an elastic diaphragm. In turn, this deflection causes the capacitance of the system to change. In order to build such a device it is necessary to understand the relationship between capacitance and applied pressure. We compute the capacitance-pressure curve for a disk-shaped sensor with a deformable elastic membrane as its central component.

5.2 Examples of Elastic Structures in MEMS/NEMS

In this section we take a brief look at seven microsystems that demonstrate the range of elastic structures used in MEMS and NEMS. Our selections are representative rather than comprehensive. Researchers are continuously attempting to fabricate and use increasingly complex elastic structures in MEMS devices. Many examples appear in the bibliography; the reader is directed to the literature for other examples of elastic structures in MEMS and NEMS.

5.2.1 Micromirrors

The torsion micromirror shown in Figure 5.1 illustrates a simple elastic structure often found in MEMS. In this system the goal is to change the

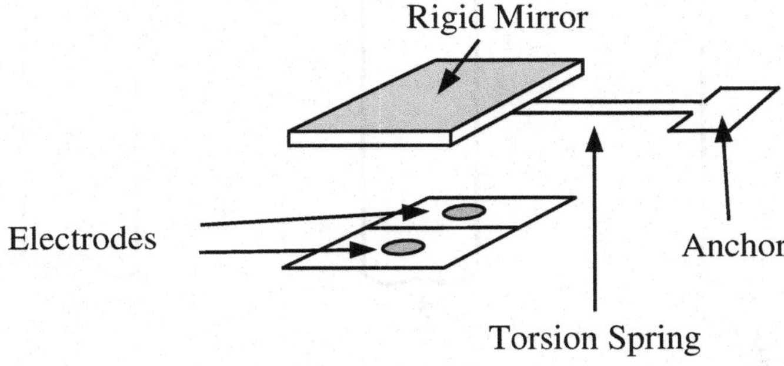

FIGURE 5.1: Sketch of a torsion micromirror.

angle of the rigid mirror in order to control the location of a reflected beam of light. This allows the mirror to act as an optical switch for optical fiber networks. The mirror is made of a material with sufficient stiffness to prevent bending. The mirror is attached to a thin beam that is anchored in place at the opposite end. The beam is free to twist and acts as a torsion spring.[1] Often the mirror is attached to two beams, one on each side rather than a single beam as shown in the figure. Thermopneumatic, thermoelastic, electrostatic, magnetic, or piezoelectric forces may be used to make the mirror move. In the structure shown in Figure 5.1, the use of electrostatic force is imagined. A pair of electrodes reside below the mirror. When a potential difference is applied between either of the electrodes and the mirror, a torque is exerted causing the desired rotation. A sample of other micromirror structures and actuation methods can be found in [22, 43, 122, 142, 147, 200, 209, 220].

5.2.2 Carbon Nanotube Caps

The discovery of C_{60}, the carbon "Buckeyball" structure, by Curl, Kroto, and Smalley, surely ranks as one of the most significant scientific achievements of the late twentieth century. In fact in 1996 the trio received the Nobel prize for their efforts. The follow-up discovery of carbon nanotubes, sheets of carbon wrapped into a nanoscale cylinder, promises to be one of the technologically most important advances of the late twentieth and early twenty-first centuries. In Figure 5.2 we illustrate one of the possible myriad uses of carbon nanotubes.

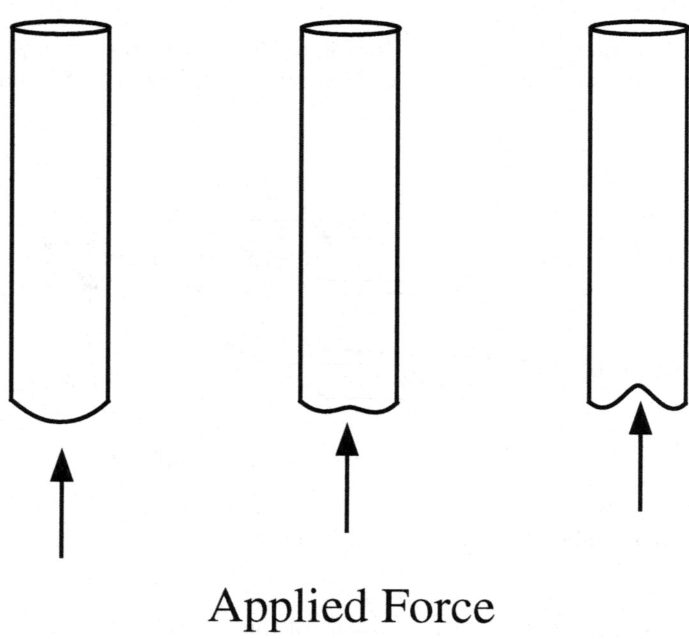

FIGURE 5.2: The carbon nanotube as a Hookean spring.

The idea shown in Figure 5.2 was explored by Yao and Lordi using molecular dynamics simulations [210]. They showed that a carbon nanotube capped with a hemispherical carbon structure can be used as a nanoscale Hookean spring. That is, if the cap is indented as shown in Figure 5.2 the nanotube structure exerts a restoring force that is linearly proportional to the size of this indentation. This is precisely the same as the restoring force exerted by a Hookean spring. Recent work [146] has shown that this effect may depend on temperature and can also be modified by filling the nanotube with different materials.

5.2.3 Pressure Sensors

The pressure sensor is a staple of MEMS engineering. The demand for inexpensive, high-quality, miniature pressure sensors comes from a variety of industries, the automotive, aeronautic, and medical being a few of the most important. The basic pressure sensor design is shown in Figure 5.3. The system consists of a deformable diaphragm over a sealed cavity. The cavity may contain a gas at a known reference pressure or may contain a vacuum. The opposite side of the diaphragm is exposed to the atmosphere whose pressure one wishes to sample. The difference in pressure between opposing sides of the diaphragm causes a stress and possibly a displacement of the diaphragm. A variety of methods have been developed to detect the stress or displacement. Capacitive pressure sensors detect the displacement by measuring the change in capacitance of the system composed of the diaphragm and a ground plate. Piezoresistive pressure sensors make use of the piezoresistive effect, where the electrical resistance of the diaphragm changes in response to a stress. In the appropriate circuit the resistance can be measured and pressure inferred. Numerous groups have fabricated pressure sensors and tested them in a wide variety of environments. A sample of different pressure sensor designs may

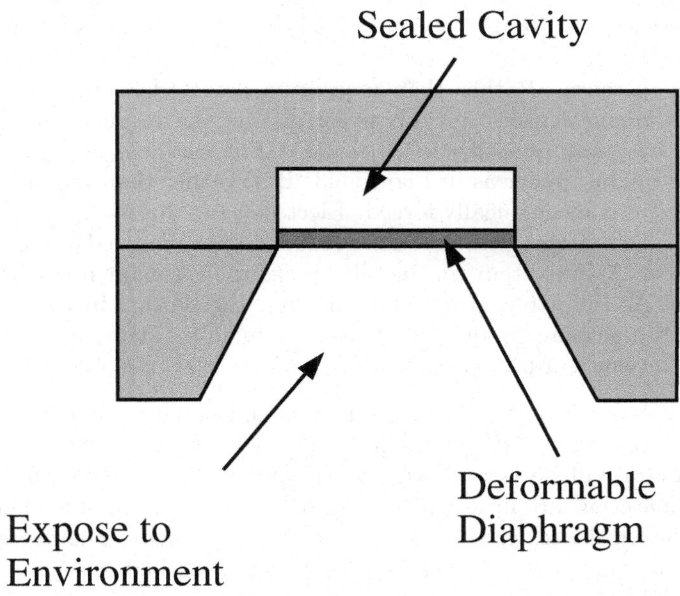

FIGURE 5.3: The basic pressure sensor design.

be found in [28, 29, 45, 116, 133, 203, 207]. Understanding the relationship between pressure and elastic deflections is an important part of designing any MEMS pressure sensor.

5.2.4 Resonant Sensors

A class of MEMS and NEMS sensors similar to the pressure sensor are the so-called *resonant sensors.* As with the pressure sensors, the resonant sensor correlates a change in the elastic behavior of the system with the quantity one wishes to sense. Rather than measuring device deflections or stresses, however, resonant sensors rely upon correlating the resonant frequency of the device with the quantity to be sensed. For example, a resonant pressure sensor might appear as in Figure 5.3. But, rather than remaining fixed, the diaphragm is mechanically forced. Electrostatic, thermoelastic, thermopneumatic, magnetic, or piezoelectric actuation may be used to provide the driving force. A subsystem continually tracks the resonant frequency of the membrane. As the external pressure changes, the tension in the membrane and hence the resonant frequency of the system shifts; detection of this shift provides a measure of pressure.

The same basic system may be used to make resonant chemical or biosensors. In this case the structure again might appear as in Figure 5.3; however, the diaphragm will be coated with a polymer layer. Specific chemicals or biological materials are absorbed by the polymer. In turn, this changes the mass and hence the resonant frequency of the diaphragm. The measurement of the resonant frequency is correlated with the absorbed mass to provide a measure of the chemical or biological material in the environment.

5.2.5 Micromachined Switches

A system that illustrates the role of various elastic structures in microscale engineering is the micromachined switch. The basic structure is quite simple. A pair of electrodes comprise the switch. One electrode is typically rigid and fixed in space. The other electrode is a deformable elastic structure. This electrode may be fabricated in a variety of shapes and resemble an elastic membrane, an elastic beam, or an elastic plate. The switch is closed by applying a potential difference between the two electrodes. This creates an electrostatic force deflecting the deformable electrode and causing the desired contact between the electrodes. Numerous groups have fabricated and tested such structures. The mathematical modeling of these devices has ranged from simple mass-spring-based models to full-blown, three-dimensional finite element simulations. A sample of both experimental and theoretical studies of micromachined switches may be found in [26, 136].

5.2.6 Micro- and Nanotweezers

A particularly novel application of NEMS technology is the *nanotweezer* developed by Philip Kim and Charles Lieber at Harvard University [103]. In their design a pair of carbon nanotubes are attached to gold electrodes, which in turn, are fastened to a tapered glass micropipette. The basic design is shown

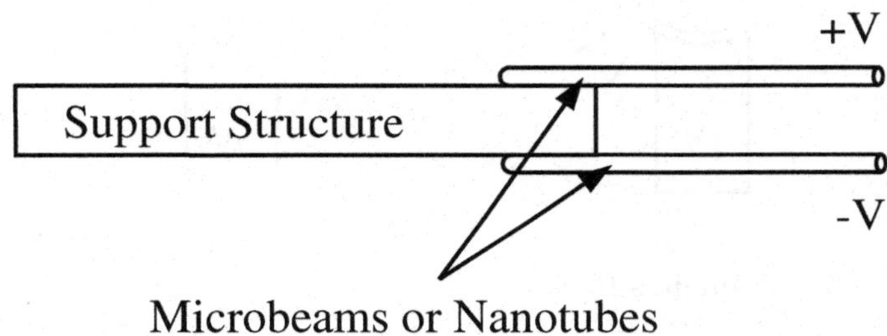

FIGURE 5.4: The basic micro- or nanotweezer design.

in Figure 5.4. A potential difference is then applied between the nanotubes, creating an attractive electrostatic force. Altering the applied voltage changes the force on the nanotubes and hence the distance between the tubes. The nanotweezers can then be used like ordinary tweezers. Nanoscale objects—cells, nanomachines, etc.—can be manipulated. The microscale analog of nanotweezers, the microtweezer, has also been fabricated. The joint effort by Xerox Corporation and Cornell University was perhaps the earliest effort in this direction [176, 177]. In both the micro- and nanotweezer designs the elastic beam plays a central role. Understanding the deformation of elastic beams subjected to various loads is essential for understanding and designing micro- and nanotweezers.

5.2.7 The Nanomechanical Resonator

> In September 1752, I erected an Iron Rod to draw the Lightning down into my House, in order to make some Experiments on it, with two Bells to give Notice when the Rod should be electrified. A contrivance obvious to every Electrician.

This excerpt from the papers of Benjamin Franklin [58] describes the electromechanical resonator invented by Franklin and recently replicated on the nanoscale as the centerpiece of a charge detection device. At the macroscale Franklin's bells ring due to a potential difference between the two bells. The potential difference induces charge on the two bells; the charge is equal but

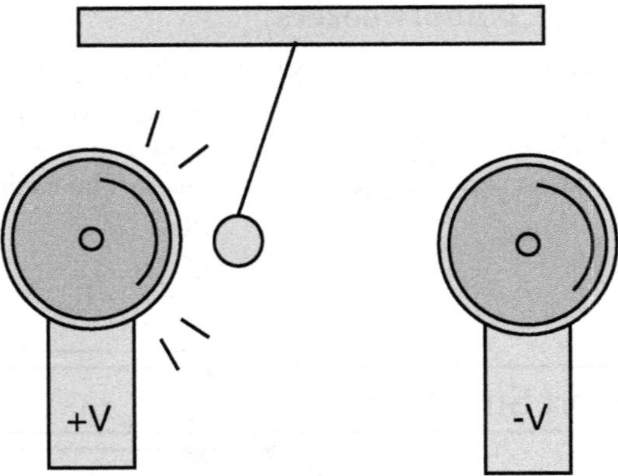

FIGURE 5.5: Franklin's Bells.

of opposite sign. A small charge on the ball suspended between the bells causes it to be attracted to one of the two bells. Upon making contact with a bell, the ball deposits its charge, picks up the opposite charge, and now being charged in the same manner as the bell it is contacting is violently repelled. The ball then makes contact with the opposite bell and the cycle repeats.

At the nanoscale the driving force is the same. However, rather than a ball suspended between two bells, a stiff silicon beam supports a nanoscale "island" between a pair of electrodes. The nanoscale island has a characteristic length of 100nm. This basic structure can then be used as a charge detection device, a parametric amplifier, a nanoscale clock, or a resonant sensor. Nanoscale resonators operating at frequencies above 500MHz and with quality factors as large as $250,000$ have been demonstrated. Understanding the deformation of elastic beams and the phenomenon of resonance is central to understanding and designing these devices. An introduction to the rapidly changing literature on nanomechanical resonators can be found in [15, 48, 49, 50].

5.3 The Mass on a Spring

The mass on a spring—every student of physics spends time contemplating this system. Before long, simple harmonic motion, Hooke's law and resonance have been added to the student's lexicon. And for good reason. Fields as seemingly disparate as elasticity, electric circuits, acoustics, statistical mechanics, chemistry, and even population biology all have the harmonic oscillator as one of their basic models. The study of microsystems is no exception. As we shall see in this and later chapters the mass-spring oscillator is often a key component of a "first-cut" model of a MEMS or NEMS device. In preparation we review the harmonic oscillator and study the motion of a mass on a spring. We examine damped unforced oscillations and introduce the ideas of energy and the phase plane. We review the notion of resonance through the study of the forced harmonic oscillator. We briefly discuss the origin of nonlinear terms in mass-spring models and their relationship to MEMS and NEMS. We also indicate how parameters from a real system may be related to a mass-spring model through the use of an effective spring constant.

5.3.1 Governing Equation and Scaling

We begin by analyzing the balance of forces for the system pictured in Figure 5.6. The governing equation for our mass-spring system follows directly

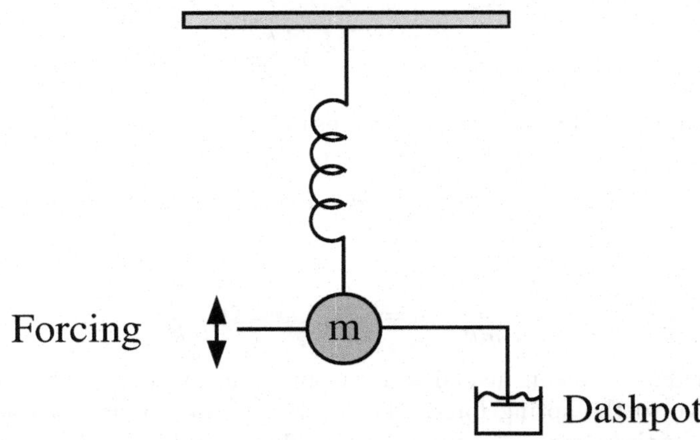

FIGURE 5.6: The mass on a spring with dashpot and forcing.

from Newton's second law. That is,

$$m\frac{d^2x}{dt'^2} = \sum \text{forces}. \tag{5.1}$$

Here, x is the displacement of the mass, m, from the top support. The forces acting on our system are the spring force, F_s, a damping force represented by the dashpot in Figure 5.6, F_d, and the imposed forcing, F_e. We assume that the spring is a linear spring and follows Hooke's law,

$$F_s = -k(x - l), \tag{5.2}$$

where l is the rest length of the spring and k is the spring constant. We assume that damping is linearly proportional to the velocity, that is,

$$F_d = -a\frac{dx}{dt'} \tag{5.3}$$

and we assume the imposed forcing is time harmonic,

$$F_e = A\cos(\Omega t'). \tag{5.4}$$

Inserting equations (5.2), (5.3), and (5.4) into equation (5.1) yields

$$m\frac{d^2x}{dt'^2} + a\frac{dx}{dt'} + k(x-l) = A\cos(\Omega t'). \tag{5.5}$$

It is convenient to rescale equation (5.5) and rewrite in terms of dimensionless variables. We introduce the scalings,

$$u = \frac{x-l}{B}, \quad t = \sqrt{\frac{k}{m}}\,t', \tag{5.6}$$

into equation (5.5). This yields the dimensionless equation

$$\frac{d^2u}{dt^2} + \gamma\frac{du}{dt} + u = \delta\cos(\omega t), \tag{5.7}$$

where

$$\gamma = \frac{a}{\sqrt{mk}}, \quad \delta = \frac{A}{kB}, \quad \omega = \Omega\sqrt{\frac{m}{k}}. \tag{5.8}$$

The parameter γ is a dimensionless measure of the strength of the damping force relative to the spring force. The reciprocal of γ is called the *quality factor* or Q of the system. As we shall see below γ and hence Q controls the rate of decay of oscillations; roughly it tells us how often the system oscillates before being completely damped. The parameter ω is a ratio of timescales; the natural timescale of oscillation to the forcing timescale. The parameter δ is a dimensionless measure of amplitudes in the system. Notice we have not specified B. We have several choices for B. For example, we could choose B to be the initial displacement of the mass, or we could choose $B = A/k$. The proper choice of B will depend on the problem under consideration.

5.3.2 Unforced Oscillations

Next, we review unforced or free oscillations of the mass-spring system. That is, we set $\delta = 0$ and study

$$\frac{d^2u}{dt^2} + \gamma\frac{du}{dt} + u = 0. \tag{5.9}$$

This is a linear, constant coefficient, second-order ordinary, differential equation. We can solve exactly by substituting the trial solution $e^{\lambda t}$, into equation (5.9) to obtain the quadratic equation for λ

$$\lambda^2 + \gamma\lambda + 1 = 0. \tag{5.10}$$

This equation has two roots that we label λ_1 and λ_2. In terms of these roots the general solution to equation (5.9) can be written as

$$u(t) = c_0 e^{\lambda_1 t} + c_1 e^{\lambda_2 t}. \tag{5.11}$$

The behavior of this solution depends on the exact nature of the λ_i. We solve the quadratic equation for λ and find

$$\lambda = -\frac{\gamma}{2} \pm \frac{1}{2}\sqrt{\gamma^2 - 4}. \tag{5.12}$$

Physically we are interested in the case $\gamma \geq 0$. It is easy to see that if $\gamma = 0$, then λ_1 and λ_2 are a pure imaginary complex conjugate pair. In particular, $\lambda_{1,2} = \pm i$. Thus, the solution for $u(t)$, equation (5.11), is purely oscillatory. Physically this is obvious, $\gamma = 0$ corresponds to no damping. If $0 < \gamma < 2$, the roots λ_1 and λ_2 are a complex conjugate pair with negative real part and nonzero imaginary part. Hence the solution for $u(t)$ consists of damped oscillations. Notice the rate of decay is governed by the dimensionless parameter γ. In particular, the "decay envelope" is the curve $e^{-\gamma t/2}$. The damped oscillation and the decay envelope are shown in Figure 5.7. If $\gamma > 2$ the roots λ_1 and λ_2 are purely real and negative. Hence the solution for $u(t)$ is strictly decaying with no oscillations.

It is worth investigating an alternate perspective. If we multiply equation (5.9) by du/dt, we can rewrite as[2]

$$\frac{d}{dt}\left[\frac{1}{2}\left(\frac{du}{dt}\right)^2 + \frac{u^2}{2}\right] = -\gamma\left(\frac{du}{dt}\right)^2. \tag{5.13}$$

We denote the term in brackets by $E(t)$, that is,

$$E(t) = \frac{1}{2}\left(\frac{du}{dt}\right)^2 + \frac{u^2}{2}. \tag{5.14}$$

The function $E(t)$ may be interpreted as an *energy* for the system. The first term in $E(t)$ is proportional to the square of the velocity and represents the kinetic energy. The second term is proportional to the square of the displacement and represents the potential energy stored in the spring. With this interpretation, since the right-hand side of equation (5.13) is negative for $\gamma > 0$, we see that equation (5.13) is simply the statement that the energy in the system is decreasing. Physically this is consistent since when γ is positive, the dashpot dissipates the kinetic and potential energy. Also, when $\gamma = 0$

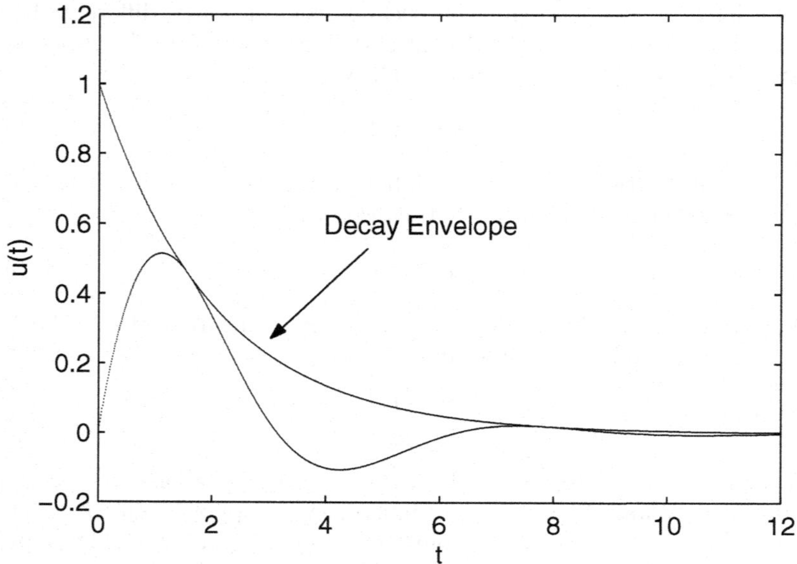

FIGURE 5.7: Unforced damped oscillations of the mass-spring system. The decay envelope is also shown.

equation (5.13) says that energy is conserved. Again this is physically realistic, $\gamma = 0$ corresponds to no damping. Notice in this special case equation (5.13) becomes

$$\frac{dE}{dt} = 0, \tag{5.15}$$

which immediately implies that E is a constant. We can imagine E being determined by initial conditions so that

$$E = \left[\frac{1}{2} \left(\frac{du}{dt} \right)^2 + \frac{u^2}{2} \right]_{t=0}. \tag{5.16}$$

Now, imagining E is constant as determined by initial conditions we can rewrite equation (5.14) as

$$E - \frac{u^2}{2} = \frac{1}{2} \left(\frac{du}{dt} \right)^2. \tag{5.17}$$

Since the right-hand side of this equation is always positive, the left-hand side must also be. This means that motion can be envisioned as restricted to the potential well defined by

$$V(u) = \frac{u^2}{2}. \tag{5.18}$$

That is, if we plot $V(u)$ and then draw a line representing the fixed energy E, motion can only occur in a way consistent with the condition that $E - V(u) > 0$. This immediately implies that $u(t)$ is bounded and periodic. A plot of this potential well is shown in Figure 5.8. In Chapter 6 we shall see yet another

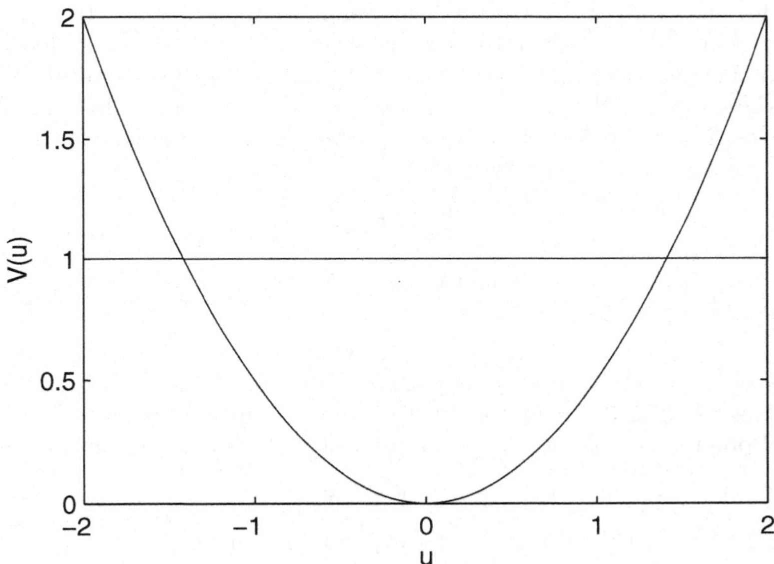

FIGURE 5.8: The potential well for the undamped unforced mass-spring system. The horizontal line indicates a typical constant value of the energy. Motion is restricted to the region between the points of intersection of the line and the parabola.

interpretation of the function $E(t)$ as a Lyapunov function for our system. Its existence implies that $u = 0$ is a stable solution when $\gamma > 0$. That is, if we perturb the system from the rest state, it returns to the state where both u and du/dt are equal to zero; the mass-spring oscillator comes to rest at the rest length of the spring.

5.3.3 The Phase Plane

Yet another perspective on our unforced mass-spring system is worth investigating. The stable rest state of the mass-spring oscillator for the case $\gamma > 0$ can be viewed as a point in the plane. This plane is called the *phase plane* for the system and the axes are the position of the mass, $u(t)$, and the velocity of the mass, $v(t) = du/dt$. If we rewrite our unforced model as a first-order system,

$$\frac{d}{dt}\begin{bmatrix} u \\ v \end{bmatrix} = \begin{bmatrix} 0 & 1 \\ -1 & -\gamma \end{bmatrix}\begin{bmatrix} u \\ v \end{bmatrix}, \tag{5.19}$$

this becomes immediately clear. The point $u = 0$, $v = 0$, is called a *critical point* for our problem. If we start the system at the point $u = 0$, $v = 0$, we stay at that point for all time. This is clear from rewriting as a first-order system since we see immediately that both du/dt and dv/dt are identically zero when $u = 0$ and $v = 0$. No change occurs at this point. In general, we can investigate the behavior of a second-order nonlinear system near a critical point. That is, a system of the form[3]

$$\frac{d\mathbf{u}}{dt} = F(\mathbf{u}), \tag{5.20}$$

where $\mathbf{u}$ is the 2 component vector $[u, v]$. A critical point for this system is simply a point such that

$$F(\mathbf{u}) = 0. \tag{5.21}$$

Where do solutions started nearby the critical point of equation (5.20) end up? This information is extracted from the system by *linearizing* near the critical point. If we denote a critical point of (5.20) by $\mathbf{u}^*$ and introduce

$$\mathbf{u} = \mathbf{u}^* + \epsilon\mathbf{w} \tag{5.22}$$

into equation (5.20) and then Taylor expand the function F about the point $\epsilon = 0$ we obtain

$$\epsilon\frac{d\mathbf{w}}{dt} = F(\mathbf{u}^*) + \epsilon DF(\mathbf{u}^*)\mathbf{w} + \cdots \tag{5.23}$$

The terms suppressed in this expression are all $O(\epsilon^2)$ and may be ignored for small ϵ. That is, nearby the critical point we retain only linear terms. If we disregard these higher-order terms and recall that $F(\mathbf{u}^*) = 0$ our linearized problem becomes

$$\frac{d\mathbf{w}}{dt} = DF(\mathbf{u}^*)\mathbf{w}. \tag{5.24}$$

The DF term is the *Jacobian matrix* of the function F at the point $\mathbf{u}^*$. If we denote the components of the function F by F_1 and F_2, then DF is[4]

$$DF(\mathbf{u}^*) = \begin{bmatrix} \dfrac{\partial F_1}{\partial u} & \dfrac{\partial F_1}{\partial v} \\ \\ \dfrac{\partial F_2}{\partial u} & \dfrac{\partial F_2}{\partial v} \end{bmatrix}_{\mathbf{u} = \mathbf{u}^*}. \tag{5.25}$$

So, the linearization near a critical point of the second-order nonlinear system equation (5.20) results in a linear system of the form

$$\frac{d}{dt}\begin{bmatrix} u \\ v \end{bmatrix} = \begin{bmatrix} a & b \\ c & d \end{bmatrix}\begin{bmatrix} u \\ v \end{bmatrix}. \tag{5.26}$$

Our damped unforced harmonic oscillator equation is simply a special case of this general linear system. For this linear system, equation (5.26), finding the critical points is equivalent to finding all solutions of

$$A\mathbf{u} = 0, \tag{5.27}$$

where

$$A = \begin{bmatrix} a & b \\ c & d \end{bmatrix}. \tag{5.28}$$

It is easy to see that if $ad - bc \neq 0$, then $\mathbf{u} = 0$ is the only critical point. The quantity $ad - bc$ is called the *determinant* of A. For our unforced harmonic oscillator, the determinant is equal to one and hence the origin is the only critical point for our oscillator. We know from the previous subsection that a solution starting with any initial conditions, i.e., at any point in the phase plane, decays toward the origin as time increases if $\gamma > 0$. If $\gamma = 0$ the solution oscillates for all time. If we sketch these possible behaviors in the phase plane, we obtain the phase portraits shown in Figure 5.9. In the case where $\gamma = 0$, all orbits in the phase plane are closed curves. In this case, the critical point at the origin is called a *center.* When $0 < \gamma < 2$, the phase plane appears as in Figure 5.9(a). In this case the critical point at the origin is called a *stable spiral.* In the case when $\gamma > 2$, the phase plane appears as in Figure 5.9(c). In this case the critical point at the origin is called a *stable node.* The remaining possible behaviors of the critical point at the origin are not realized by our unforced harmonic oscillator. However, in general, equation (5.26) could have another of the phase portraits shown in Figure 5.9. The phase portrait of Figure 5.9(e) is called a *saddle* while Figure 5.9(f) and (d) are called an *unstable spiral* and *unstable node*, respectively. The reader is referred to the exercises and the related reading at the end of the chapter for more information on the phase plane.

5.3.4 Forced Oscillations and Resonance

The forced second-order harmonic oscillator equation,

$$\frac{d^2u}{dt^2} + \gamma\frac{du}{dt} + u = \delta\cos(\omega t), \tag{5.29}$$

is also linear and may be solved exactly.[5] We can write the solution as

$$u(t) = c_0 e^{-\frac{\gamma t}{2}}\sin(\omega_0 t) + c_1 e^{-\frac{\gamma t}{2}}\cos(\omega_0 t) + A\cos(\omega t - \phi), \tag{5.30}$$

where

$$\omega_0 = \sqrt{1 - \frac{\gamma^2}{2}} \tag{5.31}$$

$$\phi = \tan^{-1}\left(\frac{\gamma\omega}{\sqrt{1-\omega^2}}\right) \tag{5.32}$$

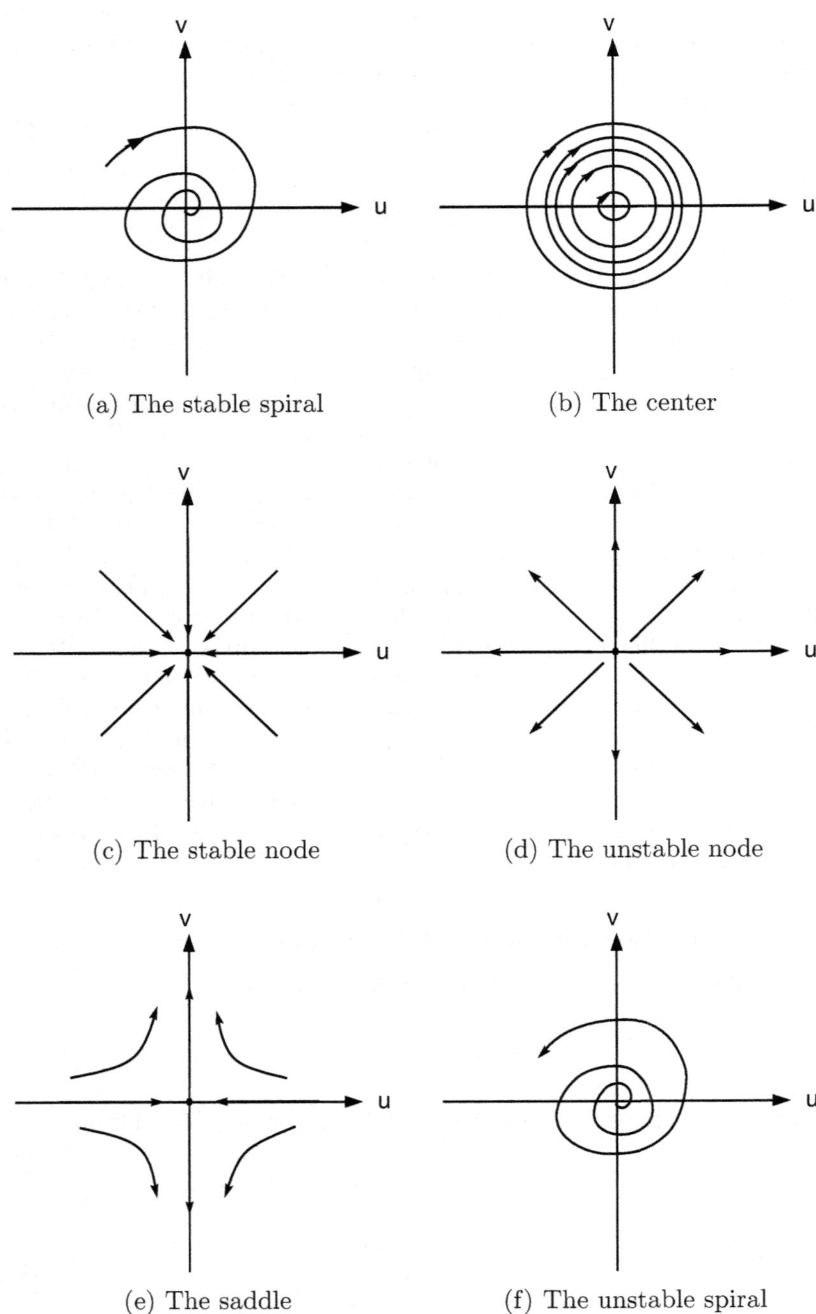

(a) The stable spiral

(b) The center

(c) The stable node

(d) The unstable node

(e) The saddle

(f) The unstable spiral

FIGURE 5.9: Phase plane portraits.

and

$$A = \frac{\delta}{\sqrt{(1-\omega^2)^2 + \gamma^2\omega^2}}. \tag{5.33}$$

The first two terms in the solution may be thought of as the *transient response*. If $\gamma > 0$ these terms decay as time increases. The third term is purely oscillatory, is present for all time, and may be thought of as the response of the system to the forcing function. Notice that this term oscillates with the same frequency as the forcing function, but with a phase ϕ that is generally different from the forcing term. After all transient terms have died out, the system is oscillating at the forcing frequency but with a phase shift. The amplitude of the solution after the transient effect has passed is contained in the A term. Notice this term is proportional to the magnitude of the forcing term through δ but also depends on the dimensionless measure of damping, γ, and the dimensionless forcing frequency, ω. If we plot A versus ω for fixed γ and δ, we obtain what is called the frequency-response diagram for the system. See Figure 5.10. At a glance it shows the magnitude of the response to different

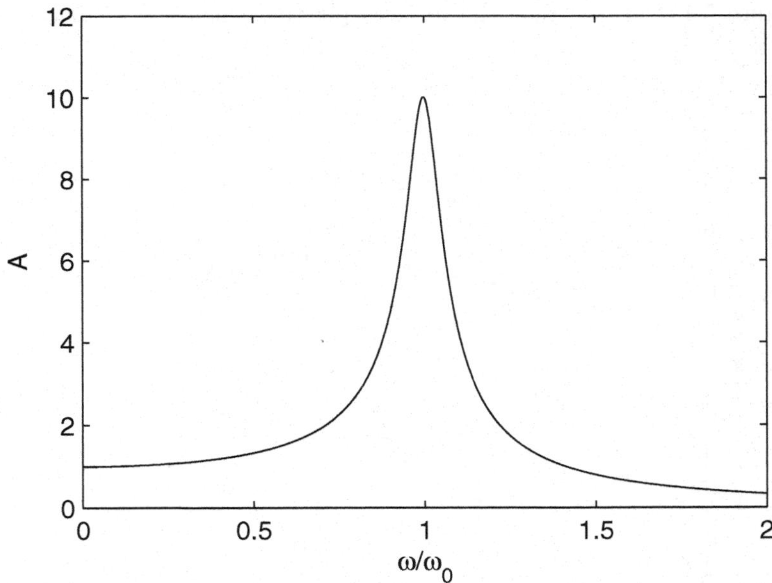

FIGURE 5.10: The frequency-response diagram for the mass-spring system. Note the peak at $\omega = \omega_0$, the scaled resonant frequency.

forcing frequencies. Notice that this plot is peaked at the point $\omega = \omega_0$, the natural or *resonant frequency* of the system. This is the phenomenon known as *resonance*, that is, the amplitude of the response of the system to forcing at the natural frequency is larger than the amplitude of the response to forcing at any other frequency. Finally, we note that we can easily compute the resonant frequency in the familiar units of inverse time. The argument of our solution is of the form $\omega_0 t$, which in terms of dimensional time, is

$$\omega_0 t' \sqrt{\frac{k}{m}} \tag{5.34}$$

and hence the natural frequency of the harmonic oscillator is

$$f_0 = \sqrt{\frac{k}{m}\left(1 - \frac{\gamma^2}{2}\right)}. \tag{5.35}$$

5.3.5 Nonlinear Springs

Real springs and real elastic systems often have nonlinear behavior. One way to include nonlinear terms in the mass-spring system is to return to Hooke's law and make a modification. The spring force modeled by Hooke's law is assumed to be a restoring force. If we plot the force F_s versus x as in Figure 5.11, we see that the force is positive when the displacement is

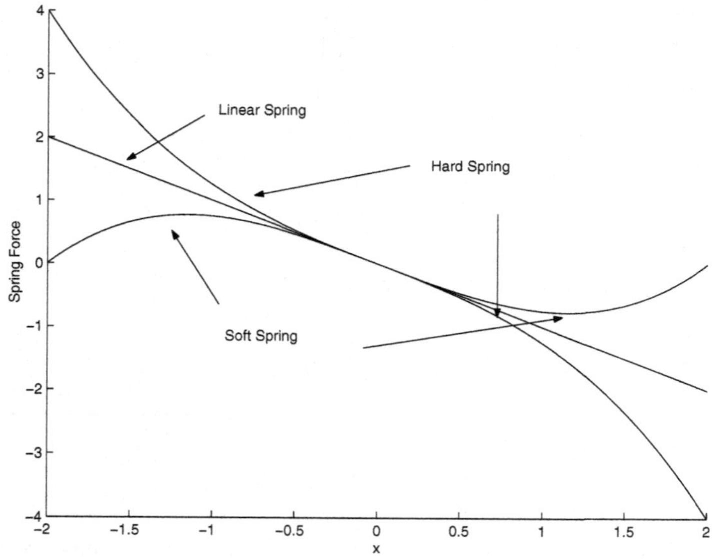

FIGURE 5.11: Linear and nonlinear spring force versus deflection.

negative and vice versa when the displacement is positive. That is, the force acts to oppose the motion. We can preserve this restorative behavior and make Hooke's law nonlinear by adding a cubic term. That is, by introducing a modified spring force of the form

$$F_m = -kx + k_1 x^3. \tag{5.36}$$

For small displacements this preserves the restoring force behavior as desired. Notice that according to the sign of k_1 two types of behavior are possible. If $k_1 > 0$ then F_m is always less than F_s. In this case we say the spring is a weak or soft spring. The nonlinear restoring force is always less than the linear case. If $k_1 < 0$ then F_m is always greater than F_s. In this case we say the spring is a strong or hard spring. The nonlinear restoring force is always greater than in the linear case. In microsystems these nonlinear terms often become important in the case of strain-hardening or strain-softening. This is the tendency of materials to become harder or softer when subjected to large displacements. Studying a mass-spring system with the appropriate nonlinearity can lend insight into device design. The addition of nonlinear terms to a mass-spring model of a MEMS or NEMS device may also be appropriate when displacements in the system are expected to be large. As was mentioned in Chapter 2, the linear elastic theory presented assumed small displacements and small gradients. Real systems may or may not abide by these assumptions. Taking a first crack at estimating the effect of large displacements by adding nonlinear terms to a mass-spring model is often useful.

5.3.6 Effective Spring Constants

Few MEMS devices will have the simple structure pictured in Figure 5.6. One way to view our study of the mass-spring model is simply as an abstraction of the essential features of elastic devices. However, we may wish to go further. In fact, one might attempt to relate the parameters in our mass-spring model to material parameters for a real device. But few real devices have a simple spring with known spring constant as their key mechanical element. It is possible and does happen that the spring in Figure 5.6 is replaced by a beam or by several beams. In this case, we can write Hooke's law in 1-D for a beam as

$$\sigma = E\frac{\partial u}{\partial x}, \tag{5.37}$$

where E is Young's modulus, σ is the stress in the axial direction, u the strain in the axial direction, and x a coordinate measuring distance along the axis of the beam. In this way, we see that k can be related to the Young's modulus, E, for beam-like springs. The Young's modulus is tabulated[6] for many materials used in MEMS construction, and hence this is particularly convenient. A typical computation of such an *effective spring constant* would proceed as follows. First, we note that Hooke's law for a spring states

$$F_s = -kx. \tag{5.38}$$

Now, if the we apply a force to the right end of the beam in Figure 5.12 of

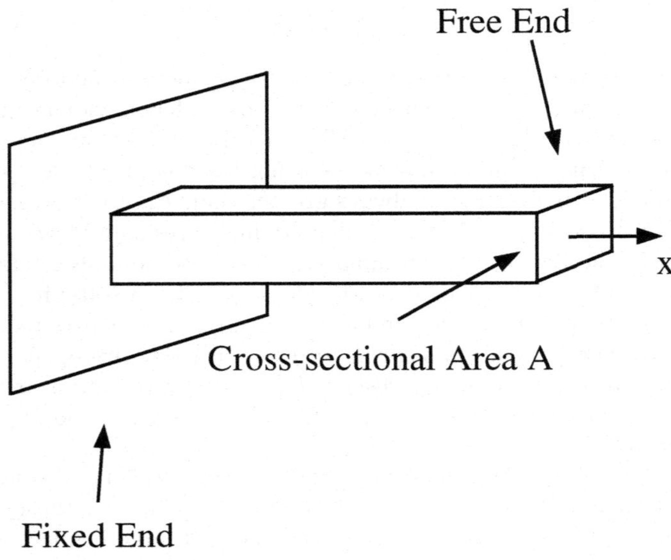

FIGURE 5.12: The spring-like beam for computing an effective spring constant.

magnitude F, the resulting stress on the beam has magnitude

$$\sigma = \frac{F}{A}, \tag{5.39}$$

where A is the cross-sectional area of the beam. From equation (5.37) we can find the displacement of the beam by integrating. We find

$$u(x) = \frac{F}{AE}x + c_0. \tag{5.40}$$

Requiring that the beam be fixed at $x = 0$ implies that $c_0 = 0$. So, the displacement of the right end of the beam is

$$u(L) = \frac{FL}{AE}. \tag{5.41}$$

But $u(L)$ is simply the change in length of the beam, or in terms of Hooke's law, x. Equating our two expressions for the change in length we arrive at

$$k_{\text{eff}} = \frac{AE}{L} \tag{5.42}$$

as our effective spring constant for the beam system.[7]

5.4 Membranes

Pressure sensors, micropumps, thermopneumatic actuators, electrostatic display systems, and numerous other MEMS and NEMS devices make use of thin membranes held under tension. Consequently, the elastic membrane will be central to many of the models studied in the remainder of this text. In Chapter 6 we will study a thermopneumatic actuator that consists of an elastic membrane capping a gas-filled cylinder. In Chapter 7, the elastic membrane will be at the heart of our investigation into electrostatic actuation. In Chapter 8, the elastic membrane will be an important component in modelling magnetic actuation. Here we study the deformation and vibration of membranes with known loading.

5.4.1 Governing Equation

In Chapter 2 we wrote down the equation governing an elastic membrane as a simplification of the Navier equations of elasticity. In an exercise for Chapter 2, the membrane equation was rederived from a variational principle. Recall that an elastic membrane is a body with tension. It has no resistance to bending, but resists stretching. We modify the equation of Chapter 2 to allow for a load and write our membrane equation as

$$\rho \frac{\partial^2 u'}{\partial t'^2} = \mu \nabla^2 u' + f(x', y', t'). \tag{5.43}$$

We are assuming the coordinate system shown in Figure 2.3. The function $u(x', y', t')$ is the height of the membrane at location (x', y') and time t' above some reference position. The function $f(x', y', t')$ is the load on the membrane. Recall that ρ is the density of the membrane and μ is the tension in the membrane. As with the mass on a spring, it is convenient to rescale our governing equations to reduce to a problem with fewer dimensionless parameters. We scale u' with some characteristic displacement, x' and y' with some characteristic length in the system, and time with a natural time scale of vibration by introducing

$$u = \frac{u'}{A}, \quad x = \frac{x'}{L}, \quad y = \frac{y'}{L}, \quad t = \sqrt{\frac{\mu}{\rho}} \frac{t'}{L} \tag{5.44}$$

into equation (5.43), which yields

$$\frac{\partial^2 u}{\partial t^2} = \nabla^2 u + F(x, y, t). \tag{5.45}$$

The function F is a scaled form of the forcing term f.

5.4.2 One-Dimensional Membrane with Uniform Load

As a first example of membrane mechanics, we look at a one-dimensional membrane. We work with the scaled equation obtained above. The one-dimensional membrane may be thought of as an infinite strip or as a simple string. Recall from Chapter 2 that the string equation is just the membrane equation with u a function of one Cartesian coordinate only, which here we take to be x. Let us examine the steady-state deflection of such a string with a constant load of magnitude p. The governing equation is then

$$\frac{d^2u}{dx^2} + p = 0. \tag{5.46}$$

We assume the string is held fixed at two endpoints. These may be taken to be $x = \pm 1/2$ as the string length has been scaled to one. We impose the boundary conditions,

$$u(-1/2) = u(1/2) = 0. \tag{5.47}$$

Equation (5.46) is easy to integrate. We find

$$u(x) = -\frac{px^2}{2} + c_0 x + c_1 \tag{5.48}$$

and applying the boundary conditions, we find the parabolic profile,

$$u(x) = -\frac{p}{2}\left(x^2 - \frac{1}{4}\right). \tag{5.49}$$

Or, returning to dimensional variables,

$$u'(x') = -\frac{PL^2}{2\mu}\left[\left(\frac{x'}{L}\right)^2 - \frac{1}{4}\right], \tag{5.50}$$

where P is the dimensional load on the string.

As a slightly more challenging example, let us compute the deflection of a circular elastic membrane with constant load of magnitude p. Here, again the problem may be assumed to be one-dimensional with the deflection a function of the radial variable r only. Recalling the Laplace operator[8] in cylindrical coordinates, we may write the governing equation as

$$\frac{d^2u}{dr^2} + \frac{1}{r}\frac{du}{dr} + p = 0. \tag{5.51}$$

We assume the membrane is held fixed on its outer boundary. This may be taken to be at $r = 1$ as the membrane radius has been scaled to one. We impose the boundary condition

$$u(1) = 0. \tag{5.52}$$

Notice that this is only one condition while our governing equation is of second order. If we integrate our governing equation we find

$$u(r) = -\frac{pr^2}{4} + c_0 \log(r) + c_1. \tag{5.53}$$

The second term in this expression becomes infinite at $r = 0$. This tells us that as a second condition we should require that $u(r)$ be bounded at the origin and take $c_0 = 0$. Imposing the condition at $r = 1$ we find

$$u(r) = -\frac{p}{4}\left(r^2 - 1\right). \tag{5.54}$$

Or, returning to dimensional variables,

$$u'(r') = -\frac{PL^2}{4\mu}\left[\left(\frac{r'}{L}\right)^2 - 1\right], \tag{5.55}$$

where P is the pressure on the membrane and L is the membrane radius.

5.4.3 One-Dimensional Membrane with Concentrated Load

Understanding the response of a system to a concentrated load or point force is useful for both theoretical and practical reasons. As we will see in Section 5.4.4, from the theoretical point of view it is useful since the solution for an arbitrary load can be constructed using only the known load and the solution for a point source. From a practical point of view, a point source is a useful model of the behavior of many real systems. An electrostatic actuator with a small electrode or a magnetic actuator with a small permanent magnet are a few examples of MEMS systems where modeling with a concentrated load may be appropriate.

We model the point source by using the *Dirac delta function* and consider

$$\frac{d^2g}{dx^2} + \delta(x - x_0) = 0. \tag{5.56}$$

The Dirac delta function[9] is zero at all points except $x = x_0$ and has the property that

$$\int_{-\infty}^{\infty} \delta(x - x_0)dx = 1. \tag{5.57}$$

Equation (5.56) models a one-dimensional string with point source at $x = x_0$. We are using g rather than u for the solution. The solution to this problem is called the *Green's function*. As we will see in the next section, it is convenient to have a separate notation for g. We assume the ends of the string are held fixed and impose

$$g(-1/2) = g(1/2) = 0. \tag{5.58}$$

We solve this problem by using the fact that the delta function vanishes away from the point $x = x_0$. Hence, to the left and right of x_0, the solution is simply a linear function. That is,

$$g(x) = \begin{cases} ax + b \;, -1/2 \le x < x_0 \\ cx + d \;, x_0 \le x \le 1/2. \end{cases} \tag{5.59}$$

Imposing the boundary conditions at $x = \pm 1/2$ reduces this to

$$g(x) = \begin{cases} a(x + 1/2) \;, -1/2 \le x < x_0 \\ c(x - 1/2) \;, x_0 \le x \le 1/2. \end{cases} \tag{5.60}$$

Now, we still have two unknown constants, a and c. In order to determine them, we require first that g be continuous at the point x_0, i.e.,

$$g(x_0-) = g(x_0+), \tag{5.61}$$

where the plus and minus indicate solutions to the right and left of x_0, respectively. The final condition comes from integrating the governing equation across the point x_0 and using the properties of the Dirac delta function. This says

$$\frac{dg}{dx}(x_0+) - \frac{dg}{dx}(x_0-) = -1. \tag{5.62}$$

Applying these two conditions we find

$$g(x) = -\begin{cases} (x + 1/2)(x_0 - 1/2) \;, -1/2 \le x < x_0 \\ (x - 1/2)(x_0 + 1/2) \;, x_0 \le x \le 1/2 \end{cases} \tag{5.63}$$

as the deflection of an elastic membrane with concentrated load of unit strength at the point x_0.

5.4.4 One-Dimensional Membrane with Arbitrary Load

As a final example in static membrane mechanics, we compute the deflection of a one-dimensional membrane subjected to an arbitrary load. In addition to intrinsic interest, this example is interesting because of the method used to construct the solution. In particular, we make use of the Green's function derived in the previous subsection. The technique introduced is powerful and generalizes to allow construction of representations of the solution to linear problems in both ordinary and partial differential equations. Further, this technique is useful for constructing integral representations of nonlinear differential equations. In Chapter 7 we will work out an example utilizing this technique for a nonlinear equation modeling an electrostatically deflected membrane. Properties of the solution will follow directly from this representation as will insight into iterative numerical methods.

We consider

$$\frac{d^2u}{dx^2} + p(x) = 0, \tag{5.64}$$

where $p(x)$ is a known arbitrary load. We assume the ends of the string are held fixed and impose

$$u(-1/2) = u(1/2) = 0. \tag{5.65}$$

Now, we use the Green's function of the previous subsection to derive a solution for this problem. We start by multiplying equation (5.56) by u, equation (5.64) by g, subtracting the two equations, and integrating from $x = -1/2$ to $x = 1/2$ to obtain

$$\int_{-1/2}^{1/2} \left(g(x)\frac{d^2u}{dx^2} - u(x)\frac{d^2g}{dx^2} + p(x)g(x,x_0) - u(x)\delta(x-x_0) \right) dx = 0. \tag{5.66}$$

Next, we integrate the first two terms by parts one time and apply the boundary conditions on both u and g. All contributions from these first two terms vanish and we are left with

$$u(x_0) = \int_{-1/2}^{1/2} p(x)g(x,x_0)dx. \tag{5.67}$$

Noting that g is a symmetric function of x and x_0 we may rename the variables and write

$$u(x) = \int_{-1/2}^{1/2} p(x_0)g(x,x_0)dx_0. \tag{5.68}$$

This is the Green's function representation of the solution.

5.4.5 Free Vibrations of a One-Dimensional Membrane

Next we investigate vibrations of a one-dimensional elastic membrane. We work with the unforced one-dimensional membrane equation in dimensionless form, i.e.,

$$\frac{\partial^2 u}{\partial t^2} = \frac{\partial^2 u}{\partial x^2}. \tag{5.69}$$

We should think of this as the governing equation for the motion of a string with x the coordinate along the string. We assume the string is held fixed at two endpoints. These may be taken to be $x = 0$ and $x = 1$ as the string length has been scaled to one. We impose the boundary conditions,

$$u(0,t) = u(1,t) = 0. \tag{5.70}$$

We seek solutions of this equation via the method of *separation of variables.* In this section we work through the method in detail. In the remainder of the text, we assume the reader is conversant with this method. For more practice with separation of variables, consult the exercises and related reading at the end of the chapter. To begin, we assume the solution $u(x,t)$ can be written as the product of two functions, one a function of x alone, the other a function of t alone. In particular we take

$$u(x,t) = A(t)\phi(x). \tag{5.71}$$

Substituting into equation (5.69) we obtain

$$\ddot{A}\phi = A\phi'', \tag{5.72}$$

where the dots indicate differentiation with respect to time and the primes differentiation with respect to x. Now, dividing by $A\phi$ this may be rewritten as

$$\frac{\ddot{A}}{A} = \frac{\phi''}{\phi}. \tag{5.73}$$

The key observation of separation of variables is that the left-hand side of this expression is a function of t alone while the right-hand side is a function of x alone. This can only occur if both sides are equal to a constant. We denote this *separation constant* by $-\lambda^2$ and write

$$\frac{\ddot{A}}{A} = \frac{\phi''}{\phi} = -\lambda^2. \tag{5.74}$$

The second of these two equations can be written as

$$\phi'' + \lambda^2\phi = 0 \tag{5.75}$$

and the boundary conditions on $u(x,t)$ immediately imply the conditions

$$\phi(0) = \phi(1) = 0. \tag{5.76}$$

Equations (5.75) and (5.76) constitute an *eigenvalue problem.* That is, this system only has nontrivial solutions for certain characteristic values of λ, called *eigenvalues.* We see this by first writing the general solution of equations (5.75) and (5.76) as

$$\phi(x) = c_0\sin(\lambda x) + c_1\cos(\lambda x). \tag{5.77}$$

Imposing the boundary condition at $x = 0$ implies that $c_1 = 0$ while the boundary condition at $x = 1$ says that either $c_0 = 0$ or

$$\sin(\lambda) = 0. \tag{5.78}$$

Since $c_0 = 0$ yields only the trivial solution we must require (5.78). This implies that $\lambda = n\pi$ where $n = 1, 2, 3, \ldots$. These values of λ are the *eigenvalues* or natural frequencies of the system. They are the analog of the natural frequency we found in the mass-spring system. The difference is that the one-dimensional membrane has a countable infinity of natural frequencies.

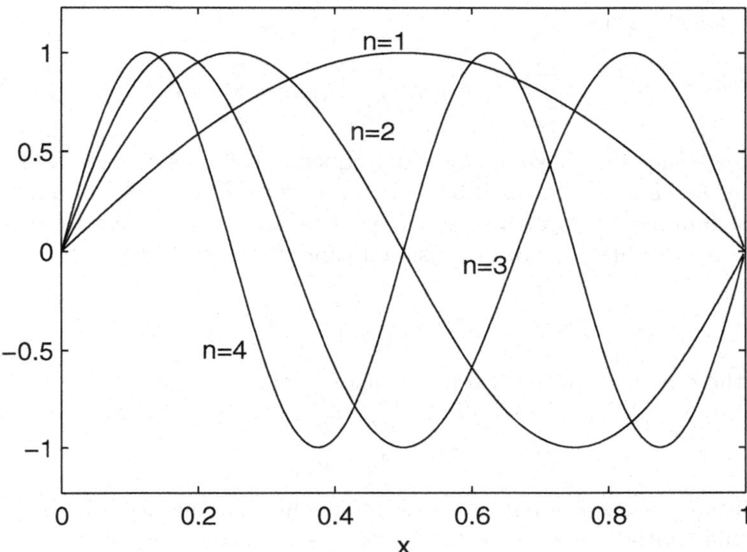

FIGURE 5.13: The first four normal modes of the vibrating string. We use the normalization $c_n = 1$.

The corresponding solutions to the spatial eigenfunction problem, called the *eigenfunctions* or *normal modes*, are

$$\phi_n(x) = c_n \sin(n\pi x). \tag{5.79}$$

The first few normal modes are plotted in Figure 5.13. We use the normalization $c_n = 1$. Notice that the normal modes corresponding to even values of n are symmetric about $x = 1/2$ while the normal modes corresponding to odd values of n are antisymmetric. Next, we use the values $\lambda_n = n\pi$ to solve the time-dependent equation in equation (5.74). That is,

$$\ddot{A} + n^2\pi^2 A = 0. \tag{5.80}$$

This equation has the general solution

$$A_n(t) = a_n \cos(n\pi t) + b_n \sin(n\pi t). \tag{5.81}$$

Hence we may construct product solutions for $u(x,t)$,

$$A_n(t)\phi(x) = (a_n \cos(n\pi t) + b_n \sin(n\pi t)) \sin(n\pi x), \tag{5.82}$$

where we have absorbed the constant in the spatial eigenfunction into the a_n and b_n. The general solution to equation (5.69) is then a summation of these product solutions, i.e.,

$$u(x,t) = \sum_{n=1}^{\infty} (a_n \cos(n\pi t) + b_n \sin(n\pi t)) \sin(n\pi x). \tag{5.83}$$

It is also instructive to take the natural frequencies of our system in dimensionless form, $n\pi$, and relate them back to the natural frequencies of our system with the familiar units of inverse time. The argument of our time-varying solutions is $n\pi t$, in terms of dimensional time t', this is

$$n\pi t' \sqrt{\frac{\mu}{\rho}} \frac{1}{L}. \tag{5.84}$$

Hence, the dimensional natural frequencies of our vibrating string are

$$f_n = \frac{n\pi}{L}\sqrt{\frac{\mu}{\rho}}. \tag{5.85}$$

Finally we comment on using the general solution, equation (5.83), to solve a particular initial value problem. Usually, we impose conditions on the displacement and velocity of $u(x,t)$ at time $t = 0$. For example,

$$u(x,0) = g(x) \tag{5.86}$$

and

$$\frac{\partial u}{\partial t}(x,0) = h(x). \tag{5.87}$$

The a_n's and b_n's are then determined by the initial conditions. In particular,

$$g(x) = \sum_{n=1}^{\infty} a_n \sin(n\pi x) \tag{5.88}$$

and

$$h(x) = \sum_{n=1}^{\infty} n\pi b_n \sin(n\pi x). \tag{5.89}$$

To extract the individual values of the a_n's and b_n's, we use the *orthogonality* property of the eigenfunctions. That is,

$$\int_0^1 \phi_n(x)\phi_m(x)dx = 0 \tag{5.90}$$

if $n \neq m$ and

$$\int_0^1 \phi_n(x)\phi_n(x)dx = \frac{1}{2}. \tag{5.91}$$

So, we multiply equation (5.88) or equation (5.89) by $\sin(m\pi x)$, integrate from 0 to 1 and solve for a_n and b_n to find

$$a_n = 2\int_0^1 g(x)\sin(n\pi x)dx \tag{5.92}$$

$$b_n = \frac{2}{n\pi}\int_0^1 h(x)\sin(n\pi x)dx. \tag{5.93}$$

5.4.6 Forced Vibrations and Resonance

Our final discussion of membranes in this chapter concerns the forced vibration of a one-dimensional elastic membrane. In particular, we consider

$$\frac{\partial^2 u}{\partial t^2} = \frac{\partial^2 u}{\partial x^2} - \delta(x - x_0)e^{-i\omega t}. \tag{5.94}$$

That is, we are looking at the response of the membrane to time harmonic forcing of frequency ω concentrated at the point x_0. As with the steady-state above, this is actually the Green's function problem for the membrane. A solution for any time harmonic load can be synthesized from the solution we construct here. The exponential form of the forcing is chosen for convenience, a real solution may be obtained by taking the real part of $u(x,t)$ at the end of the analysis. We assume fixed boundary conditions

$$u(0,t) = u(1,t) = 0 \tag{5.95}$$

and we seek a solution of the form

$$u(x,t) = \phi(x)e^{-i\omega t}. \tag{5.96}$$

Substituting this into equation (5.94) and canceling the exponential terms yields

$$\frac{d^2\phi}{dx^2} + \omega^2\phi = \delta(x - x_0). \tag{5.97}$$

This is very similar to the Green's function problem we studied earlier. We can construct a solution in the same way. That is, to the left and right of x_0 we solve the equation taking the right-hand side to be zero. This yields

$$\phi(x) = \begin{cases} a\sin(\omega x) + b\cos(\omega x) & ,0 \le x < x_0 \\ c\sin(\omega x) + d\cos(\omega x) & ,x_0 \le x \le 1. \end{cases} \tag{5.98}$$

The boundary conditions on u imply that

$$\phi(0) = \phi(1) = 0. \tag{5.99}$$

Applying these to the solution for ϕ yields

$$\phi(x) = \begin{cases} a\sin(\omega x) & ,0 \le x < x_0 \\ c(\sin(\omega x) - \tan(\omega)\cos(\omega x)) & ,x_0 \le x \le 1. \end{cases} \tag{5.100}$$

Once again we are left with two unknowns and must find two additional conditions to impose. The first of these is continuity of ϕ at the point x_0, i.e.,

$$\phi(x_0-) = \phi(x_0+), \tag{5.101}$$

and the second is obtained by integrating equation (5.97) across the point x_0 to obtain

$$\frac{d\phi}{dx}(x_0+) - \frac{d\phi}{dx}(x_0-) = 1. \tag{5.102}$$

Applying these two conditions we find

$$\phi(x) = \begin{cases} -\dfrac{\sin(\omega x)\sin(\omega(1-x_0))}{\omega \sin(\omega)}, & 0 \leq x < x_0 \\ -\dfrac{\sin(\omega x_0)\sin(\omega(1-x))}{\omega \sin(\omega)}, & x_0 \leq x \leq 1. \end{cases} \tag{5.103}$$

Notice that the amplitude of $\phi(x)$ becomes infinite when the forcing frequency ω equals $n\pi$. These values of ω correspond exactly to the natural frequencies of the system we uncovered in our study of free vibrations of the string. Again we have the phenomenon of resonance occurring in our system. The amplitude of the response at and near the resonant frequencies is larger than at any other forcing frequency. Unlike our mass-spring system, here the amplitude becomes infinite at the resonant frequencies. This is due to the fact that we have not included damping in our system. If a damping term were included, we would simply see that the amplitude of $\phi(x)$ became large but remained finite at the natural frequencies.

5.5 Beams

The elastic beam is a ubiquitous structure in MEMS and NEMS engineering. Resonant sensors, electrostatic comb drives, thermal bimorph actuators, and other devices make use of beam-like mechanisms. The beam structure will appear repeatedly throughout this text. For example, in Chapter 6, while studying thermal actuation we construct a model of a thermal bimorph actuator composed of elastic beams. Here we study the deformation and vibration of beams with known loading.

5.5.1 Governing Equation

In Chapter 2 the equation governing the motion of an elastic beam was derived from Hamilton's principle of least action. Recall that an elastic beam is an idealized body that exhibits resistance to both bending and stretching. We modify the equation of Chapter 2 to allow for a load and write our beam equation as

$$\rho A \frac{\partial^2 u'}{\partial t'^2} - \mu \frac{\partial^2 u'}{\partial x'^2} + \frac{\partial^2}{\partial x'^2}\left(EI \frac{\partial^2 u'}{\partial x'^2}\right) = f(x', t'). \tag{5.104}$$

We are assuming the coordinate system shown in Figure 5.12. That is, x' is an axial coordinate while the u' captures deflections in the $x' - y'$ plane. The function $f(x', t')$ is the load on the beam. Recall ρ is the beam density, μ the tension in the beam, and EI the flexural rigidity. We assume that EI is constant in the remainder of this section. It is convenient to rescale our governing equation to reduce to dimensionless form. We scale x' with the length of the beam, time with a natural time scale of vibration, and u' with some characteristic displacement by introducing

$$x = \frac{x'}{L}, \quad u = \frac{u'}{B}, \quad t = \frac{t'}{L^2}\sqrt{\frac{EI}{A\rho}} \tag{5.105}$$

into equation (5.104), which yields

$$\frac{\partial^2 u}{\partial t^2} - \alpha\frac{\partial^2 u}{\partial x^2} + \frac{\partial^4 u}{\partial x^4} = F(x, t). \tag{5.106}$$

The parameter α is given by

$$\alpha = \frac{\mu A}{L^4 EI} \tag{5.107}$$

and is a dimensionless measure of the relative strengths of resistance to stretching and bending in the problem. The function F is a dimensionless form of the forcing function f.

5.5.2 Cantilevered Beam with Uniform Load

As a first example of beam mechanics, we examine steady-state deflections of the elastic beam with constant load. We assume that $\alpha = 0$, i.e., we neglect the effects of tension in the beam, and study

$$\frac{d^4 u}{dx^4} = -p, \tag{5.108}$$

where p is a constant. We consider the case of a cantilevered beam where the end at $x = 0$ is fixed and the end at $x = 1$ is free. From Chapter 2 we choose the appropriate boundary conditions and impose

$$u(0) = \frac{du}{dx}(0) = \frac{d^2u}{dx^2}(1) = \frac{d^3u}{dx^3}(1) = 0. \tag{5.109}$$

Equation (5.108) is easy to integrate. We find

$$u(x) = -\frac{px^4}{24} + \frac{c_0x^3}{6} + \frac{c_1x^2}{2} + c_2x + c_3, \tag{5.110}$$

and applying the boundary conditions we find the profile

$$u(x) = -\frac{px^2}{2}\left(\frac{x^2}{12} - \frac{x}{3} + \frac{1}{2}\right). \tag{5.111}$$

Or, in terms of dimensional variables,

$$u'(x') = -\frac{PL^4}{2EI}\left(\frac{x'}{L}\right)^2\left[\frac{1}{12}\left(\frac{x'}{L}\right)^2 - \frac{x'}{3L} + \frac{1}{2}\right], \tag{5.112}$$

where P is the dimensional load on the beam.

5.5.3 Cantilevered Beam with Concentrated Load

Next, we consider the response of the cantilevered beam to a concentrated load at the point x_0. As with the membrane, we use the Dirac delta function to model the point source. We take $\alpha = 0$ and consider

$$\frac{d^4u}{dx^4} = \delta(x - x_0). \tag{5.113}$$

We again consider the case of a cantilevered beam where the end at $x = 0$ is fixed and the end at $x = 1$ is free. We impose the boundary conditions,

$$u(0) = \frac{du}{dx}(0) = \frac{d^2u}{dx^2}(1) = \frac{d^3u}{dx^3}(1) = 0. \tag{5.114}$$

Now, using the fact that the Dirac delta function vanishes away from x_0, we solve for x less than and greater than x_0 to find

$$u(x) = \begin{cases} a_0x^3 + a_1x^2 + a_2x + a_3 & , 0 \le x < x_0 \\ b_0x^3 + b_1x^2 + b_2x + b_3 & , x_0 \le x \le 1. \end{cases} \tag{5.115}$$

Applying the boundary conditions at $x = 0$ and $x = 1$ reduces this to

$$u(x) = \begin{cases} a_0x^3 + a_1x^2 & , 0 \le x < x_0 \\ b_2x + b_3 & , x_0 \le x \le 1. \end{cases} \tag{5.116}$$

Notice that we still have four unknown constants to determine. We require that the solution, its first derivative, and its second derivative all be continuous at x_0. That is, we impose

$$u(x_0-) = u(x_0+) \tag{5.117}$$

$$\frac{du}{dx}(x_0-) = \frac{du}{dx}(x_0+) \tag{5.118}$$

$$\frac{d^2u}{dx^2}(x_0-) = \frac{d^2u}{dx^2}(x_0+). \tag{5.119}$$

The final condition comes from integrating equation (5.113) across the point x_0 to obtain

$$\frac{d^3u}{dx^3}(x_0+) - \frac{d^3u}{dx^3}(x_0-) = 1. \tag{5.120}$$

Applying these four conditions we find

$$u(x) = \begin{cases} -\frac{x^2}{2}\left(x_0 - \frac{x}{3}\right), 0 \le x < x_0 \\ -\frac{x_0^2}{2}\left(x - \frac{x_0}{3}\right), x_0 \le x \le 1 \end{cases} \tag{5.121}$$

as the deflection of a cantilevered elastic beam with concentrated load of unit strength at the point x_0.

5.5.4 Free Vibrations of a Beam

Next, we investigate the free vibrations of an elastic beam. We work with our equation in dimensionless form and take $\alpha = 0$. That is, we consider

$$\frac{\partial^2 u}{\partial t^2} + \frac{\partial^4 u}{\partial x^4} = 0. \tag{5.122}$$

This time we consider the case where both ends of the beam are pinned. So, we impose the boundary conditions

$$u(0,t) = \frac{\partial^2 u}{\partial x^2}(0,t) = 0 \tag{5.123}$$

$$u(1,t) = \frac{\partial^2 u}{\partial x^2}(1,t) = 0. \tag{5.124}$$

Using the method of separation of variables, we seek product solutions in the form $u(x,t) = A(t)\phi(x)$ and obtain

$$\frac{d^4\phi}{dx^4} - \lambda^2\phi = 0, \tag{5.125}$$

where λ^2 is the separation constant. The boundary conditions on u imply

$$\phi(0) = \frac{d^2\phi}{dx^2}(0) = 0 \tag{5.126}$$

$$\phi(1) = \frac{d^2\phi}{dx^2}(1) = 0. \tag{5.127}$$

Equation (5.125) has the general solution,

$$\phi(x) = a_0 \sin(\lambda x) + a_1 \cos(\lambda x) + a_2 \sinh(\lambda x) + a_3 \cosh(\lambda x). \tag{5.128}$$

Applying the boundary conditions we find that $a_1 = a_2 = a_3 = 0$, a_0 is arbitrary and λ satisfies

$$\sin(\lambda) = 0. \tag{5.129}$$

Hence, $\lambda = n\pi$, where $n = 1, 2, 3, \ldots$ Solving the problem for $A(t)$ yields

$$A_n(t) = a_n \cos(n\pi t) + b_n \sin(n\pi t), \tag{5.130}$$

so that we have product solutions,

$$\phi(x)A_n(t) = (a_n \cos(n\pi t) + b_n \sin(n\pi t))\sin(n\pi x), \tag{5.131}$$

and the general solution,

$$u(x,t) = \sum_{n=1}^{\infty}(a_n \cos(n\pi t) + b_n \sin(n\pi t))\sin(n\pi x). \tag{5.132}$$

The general solution can be used to solve an initial value problem in the same way in which the initial value problem for the vibrating membrane example was solved. We note that here our natural frequencies are $n\pi$ and that these may be related back to beam parameters. The natural frequencies of our vibrating beam are

$$f_n = \frac{n\pi}{L^2}\sqrt{\frac{EI}{\rho A}}. \tag{5.133}$$

5.6 Plates

The elastic plate is the two-dimensional analog of the one-dimensional elastic beam. The plate is an idealized elastic structure that resists both bending and stretching, but occupies a two-dimensional rather than one-dimensional region. Consequently, the elastic plate appears in MEMS and NEMS structures such as micropumps and microvalves where planar structure is essential. Here we study the free vibration and deformation of elastic plates.

5.6.1 Governing Equation

In Chapter 2 we indicated how the plate equation may be derived using Hamilton's principle and the calculus of variations. We modify the plate from Chapter 2 to allow for a load and study

$$\rho h \frac{\partial^2 w'}{\partial t^2} - T\nabla^2 w' + D\nabla^4 w' = f(x', y', t'). \tag{5.134}$$

The x' and y' coordinates are in the plane of the un-deformed plate while w' represents the deflection of the plate in the z' direction. The function f represents a load on the plate. Recall that ρ is plate density, h is plate thickness, T is the tension in the plate, and D is the flexural rigidity. We have assumed that D is a constant. It is convenient to rescale our governing equation and rewrite in dimensionless form. We scale the x' and y' coordinates with a characteristic length of the system, w', with a characteristic deflection, and time with the natural timescale of the system. We introduce

$$x = \frac{x'}{L}, \quad y = \frac{y'}{L}, \quad w = \frac{w'}{A}, \quad t = \frac{t'}{L^2}\sqrt{\frac{D}{\rho h}} \tag{5.135}$$

into equation (5.134) and obtain

$$\frac{\partial^2 w}{\partial t^2} - \alpha \nabla^2 w + \nabla^4 w = F(x, y, t), \tag{5.136}$$

where

$$\alpha = \frac{TL^2}{D}. \tag{5.137}$$

As with the elastic beam the dimensionless parameter α is a measure of the relative effects of tension and flexural rigidity in the plate. The function F is a scaled version of the load function f.

5.6.2 Circular Plate with Uniform Load

As a first example of plate mechanics, we consider steady-state deflections of a circular elastic plate subjected to a uniform load. We assume $\alpha = 0$, i.e., we ignore tension in the plate, and study

$$\nabla^4 w = p. \tag{5.138}$$

Further, we assume that solutions are radially symmetric. That is, we assume there is no variation in the θ direction. Hence, we rewrite the bi-Laplacian in equation (5.138), keeping only terms depending on the radial coordinate r,

$$\left(\frac{d^2}{dr^2} + \frac{1}{r}\frac{d}{dr}\right)\left(\frac{d^2 w}{dr^2} + \frac{1}{r}\frac{dw}{dr}\right) = p. \tag{5.139}$$

We assume that the edge of the plate is clamped or fixed; this implies the boundary conditions,

$$w(1) = \frac{dw}{dr}(1) = 0. \tag{5.140}$$

Equation (5.139) is easily integrated. We find

$$w(r) = \frac{pr^4}{64} + c_1 + c_2 r^2 + c_3 \log(r) + c_4 r^2 \log(r). \tag{5.141}$$

Notice that we have four unknown constants and thus far have only imposed two boundary conditions. However, as with the circular membrane, we must require that the solution be bounded at $r = 0$. This immediately implies that $c_3 = c_4 = 0$. Imposing the remaining two boundary conditions, we find

$$w(r) = \frac{p}{64}\left(r^2 - 1\right)^2 \tag{5.142}$$

as the deflection of a circular elastic plate subjected a uniform load. Or, in terms of dimensional variables,

$$w'(r') = \frac{PL^4}{64D}\left[\left(\frac{r'}{L}\right)^2 - 1\right]^2, \tag{5.143}$$

where P is the pressure on the plate.

5.6.3 Free Vibrations of a Circular Plate

Next we investigate free vibrations of a circular elastic plate. We work with our equation in dimensionless form and ignore tension by taking $\alpha = 0$. Hence we study

$$\frac{\partial^2 w}{\partial t^2} + \nabla^4 w = 0. \tag{5.144}$$

We consider the case where the outer edges of the plate are held fixed or clamped and impose the boundary conditions,

$$w(1, \theta, t) = 0 \tag{5.145}$$

$$\frac{\partial w}{\partial r}(1, \theta, t) = 0. \tag{5.146}$$

Throughout this section we will use circular polar coordinates. That is, r will be a radial coordinate and θ an angular coordinate. Notice that unlike our investigations of beams and membranes, the solution here is a function of three variables. So, while we use separation of variables, we will need to proceed with care. First, we separate out the time behavior by seeking a solution in the form

$$w(r, \theta, t) = \psi(r, \theta)e^{-i\omega t}. \tag{5.147}$$

This yields

$$\nabla^4\psi - \omega^2\psi = 0. \tag{5.148}$$

Next, we note that this equation has the solution $\psi(r, \theta) = M(r, \theta) + N(r, \theta)$ where M and N satisfy

$$\nabla^2 M - \omega M = 0 \tag{5.149}$$

and

$$\nabla^2 N + \omega N = 0. \tag{5.150}$$

Now, we separate variables in each of these equations. First, seek solutions to the N equation in the form

$$N(r, \theta) = A(\theta)\phi(r). \tag{5.151}$$

Denoting the separation constant by λ^2, this yields the two equations,

$$\frac{d^2 A}{d\theta^2} + \lambda^2 A = 0 \tag{5.152}$$

and

$$\frac{d^2\phi}{dr^2} + \frac{1}{r}\frac{d\phi}{dr} + \left(\omega - \frac{\lambda^2}{r^2}\right)\phi = 0. \tag{5.153}$$

Equation (5.152) is easily solved. We find

$$A(\theta) = a_0 \sin(\lambda\theta) + b_0 \cos(\lambda\theta). \tag{5.154}$$

Since our plate is a disk, we require that $A(\theta) = A(\theta + 2\pi)$, which immediately implies that $\lambda = n = 1, 2, 3, \ldots$ Hence our solutions in the θ direction are

$$A_n(\theta) = a_n \sin(n\theta) + b_n \cos(n\theta). \tag{5.155}$$

Our equation for $\phi(r)$ then becomes

$$\frac{d^2\phi}{dr^2} + \frac{1}{r}\frac{d\phi}{dr} + \left(\omega - \frac{n^2}{r^2}\right)\phi = 0, \tag{5.156}$$

which is Bessel's equation and has solution

$$\phi_n(r) = c_n J_n(\sqrt{\omega}r) + d_n Y_n(\sqrt{\omega}r). \tag{5.157}$$

The functions J_n and Y_n are known as the *Bessel functions.* Properties of the Bessel functions may be found in any text on mathematical physics. The function Y_n is unbounded at $r = 0$; hence we require $d_n = 0$. So, our solutions for $N(r, \theta)$ have the form

$$N_n(r,\theta) = J_n(\sqrt{\omega}r)(a_n \sin(n\theta) + b_n \cos(n\theta)) \tag{5.158}$$

where the c_n have been absorbed into the a_n and b_n. In the same way, we can construct solutions for M and find

$$M_n(r,\theta) = I_n(\sqrt{\omega}r)(c_n \sin(n\theta) + d_n \cos(n\theta)) \tag{5.159}$$

where the I_n are Bessel functions of pure imaginary argument. Our solutions for ψ then have the form

$$\begin{aligned}\psi_n(r,\theta) = {} & (a_n J_n(\sqrt{\omega}r) + c_n I_n(\sqrt{\omega}r)) \sin(n\theta) \\ & + (b_n J_n(\sqrt{\omega}r) + d_n I_n(\sqrt{\omega}r)) \cos(n\theta).\end{aligned} \tag{5.160}$$

Finally, we apply the boundary conditions and find that the normal mode frequencies are ω's, which solve

$$J_n(\sqrt{\omega})I_n'(\sqrt{\omega}) - J_n'(\sqrt{\omega})I_n(\sqrt{\omega}) = 0. \tag{5.161}$$

For each value of n this equation has a countable infinity of solutions. We denote the solutions by ω_{nm}. Returning to dimensional variables the natural frequencies of our circular elastic plate are

$$f_n = \frac{\omega_{nm}}{L^2}\sqrt{\frac{D}{\rho h}}, \tag{5.162}$$

where the roots of equation (5.161), i.e., the ω_{nm}, are to be determined numerically.

5.7 The Capacitive Pressure Sensor

In this section we outline how the results of this chapter may be applied to the study of the capacitive pressure sensor discussed in Section 5.2. In this system it is essential to understand the relationship between the capacitance of the device, deflection of the elastic component, and the pressure exerted by the environment. In particular, since capacitance is the quantity that will be measured, a relationship between pressure and capacitance is desirable. Recall from Section 5.2 that the basic capacitive pressure sensor consists of an elastic diaphragm over a sealed cavity. In general, the geometry of the pressure sensor appears as in Figure 5.14. The capacitance of the system

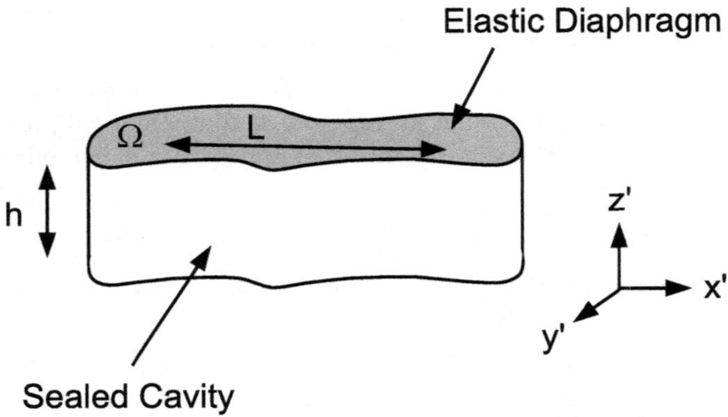

FIGURE 5.14: The geometry of the capacitive pressure sensor.

consisting of the elastic diaphragm and the bottom of the sealed cavity is approximately given by

$$C = \epsilon \int_{\Omega} \frac{dx'dy'}{h + w'(x', y')}. \tag{5.163}$$

The parameter ϵ is the permittivity of the material filling the cavity, h is the height of the cavity, and $w'(x', y')$ is the deflection of the diaphragm from the $z' = h$ position. This expression for capacitance holds in the limit where the aspect ratio of the system is small. That is, when $h/L \ll 1$. This is the limit in which fringing fields may be ignored and the electric field between the plates computed as if the plates were locally parallel. In Chapter 7 we will work this computation out in detail. Here, we are concerned with how our knowledge about elastic deflections may be employed to compute a capacitance-pressure relationship. Rather than work in dimensional variables, we scale the capacitance in the same manner we have scaled the equations of elasticity throughout this chapter. In particular, we let

$$x = \frac{x'}{L}, \quad y = \frac{y'}{L}, \quad w = \frac{w'}{h}. \tag{5.164}$$

This yields the dimensionless version of capacitance,

$$\mathcal{C} = \frac{C}{\epsilon h} = \int_{\Omega} \frac{dxdy}{1 + w(x,y)}. \tag{5.165}$$

Now, if our pressure sensor is circular, that is, if the domain Ω is a disk, and if the elastic diaphragm behaves like an elastic membrane we can use the result of Section 5.4 to compute $\mathcal{C}$. We find

$$\mathcal{C} = \int_0^{2\pi} \int_0^1 \frac{rdrd\theta}{1 - \frac{p}{4}(r^2 - 1)} \tag{5.166}$$

or

$$\mathcal{C} = -\frac{4\pi}{p} \log\left(\frac{4}{4+p}\right). \tag{5.167}$$

Note that here we are assuming that the deflection of the membrane is entirely due to pressure. In fact, the membrane will feel a force due to the voltage difference between the membrane and top plate. This is the basis of electrostatic actuation and will be discussed in Chapter 7. If the tension in the membrane is high, and the voltage difference low, the electrostatic force may be neglected in favor of the pressure force. For the geometry shown in Figure 5.14, we are interested in negative p. Negative p corresponds to a vacuum in the cavity and pressure in the external environment. Also, when $p = -4$ the membrane touches the ground plate, this is the limit of the range of the sensor. A modified design wherein the bottom plate is capped by an insulating layer would allow this range to be extended. However, computing the capacitance-pressure curve would require solving a contact problem in elasticity. While this can be done, it is considerably more complicated than the computations we have carried out in this chapter. In Figure 5.15 we plot $\mathcal{C}$ versus p for the regime, $-4 < p < 0$. This curve is the desired "look-up" table relating capacitance and pressure for the sensor. A standard modification of the sensor to improve the linearity of the capacitance-pressure curve is to use a "bossed" membrane. This idea is explored in an exercise. Finally, we note the nonlinearity of equation (5.167). Mathematical modeling is essential in order to determine the form of the nonlinearity and is a useful tool for exploring alternative designs that attempt to tailor the capacitance-pressure curve.

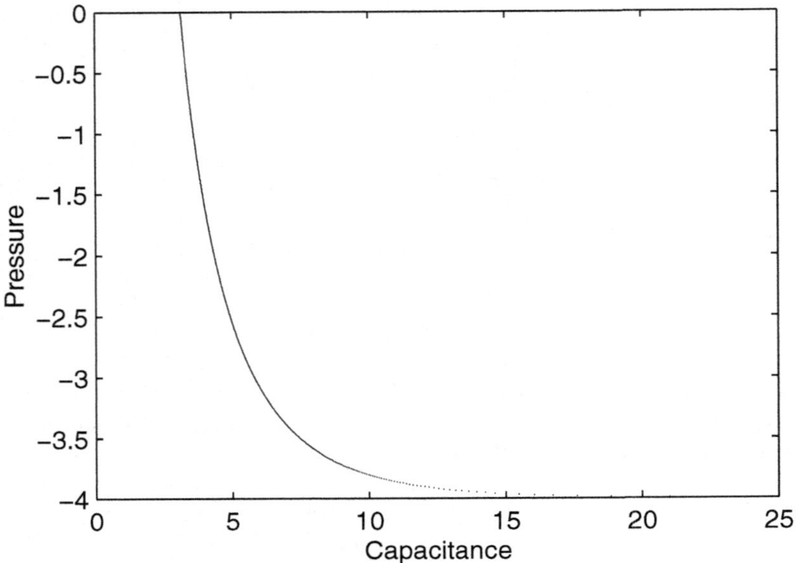

FIGURE 5.15: The capacitance-pressure relationship for the disk-shaped pressure sensor.

5.8 Chapter Highlights

- MEMS and NEMS devices make use of a wide range of elastic structures. Understanding the deformations of springs, beams, membranes, and plates is important in MEMS design.
- Mass-spring systems serve as a useful first model of many real MEMS and NEMS devices. The mass-spring system is rich enough to capture resonance and basic nonlinear behavior but lacks the effects of geometry.
- Membranes are components of real devices such as pressure sensors and micropumps. The membrane equation captures resistance to stretching and the effect of geometry. The phenomenon of resonance is present in the membrane equation. Exact solutions are possible for simple geometries.
- Beams are components of real devices such as resonant sensors and thermal bimorph actuators. The beam equation captures resistance to bending and stretching and the effect of geometry. The phenomenon of resonance is present in the beam equation. Beam boundary conditions can be used to model a variety of structures.

- Plates are components of real devices such as microvalves and micropumps. The plate equation captures resistance to bending and stretching and the effect of geometry. The phenomenon of resonance is present in the plate equation. Except in highly symmetric situations, exact solutions to the plate equation are difficult to find.

- The design of a capacitive pressure sensor relies upon knowledge of how membranes and plates deform when subjected to a load. The relationship between the deformation and capacitance is nonlinear. Mathematical modeling is a useful tool for exploring this relationship and for exploring device optimization.

5.9 Exercises

Section 5.3

1. A hollow sphere of radius R sits half submerged in a pool of water. If the sphere is slightly depressed, a restoring force equal to the weight of the displaced water pushes it upward. If it is released it will oscillate up and down in the water. Ignoring friction, find the frequency of this oscillation.

2. A mass on a rigid rod (a pendulum) will oscillate according to the nonlinear equation,

$$\frac{d^2\theta}{dt^2} + \sin(\theta) = 0,$$

where we have ignored friction. If θ is the angle the rod makes with the vertical, then compute an energy for this system and show that it is conserved.

3. Again consider the pendulum equation from Problem 2. Using the energy from Problem 2, plot the potential well for this system. Discuss the possible motions of the pendulum in terms of this potential well.

4. In Chapter 6 we will examine a mass-spring model of thermopneumatic actuation that, with an isothermal assumption, reduces to

$$\frac{d^2u}{dt^2} + u = \frac{\alpha}{1+u}.$$

Compute an energy for this system and show that it is conserved.

5. Again consider the isothermal thermopneumatic actuator from Problem 4. Using the energy from Problem 4, plot the potential well for this system. Discuss the possible motions of the pendulum in terms of this potential well.

6. Again consider the equation of motion for a pendulum from Problem 2. Rewrite this as a first-order system. Find all critical points of this system in the phase plane. Linearize near each critical point and determine the local behavior of solutions.

7. Again consider the equation of thermopneumatic actuation from Problem 4. Rewrite this as a first-order system. Find all critical points of this system in the phase plane. Linearize near each critical point and determine the local behavior of solutions.

8. Plot the frequency-response diagram for the forced mass-spring system for various values of γ. How does the resonant frequency change with γ? How does the magnitude of the response depend on γ? How does the width of the peak depend on γ? Discuss your answer in terms of the Q for the system.

9. Compute an effective spring constant for an elastic microbeam of length 1500μm, width 5μm, and height 10μm. You may assume a Young's modulus of 150GPa.

10. Compute the natural frequency of the beam system from problem 9 by modeling it as a mass-spring and using your effective spring constant. Assume the beam has density 2300kg/m^3.

11. Using the effective spring constant from Problem 9 and assuming a density of 2300kg/m^3, consider a mass-spring system where the mass is a ball 10μm in diameter. If the Q is $250,000$, what is the damping constant a?

12. Using the modified nonlinear version of Hooke's law, write down the governing equation for an undamped mass-spring oscillator. Scale your equation and interpret the dimensionless constants that arise. Compute an energy and sketch the potential well for this system.

Section 5.4

13. Consider steady-state deformations of an elastic string with load $p(x) = (x - 1/2)(x + 1/2)$. Find the shape of the deflected string using the Green's function representation of the solution.

14. Consider an elastic string with initial deformation $\sin(\pi x)$ that is released from rest. Solve for the motion of the string at all times $t > 0$. Plot your solution for various times.

15. Consider an elastic string occupying the region $[0,1]$ with initial deformation $x(1-x)$ that is released from rest. Solve for the motion of the string at all times $t > 0$. Plot your solution for various times.

16. If an elastic string is attached to a spring of spring constant k at its left end, the appropriate boundary condition becomes

$$ku'(0,t') = \mu \frac{\partial u'}{\partial x'}(0,t').$$

Consider such a string of length L that is fixed at $x = L$. Nondimensionalize this problem in the same way as the example in the text. You should find a dimensionless parameter arising in your boundary condition. Give a physical interpretation of this parameter. Find the natural frequencies of vibration for this system. How do they depend on your dimensionless parameter?

17. Use the method of eigenfunction expansion to find the steady-state deflection of a square membrane subjected to a uniform load of magnitude p. Assume the edges of the membrane are held fixed.

Section 5.5

18. Consider an elastic beam subjected to a constant load of magnitude, $-p$. Find the deflection of the beam assuming both ends are fixed. Find the deflection of the beam assuming one end is fixed and one end is pinned. Find the deflection of the beam assuming one end is free and one end is pinned.

19. Consider a cantilevered beam occupying the region $x' \in [0, L]$. Suppose a nonuniform load $p(x') = x'$ is applied. Suppose the built-in end is fixed. Solve for the deflection of the beam.

20. Consider a beam of length L that is simply supported at both ends. Suppose two concentrated loads of strength P are applied at the points $x = a$ and $x = b$ with $0 < a < b < L$. Solve for the deflection of the beam.

21. Find the natural frequencies of a beam of length L with one end clamped and the other end free. Find the natural frequencies of a beam of length L with one end clamped and the other end pinned. Find the natural frequencies of a beam of length L with one end pinned and the other end free.

22. Find the lowest natural frequency of a beam of square cross-section with length $L = 1500\mu\text{m}$, width $w = 5\mu\text{m}$, Young's modulus 150GPa, and density 2300kg/m^3.

23. Consider a beam with both ends pinned and subjected to a tensile load of magnitude μ. Find the natural frequencies of this beam. How does the lowest natural frequency change with μ?

Section 5.6

24. Find the steady-state deflection of a circular plate with clamped edges and nonzero tension subjected to a constant load. (Hint: Do not take $\alpha = 0$ in equation (5.136).)

25. Consider steady-state deflections of a circular elastic plate subjected to a constant load. Assume the edges of the plate are simply supported rather than clamped. This means the proper boundary conditions at $r = 1$ are

$$w(1) = \frac{d^2 w}{dr^2}(1) = 0.$$

Solve for the deflection.

26. Consider an annular-shaped plate, that is, a circular plate with a hole. Assume the outer and inner edges of the plate are clamped. Find the natural frequencies of this plate. How do the natural frequencies vary as a function of the ratio of the inner radius to the outer radius?

27. Consider equation (5.161), the equation that determines the natural frequencies of the circular elastic plate. Numerically compute ω_{00}, ω_{01}, ω_{10}, and ω_{11}.

Section 5.7

28. Following the example in Section 5.7, compute a capacitance-pressure curve assuming the sensor diaphragm is a circular elastic plate. Assume the edges of the plate are clamped. How does this differ from the membrane case?

29. One method used to improve the linearity of the capacitance-pressure curve for a capacitive pressure sensor is to use a *bossed* diaphragm. Model this scenario by assuming you have a circular pressure sensor where the diaphragm is an annular elastic membrane. Assume the inner part of the annulus is a circular plate that does not deform. At the boundary between the plate and the membrane, assume the membrane edge behaves as if it were free. How does the capacitive-pressure curve compare to the pure membrane case?

5.10 Related Reading

A well-written introduction to nonlinear dynamics and the phase plane is the book by Strogatz.

S.H. Strogatz, *Nonlinear Dynamics and Chaos: With Application to Physics, Biology, Chemistry, and Engineering*, Perseus Books, 2001.

Almost every text on engineering mathematics or elementary partial differential equations will discuss the method of separation of variables. One good introductory text is by Haberman.

R. Haberman, *Elementary Applied Partial Differential Equations with Fourier Series and Boundary Value Problems*, Prentice Hall, 1997.

An excellent text on the vibrations of elastic structures is the book by Graff.

K.F. Graff, *Wave Motion in Elastic Solids*, Dover, 1975.

5.11 Notes

1. A torsion spring is a spring that exerts a torque in proportion to a twist. The torque exerted is typically linearly proportional to the angle through which the spring is turned.

2. This trick of multiplying by a first derivative in order to obtain a first integral is standard in classical mechanics. It's also very useful.

3. The discussion here actually holds for an nth-order nonlinear system.

4. For an $n \times n$ system, DF is the $n \times n$ matrix whose i-jth component is the partial derivative of F_i with respect to u_j.

5. Please note that we cannot study this system in the phase plane. Since it is nonautonomous, i.e., since time appears explicitly in the equation, it is really a third-order system. The phase plane only works for second order systems.

6. It must be noted that these measurements are plagued with error and many measurements contradict one another. Please visit our website for links to typical measurements.

7. An effective spring constant for a cantilevered beam is computed in the problem section of Chapter 3.

8. The Laplace operator in various coordinate systems appears in Appendix A.

9. Actually the Dirac delta function is not a function at all. It is a *distribution.*

Chapter 6

Modeling Coupled Thermal-Elastic Systems

High technology has done us one great service: It has re-taught us the delight of performing simple and primordial tasks–chopping wood, building a fire, drawing water from a spring....

Edward Abbey

6.1 Introduction

In his classic *The Feynman Lectures on Physics* (Volume 1, Chapter 44), the physicist Richard P. Feynman describes a simple heat engine constructed from rubber bands and a light bulb. In Feynman's system, elastic bands form the spokes of a wheel as shown in Figure 6.1. When one side of the wheel is heated, the elastic bands shrink thereby shifting the center of gravity of the wheel, causing it to rotate. This is an example of a *thermal-elastic*[1] system where thermal gradients create elastic stress and this stress is used to perform work. In this chapter we investigate MEMS and NEMS devices that exploit coupled thermal-elastic effects to push, pull, pump, bend, slide, and even crawl.

Before proceeding, it is worth exploring why thermal-elastic effects are useful at the microscale. If we were to actually construct Feynman's engine from household implements, we would find that the wheel would rotate with a speed conveniently measured in revolutions per minute (RPMs). This is in agreement with our macroscopic intuition concerning heat transfer and thermoelastic behavior; macroscopic thermal-elastic systems are relatively slow. Yet in the microworld thermal-elastic effects are used to drive devices such as resonant sensors that operate in the kilohertz regime. As discussed in Chapter 3, the speed of thermal effects in the microworld is controlled by the scaling of surface area to volume with length. In Section 3.2.2 it was shown that a sphere, uniformly heated perhaps by resistive heating, would cool to the

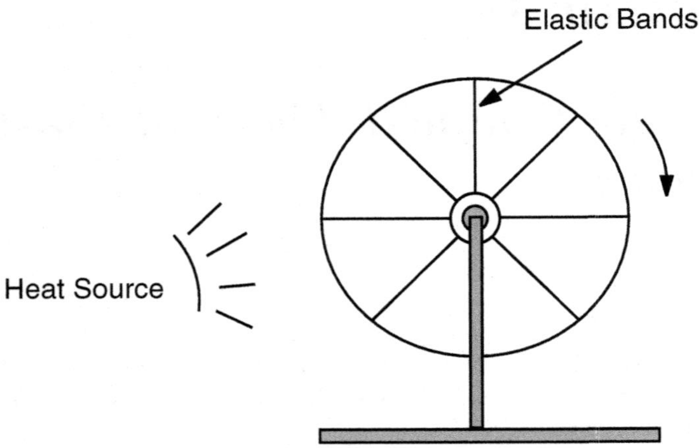

FIGURE 6.1: Feynman's thermal-elastic engine.

fraction f of the ambient environment in a time

$$t_f = -\frac{4a^2 \log f}{9\pi^2 \kappa}. \tag{6.1}$$

Recall that a is the sphere radius and κ is the thermal diffusivity. For illustrative purposes we take $a = 1\mu\text{m}$, $\kappa = 0.02 \times 10^{-4}\text{m}^2/\text{sec}$, and $f = 0.01$. This diffusivity corresponds to a typical value for silicon. Then we find $t_f \approx 10^{-7}\text{sec}$. That is, the sphere cools in a tenth of a microsecond. Assuming the sphere could be heated in the same time period, a device that heated and cooled the sphere could complete a cycle five million times each second! That is, this device could operate at 5000KHz. An actual MEMS or NEMS device may operate faster or slower, this calculation is to be taken as representative of the speeds a thermal-elastic microsystem may attain.

In addition to devices driven by thermal-elastic effects, thermal-elastic effects are important to consider while studying any type of microsystem. Thermoelastic damping plays a significant role in the microworld and strongly effects the quality factor, or Q, of resonant devices. Residual thermal stress is a byproduct of virtually every MEMS fabrication method and consequently the effects of temperature gradients on stress must always be understood. Finally, many devices operate in environments where thermal gradients are unavoidable. Such environmentally induced thermal stresses may affect the accuracy of a sensor, the efficiency of an actuator, or cause fracture and destruction of a device altogether.

In this chapter we focus on modeling devices that use thermal-elastic effects in some constructive way. We begin in the next section with an overview of devices and phenomena associated with thermal-elastic effects. We briefly describe each system, how thermal and elastic effects combine in the system, and how mathematical modeling can contribute to the design process or the understanding of the phenomena.

In Section 6.3 we turn our attention to models of thermopneumatic systems. Here the goal is to understand the volume stroke or force output of the device as a function of applied temperature. For a first model we consider a mass-spring-piston-cylinder system. This system couples together the mass-spring oscillator of mechanics and the piston-cylinder of thermodynamics to create a toy model of thermopneumatic actuation. We use the model of this system to introduce the idea of a stability analysis performed using Lyapanov functions. Next, we upgrade our model and replace the mass-spring with an elastic membrane. This immediately brings our model much closer to many systems fabricated by the MEMS community. A novel feature of this model is the *nonlocal* term that arises due to the coupling between volume change and temperature change. Such nonlocal effects are common in coupled-domain problems. We indicate how to deal with this complication and we derive the maximum displacement of our model actuator. Finally in Section 6.3, the membrane model is enhanced by replacing the membrane with an elastic plate. A set of exercises duplicating the membrane analysis for the plate model appears at the end of the chapter.

In Section 6.4 we turn our attention to the simplest conceivable thermoelastic actuator; a thermoelastic beam fixed on one end and free on the other. This system was studied in Chapter 2 as a refresher on thermoelasticity. Here we derive the displacement of the end of the rod assuming a uniform temperature increase and no lateral motion. This is used to compute the new length of the rod attained after heating. This calculation will be useful for comparison in the remaining sections of the chapter.

In Section 6.5 we consider the thermoelastic actuator based on buckling of a beam constrained axially at both ends. We find that this leads to an *eigenvalue problem* and we compute the critical temperature at which buckling occurs. In addition to modeling actual actuators, studying this system is useful for understanding unwanted thermal buckling of devices. We introduce an approximate method for computing the deflection of the beam.

In Section 6.6 we construct a model of a thermal bimorph device. We mimic the bimorph structure by constructing a model of a single beam with a thermal gradient in a direction transverse to the beam axis. This greatly simplifies the problem and allows us to avoid the complications introduced by the irregular geometry of the bimorph actuator. Nonetheless, the analysis of this system shows that the bimorph design is in fact superior to the V-beam design. We compute the maximum deflection for the case of one fixed end and one free end. This is compared to the maximum deflection for the V-beam actuator.

In Section 6.7 we discuss the bimetallic actuator. This study of the basic bimetallic structure predates the study of MEMS and NEMS by at least 45 years. In 1925, Timoshenko, published a now classic study of the buckling and bending of bimetallic structures [196]. His paper is exhaustive, covering the cases of free beams, fixed beams, beams with flexible supports, loaded beams, etc. In the past 75 years Timoshenko's analysis has found its way into most texts on elasticity and thermoelasticity. Consequently, we do not repeat this analysis, but simply discuss the principal results for bending of a free bimetallic structure.

6.2 Devices and Phenomena in Thermal-Elastic Systems

In this section we take a brief look at six devices that use thermal-elastic effects for their operation. We also briefly discuss the unwanted effects of thermoelastic damping and destructive thermal stress. Some of the systems discussed here were also mentioned in Chapter 4 during the discussion of thermally driven systems. In contrast to Chapter 4, here we are not concerned with the source of the thermal field but rather with the coupling between the thermal and elastic fields. The devices examined here are representative rather than comprehensive. We focus on the most popular and successful uses of thermal-elastic effects; however, MEMS and NEMS researchers are continuously employing these effects in MEMS and NEMS systems in new and interesting ways. The reader is directed to the references for other examples of thermal-elastic based systems and to the literature for recent developments in this area.

6.2.1 Thermoelastic Damping and Thermal Stresses

Before discussing the useful aspects of thermal-elastic effects in MEMS and NEMS, it is worth mentioning the ways in which such effects can be detrimental to MEMS and NEMS systems. There are two major phenomena to consider. The first is thermoelastic damping. Take a hammer and tap on the end of a metal rod. Elastic waves travel from the end you've tapped down the length of the rod, reflect from the far end, return, reflect, etc. After a short time, however, the rod is no longer ringing, the elastic waves have completely dissipated. The sources of this dissipation are many. One of the key players is thermoelastic damping. As the waves propagate they locally compress the material. This causes a localized increase in temperature. If there were no thermal diffusion in the rod, this thermal energy would be completely returned to elastic energy. However, in reality some of the heat diffuses away from the wavefront and hence less energy is returned to the elastic wave. If you measured the temperature of the rod before you tapped it and after letting it ring for awhile, you would notice an increase in temperature. Elastic energy is converted to thermal energy; this is thermoelastic damping. In MEMS and NEMS thermoelastic damping effects are usually discussed in the context of resonant sensors or other devices relying upon the resonant response of a structure. When a mechanical system is forced at its resonant or *natural frequency*, the amplitude of the response is larger than when the system is forced at any other frequency. The difference between the size of the response at resonance and at other nearby frequencies is captured by the quality factor or Q of the system. That is, the Q measures the selectivity of the resonant response. Viewed from another perspective, the Q measures the relative effects of damping and oscillation in the system. Hence, the quality factor of a resonant device is directly affected by thermoelastic damping. In resonant MEMS and NEMS devices, the Q is significantly lowered by thermoelastic damping [44, 86, 123, 202].

The second thermal-elastic effect detrimental to MEMS and NEMS systems is the presence of unwanted thermal stresses. These stresses may be residual stresses left in the system by the fabrication process or may be stresses created due to thermal gradients in the device or the environment during device operation. These unwanted stresses can lead to fracture of components [99], failure of electrical interconnects [180], or uncertainties in device position of systems such as hard disk drives requiring high precision [95]. The analysis of thermal stress presented in this chapter in the context of thermal-elastic actuators is applicable with small modification to the study of unwanted thermal stress.

6.2.2 Thermopneumatic Systems

In Chapter 4 we described the operation of a thermopneumatic valve (see Figure 4.3). In thermopneumatic systems such as the valve, the coupling between thermal and elastic behaviors is indirect. Resistive heating is used to heat a gas that expands and exerts a pressure force on an elastic component of the device. This stands in contrast to other thermal-elastic devices described in this chapter where the elastic component of the system is heated directly and thermal stresses in the elastic component provide the means to perform work. The indirect thermal-elastic coupling is a common feature of all thermopneumatic devices. In addition to the valve described in Chapter 4, the thermopneumatic idea has been employed in devices such as micropumps, microbellows, pressure sensor self-test systems, and thermally actuated micromirrors [19, 55, 89, 212].

An example is the microscale peristaltic pump built by J.A. Folta and his coworkers at Lawrence Livermore National Laboratory. In the peristaltic pump the thermopneumatic valve structure is repeated and the actuators used in sequence to create a pumping action. A sketch of this design is shown in Figure 6.2. Notice that the resistive heaters in Figure 6.2 do not heat the elastic membranes directly, but rather heat the gas contained in the region below the membranes. The pressure in the gas exerts the force on the membranes which in turn provide the pumping action.

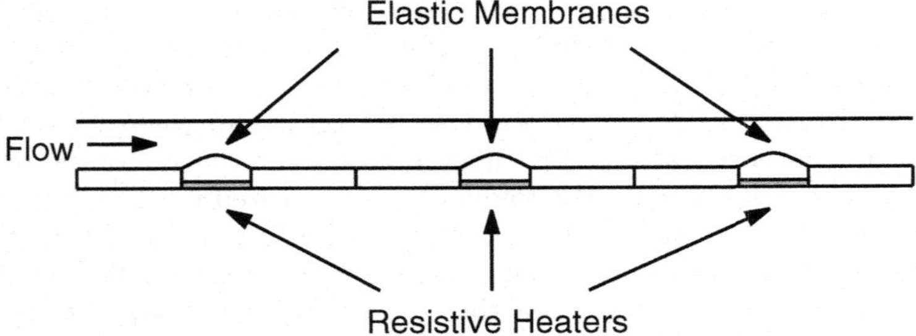

FIGURE 6.2: Sketch of a peristaltic micropump.

In the novel microbellows designed by Y.C. Tai and his group at the California Institute of Technology, an actuator was constructed based on the thermopneumatic idea, but capable of very large deflections. Similar in concept to a camera bellows, the microbellows structure uses "folded" layers of polysilicon to lessen resistance to large displacements. A sketch of this design is shown in Figure 6.3. Again notice that the resistive heater in Figure 6.3

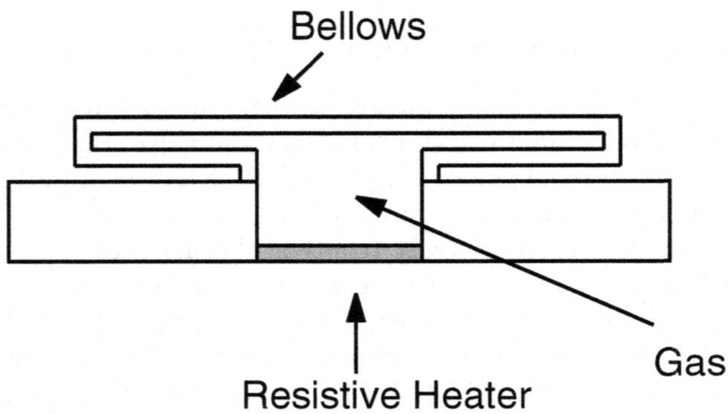

FIGURE 6.3: Sketch of a micro-bellows actuator.

does not heat the bellows directly, but rather heats the gas contained in the region below the bellows.

Other devices utilizing the thermopneumatic effect, including many examples of thermopneumatic valves, may be found in the references. In Chapter 4 we focused on modeling the thermal aspect of devices such as the microbellows. In this chapter we would like to understand how thermal effects couple to elastic effects. In the case of thermopneumatic actuation, a model relating the temperature increase in a resistive heater to the displacement of an elastic membrane is desirable.

6.2.3 The Thermoelastic Rod Actuator

The simple thermoelastic actuator consisting of a thermoelastic beam, fixed on one end and free on the other, was described in Chapter 2. A sketch of the setup is shown in Figure 2.9. We note that this design is rarely used in practice to provide actuation. As we shall see later in the chapter, the magnitude of the displacement is controlled by the coefficient of thermal expansion and the temperature increase in the rod. Given that the coefficient of thermal expansion of most MEMS materials is small, this necessitates large temperature gradients in order to achieve large displacements. A representative value of the coefficient of thermal expansion for silicon is $2 \times 10^{-6}\,\mathrm{K}^{-1}$. Large displacements forced on a device by temperature increase alone will generally lead to fracture of the system rather than a useful design. Nevertheless, this simple system is worth considering. One reason is that this setup has been used in practice to measure fracture strength! The article by Kapels et. al., describes one such study [99]. A second reason for studying this system is for purposes of comparison. Many different thermal actuator designs appear in the literature. One way to measure their efficiency is to compare with displacements achievable with the thermoelastic rod. A final reason for studying the thermoelastic rod is the ubiquity of the structure shown in Figure 2.9. Atomic force microscope tips, resonant MEMS sensors, setups for measuring material properties of DNA, and countless other systems of interest in MEMS and NEMS make use of this setup. A feel for the parameters governing thermal expansion in this system is of general interest.

6.2.4 The Thermoelastic V-Beam Actuator

The thermoelastic V-beam actuator was described in Chapter 4 (see Figure 4.1). Recall that the operating principle for this actuator is to utilize thermally induced buckling to provide a force output. As with the thermoelastic rod actuator, the V-beam actuator is not often used in practice. Still, situations occur where the V-beam design is preferable to other actuator designs. In particular, if the goal is to produce an actuator array with a high translational force output, the V-beam design may be the best choice. Also, as with the thermoelastic rod structure, the setup shown in Figure 4.1 is ever-present in MEMS and NEMS designs. Understanding when thermally induced buckling occurs is of interest and in this chapter we will compute the buckling load. In addition to being of interest from a thermomechanical point of view, this computation is of interest from a purely mechanical point of view as well. That is, in this chapter the applied load will be due to thermal stresses, but the computation of the buckling load is independent of whether the stresses are thermally induced or induced by direct mechanical loading.

6.2.5 Thermal Bimorph Actuator

The thermal bimorph actuator was designed to overcome the low force output of the thermoelastic rod actuator and thermoelastic V-beam actuator. The basic design was discussed in Chapter 4 (see Figure 4.2). Recall that the operating principle is to utilize two beams of contrasting size made of the same material and joined together mechanically. A current is passed through the entire structure, heating the small beam while leaving the large beam relatively cool. Large thermal stresses in the small beam tend to make it stretch, but this stretching is resisted by the mechanical linkage with the cool beam. This competition causes the entire structure to deflect. The wide array of devices using the thermal bimorph actuator is testimony to its efficiency. A sample of these systems may be found in the references [23, 38, 39, 88, 89, 121, 129]. It is worth noting that this actuator is particularly suitable for use in actuator arrays. For example the MEMS insect systems built by the group led by Kladitis and Bright at the University of Colorado at Boulder utilize this idea. Similarly, thermal bimorph actuator arrays have been used in self-assembly designs [162], and for actuation of micromirrors [89].

One key difference between the thermal bimorph actuator and the V-beam and rod designs is the role of geometry. In particular the question of optimizing the relative sizes of the two arms of the bimorph is of interest. On the other hand, the geometry also makes mathematical modeling of the thermal bimorph less straightforward than modeling of the V-beam and rod designs. Most researchers have turned to numerical computations to aid in the design of this actuator. In this chapter we shall examine a virtual representation of the thermal bimorph. The mechanical structure of two beams with different sizes will be replaced by a single beam with a temperature gradient perpendicular to its axis. The analysis of this system will indicate how the size of this gradient affects the deflection. In turn this suggests how the ratio of beam widths affects the operation of the thermal bimorph.[2]

6.2.6 Bimetallic Thermal Actuators

Perhaps the second most successful thermoelastic actuator is the bimetallic thermal actuator. The basic design is sketched in Figure 6.4.[3] Unlike the thermal bimorph, in the bimetallic structure, two different materials are used. The materials are chosen to have different coefficients of thermal expansion (CTE). The layers are joined along a common interface and the entire structure is heated. Since one material wants to expand more than the other but is restrained by the joint with the second material, the entire structure bends. In Figure 6.4 the structure deflection shown would occur if $\alpha_1 > \alpha_2$. We note that this bimetallic structure is not only used in MEMS but is common in many engineering designs. In particular, the typical household thermostat

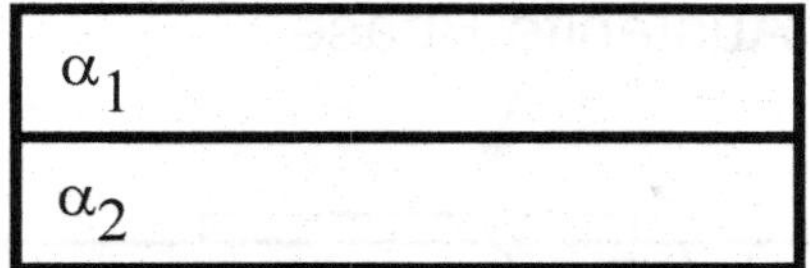

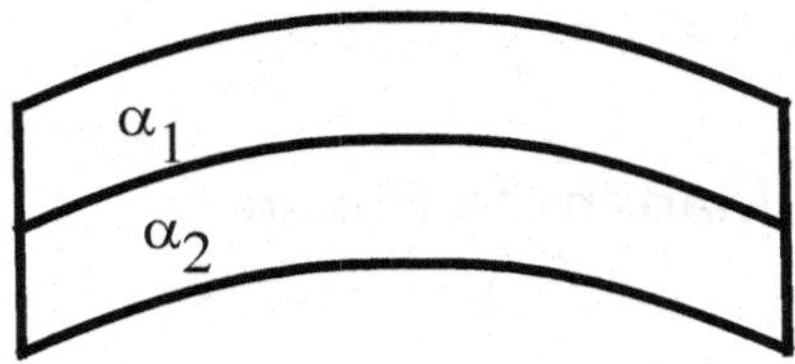

FIGURE 6.4: The basic bimetallic actuator structure. The upper structure is cool while the lower structure has been heated.

relies upon this design.

6.2.7 Shape Memory Actuators

Shape memory actuators constitute a third class of thermal-elastic systems used in MEMS. As described in Chapter 4, the shape memory effect occurs when a material undergoes a phase change in response to a temperature change. The miraculous aspect of the effect is that the material may be "trained" to return to a remembered shape upon heating. The high temperature phase, called the austenite phase, is typically stiff while the low temperature or martensite phase is soft and easily deformed. The shape memory effect has been employed in the design of a variety of MEMS devices. One example is shown in Figure 6.5, where the shape memory effect is used in a microvalve. Here, the valve is normally open. Upon heating the valve cover changes to its austenite phase sealing the opening. Heating in such a system is typically accomplished by applying an electric currect. The shape memory effect has been used in the design of micropumps [9], microactuators [67], and microrobots [90]. Shape memory effects have even been used to design microstructures for insect neural recording [186]. Several review articles on the shape memory effect in MEMS exist, in particular see [98, 114, 206].

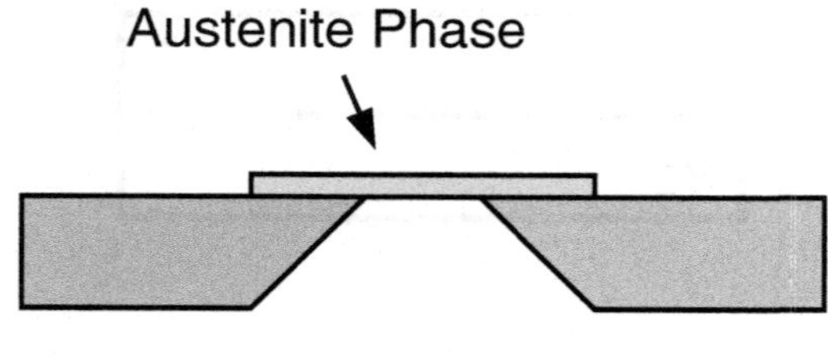

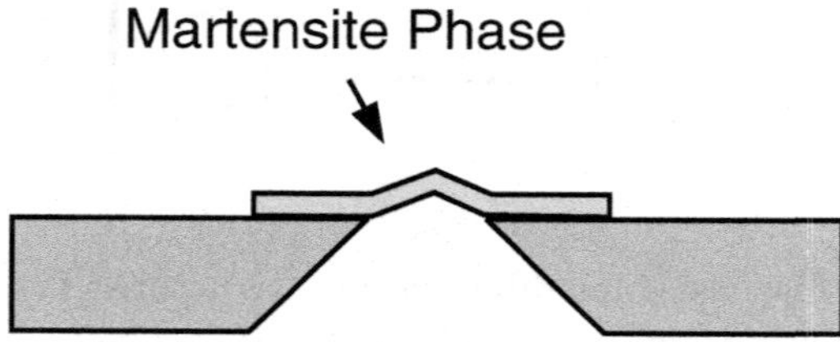

FIGURE 6.5: A normally open shape memory alloy valve.

6.3 Modeling Thermopneumatic Systems

Now that we have seen how thermal-elastic systems are used in the design of MEMS and NEMS, we would like to construct mathematical models to provide understanding and design guidelines. In this section we focus on thermopneumatic systems. We study a sequence of three simple models that allow us to relate the temperature increase in a resistive heater to the elastic displacement in the system. Our first model is a toy model that couples the canonical system of mechanics, the mass on a spring, with the canonical system of thermodynamics, the piston-cylinder, in a model of a thermopneumatic actuator.

6.3.1 A Mass-Spring-Piston Model

Consider the system pictured in Figure 6.6. We would like to formulate a model that relates the motion of the piston head to the temperature in the resistive heater at the bottom of the cylinder.[4] We assume that the piston head of mass m moves without friction and is impermeable to the ideal gas that fills the cylinder. The volume contained in the region below the piston head and above the heater is denoted by V and is related to the displacement

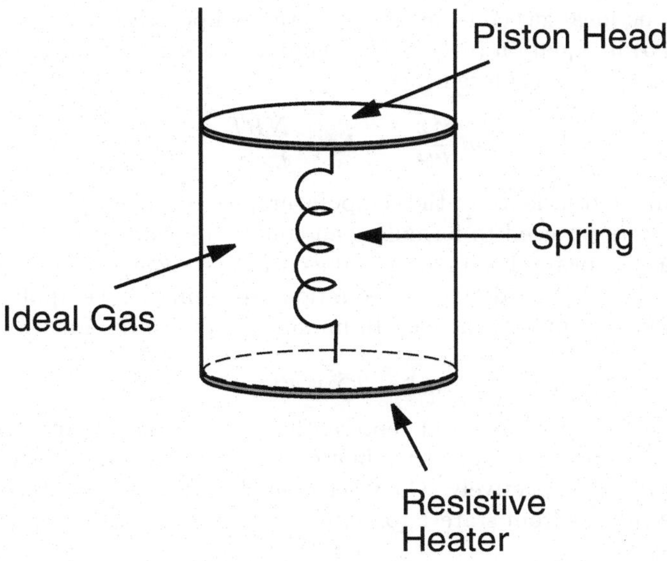

FIGURE 6.6: A mass-spring-piston system modeling thermopneumatic actuation.

of the piston head by

$$V = A(L + x(t')). \tag{6.2}$$

Here A is the cross-sectional area of the cylinder, L is the rest length of the spring, and $x(t')$ is the displacement of the piston head from its mechanical equilibrium. The equation governing the motion of the piston head follows directly from Newton's second law. That is,

$$m\frac{d^2x}{dt'^2} = \sum \text{forces}. \tag{6.3}$$

In our system the only forces are the spring force and the pressure in the gas. We assume the spring force obeys Hooke's law and apply Newton's second law to the piston head to find

$$m\frac{d^2x}{dt'^2} = -kx + Ap, \tag{6.4}$$

where k is the spring constant and p is the pressure in the gas. Note that we are assuming that the pressure above the piston head is negligible.[5] Now, we assume that the cylinder is filled with N moles of an ideal gas and apply the ideal gas law in the form,

$$pV = NRT, \tag{6.5}$$

where R is the ideal gas constant and T is the temperature of the gas. Using equation (6.5) in equation (6.4), we can eliminate the pressure in favor of the temperature, T, and displacement x,

$$m\frac{d^2x}{dt'^2} = -kx + \frac{NRT}{L+x}. \tag{6.6}$$

Having eliminated in favor of the temperature, we recognize that in order to close the system we need to formulate an equation governing T. Notice that T changes as the piston head moves and as heat is transferred to the gas from the resistive heater. To derive the equation governing T, we apply the first law of thermodynamics to the gas. In infinitesimal form the first law may be written as

$$dU = \hat{d}Q + \hat{d}W. \tag{6.7}$$

Here dU is an exact differential representing the change in internal energy, $\hat{d}Q$ and $\hat{d}W$ are process-dependent differentials representing changes in heat energy and work, respectively. Consider a small displacement of the system which takes the gas from state 1 to state 2 and integrate equation (6.7) over this process:

$$\int_1^2 dU = \int_1^2 \hat{d}Q + \int_1^2 \hat{d}W. \tag{6.8}$$

Next, we parameterize this process by assuming it begins at some time t' and ends at some times $t' + \delta t'$. Consider the first integral,

$$\int_1^2 dU = \int_{t'}^{t'+\delta t'} \frac{dU}{dt'}dt' = U(t'+\delta t') - U(t'). \tag{6.9}$$

Similarly the second integral may be written as

$$\int_1^2 \hat{d}Q = \int_{t'}^{t'+\delta t'} \frac{\hat{d}Q}{dt'}dt'. \tag{6.10}$$

We assume the sidewalls of the cylinder are perfectly insulated from the external environment but that heat is exchanged between the resistive heater and the gas. We assume that the heat capacity of the gas is so small relative to that of the heater that the heater may be considered to be held at constant temperature T_h. The model of this section may be extended by coupling to the model of Joule heating from Chapter 4 in order to include the time dependence of T_h. Here we assume T_h is known and constant. We assume that heat exchange between the gas and the heater obey's Newton's law of cooling[6],

$$\frac{\hat{d}Q}{dt'} = -hA(T - T_h), \tag{6.11}$$

and use this in equation (6.10) to obtain

$$\int_1^2 \hat{d}Q = -hA\int_{t'}^{t'+\delta t'} (T - T_h). \tag{6.12}$$

Applying the mean value theorem for integrals[7] to the right-hand side we have

$$\int_1^2 \hat{d}Q = -hA(T(\tilde{t}) - T_h)\delta t'. \tag{6.13}$$

Here $\tilde{t}$ is some time between t' and $t' + \delta t'$. Finally we deal with the third integral in equation (6.9). We make the assumption that the change of state of the system is reversible. That is, we are assuming that the change is sufficiently slow so that the system is effectively in equilibrium throughout. This allows us to express the change in work as $-pdV$. Applying the mean value theorem to the third integral in equation (6.9), we obtain

$$\int_1^2 \hat{d}W = -p(\hat{t})(V(t' + \delta t') - V(t')). \tag{6.14}$$

Collecting our three integrals together, dividing by $\delta t'$, and taking the limit as $\delta t'$ tends to zero we obtain

$$\frac{dU}{dt'} = -hA(T - T_h) - p\frac{dV}{dt'}. \tag{6.15}$$

For an ideal gas U may be expressed in terms of the temperature as

$$U = KNRT, \tag{6.16}$$

where K is a dimensionless constant related to the degrees of freedom of the gas molecules. Finally, using this expression, the ideal gas law, and the relationship between volume and piston head displacement we obtain

$$KNR\frac{dT}{dt'} + hA(T - T_h) = -\frac{NRT}{L + x}\frac{dx}{dt'}. \tag{6.17}$$

Equations (6.6) and (6.17), together with appropriate initial conditions, represent a closed system for the displacement of the piston head, x, and the temperature of the enclosed gas, T.

It is convenient to recast our model equations in dimensionless form. We scale the temperature with the heater temperature, the displacement with the rest length of the spring, and the time with the natural oscillation time of the mass-spring system by defining

$$\theta = \frac{T - T_h}{T_h}, \quad t = \sqrt{\frac{k}{m}}t', \quad u = \frac{x}{L}. \tag{6.18}$$

Inserting into equations (6.6) and (6.17), we obtain

$$\frac{d^2u}{dt^2} + u = \alpha\frac{1 + \theta}{1 + u} \tag{6.19}$$

$$\frac{d\theta}{dt} + \beta\theta = -\frac{1}{K}\frac{1 + \theta}{1 + u}\frac{du}{dt}. \tag{6.20}$$

The dimensionless parameters α and β are defined as

$$\alpha = \frac{NRT_h}{kL^2}, \quad \beta = \sqrt{\frac{m}{k}}\frac{hA}{KNR}. \tag{6.21}$$

Notice that using the ideal gas law and the definition of volume, the parameter α may be rewritten as

$$\alpha = \frac{NRT_h}{kL^2} = \frac{p_h A}{kL}, \tag{6.22}$$

where p_h is the pressure in a cylinder of length L at temperature T_h. From this representation we see that α is the ratio of a reference pressure force to a reference spring force. That is, α provides a measure of the gas's ability to stretch or compress the spring. If you wish, α may be thought of as the analog of the coefficient of thermal expansion in thermoelasticity. We also note that β has an interpretation as a ratio of timescales. We know that

$$\left[\sqrt{\frac{m}{k}}\right] = \text{time}, \quad \left[\frac{hA}{KNR}\right] = \frac{1}{\text{time}}. \tag{6.23}$$

So, β is a ratio of the characteristic oscillation time of the mass-spring to the characteristic thermal transport time. That is, β is a ratio of the rate at which mechanical oscillations occur to the rate at which thermal energy is exchanged between the gas and the heater.

Our mass-spring-piston-cylinder model of thermopneumatic actuation, equations (6.19) and (6.20), is a coupled set of nonlinear ordinary differential equations. Our first step in analyzing solutions of this system is to compute steady-state solutions by setting all time derivatives to zero. This immediately implies that in the steady-state, $\theta = 0$. This means that the temperature of the gas is the same as the temperature of the heater in the steady-state. Unsurprising! The steady-state displacement satisfies

$$u^2 + u - \alpha = 0, \tag{6.24}$$

with solutions,

$$u_1 = -\frac{1}{2} + \frac{1}{2}\sqrt{1+4\alpha}, \quad u_2 = -\frac{1}{2} - \frac{1}{2}\sqrt{1+4\alpha}. \tag{6.25}$$

Since α is a positive constant, we immediately see that $u_2 < -1$ and unphysical. That is, u_2 represents a displacement of the piston head through the resistive heater. Focusing our attention on the physical steady-state, u_1, we notice that steady-state solutions only depend on the parameter α and are independent of β. This is in accordance with our interpretation of α as a ratio of forces and β as a ratio of timescales. For small values of α we may linearize the u_1 solution about $\alpha = 0$ and recast in dimensional variables to obtain

$$x \approx \frac{NRT_h}{kL} = \frac{p_h A}{k} \tag{6.26}$$

as the approximate steady-state deflection of our piston head.

Our next step in analyzing solutions to equations (6.19) and (6.20) is to linearize about our steady-state solution and study the resulting linear equations. Assuming small α, letting $u = \alpha + w$, assuming $w \ll 1$, and $\theta \ll 1$, we obtain

$$\frac{d^2w}{dt^2} + (1+\alpha)w = \alpha\theta \tag{6.27}$$

$$\frac{d\theta}{dt} + \beta\theta = -\frac{1}{K}\frac{dw}{dt}. \tag{6.28}$$

It is convenient to recast these equations in first-order form,

$$\frac{dw}{dt} = v \tag{6.29}$$

$$\frac{dv}{dt} = -(1+\alpha)w + \alpha\theta \tag{6.30}$$

$$\frac{d\theta}{dt} = -\beta\theta - \frac{v}{K}. \tag{6.31}$$

Recasting our system in first-order form allows us to think of the system as a single vector equation,

$$\frac{d\mathbf{y}}{dt} = F(\mathbf{y}), \tag{6.32}$$

where the vector $\mathbf{y} = (w, v, \theta)$ and the function F is defined by the right-hand sides of equations (6.29)–(6.31). In general, equation (6.32) can represent any first-order system, linear or nonlinear. We always assume that $F(0) = 0$ so that $\mathbf{y} = 0$ is a steady-state solution to the system. In Chapter 4 we examined stability of steady-state solutions to such a problem using the method of linear stability analysis. Here we introduce a second method of determining stability called *Lyapunov's second method.* The result is embodied in the following theorem

THEOREM 6.1

If there exists a scalar function $V(\mathbf{y})$ that is positive definite and for which $dV/dt \leq 0$ on some region Ω containing the origin, then the zero solution of equation (6.32) is stable.

Recall that a function is positive definite on Ω if and only if $V(0) = 0$ and $V(\mathbf{y}) > 0$ for any $\mathbf{y} \neq 0$ and y in Ω. A function V satisfying the hypothesis of the theorem is called a *Lyapunov function.* For our linearized model, equations (6.29)–(6.31), it is easy to see that V defined as

$$V = \frac{w^2}{2} + \frac{v^2}{2(1+\alpha)} + \frac{\alpha K\theta^2}{2(1+\alpha)} \tag{6.33}$$

is positive definite in all of w,v,θ space. Further it is easy to compute the derivative and show that

$$\frac{dV}{dt} = -\frac{K\alpha\beta}{1+\alpha}\theta^2 \leq 0, \tag{6.34}$$

and hence V is in fact a Lyapunov function for our model. By the theorem above we see that the origin is a stable equilibrium solution of the linearized model of thermopneumatic actuation. Further steps in the analysis of this model and practice with Lyapunov functions appear in the exercises.

6.3.2 A Membrane-Based Model

A reasonable criticism of the mass-spring-piston-cylinder model is that none of the MEMS devices mentioned at the start of the chapter actually has the mass-spring structure. Rather, in each of the devices discussed, the elastic behavior was more complicated than that of a simple spring. Thermopneumatic valves in particular contain a moving part that resembles an elastic membrane rather than a mass on a spring. Hence a reasonable way to improve the model of the previous subsection is to replace the mass on a spring with an elastic membrane as in the system shown in Figure 6.7. As in the previous section we

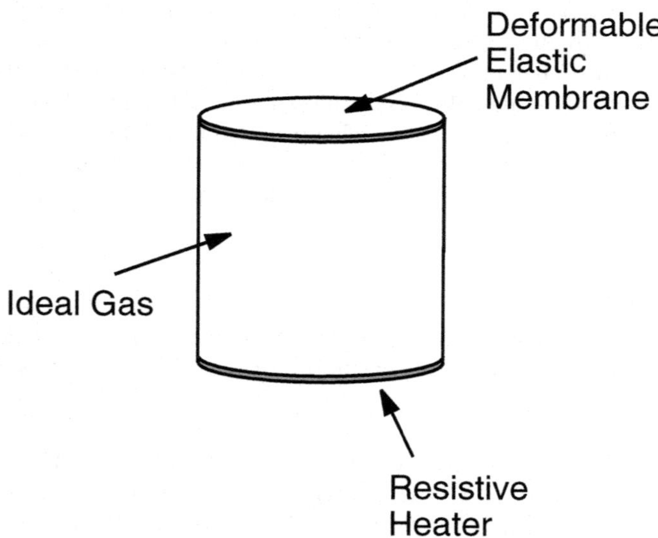

FIGURE 6.7: A cylinder-membrane system modeling thermopneumatic actuation.

would like to formulate a model that captures the motion of the membrane as a function of the temperature of the resistive heater. We assume the cylinder in Figure 6.7 has height L and radius a. We denote the deflection of the elastic membrane from the flat configuration by $u'(r',t')$. Implicit in this statement is the assumption that the membrane shape remains radially symmetric. We can express the volume, V, enclosed by the cylinder and membrane in terms of u as

$$V = \pi a^2 L + 2\pi \int_0^a u'(r',t') r' dr'. \tag{6.35}$$

Now, from Chapter 2, we know that an elastic membrane's motion is governed by the equation,

$$\rho \frac{\partial^2 u'}{\partial t'^2} = \mu \nabla^2 u' + p, \tag{6.36}$$

where ρ is the membrane density, μ is the tension in the membrane, and we have included the body force p representing the pressure in the gas. We again assume that the gas is an ideal gas and that N moles are contained in the cylinder. Then using the ideal gas law (6.5), and the expression for volume, equation (6.35) we can express the pressure in the gas in terms of temperature, T, and membrane displacement, u',

$$p = \frac{NRT}{\pi a^2 L + 2\pi \int_0^a u'(r',t') r' dr'}. \tag{6.37}$$

Hence we can eliminate the pressure in favor of temperature in equation (6.36) to obtain

$$\rho \frac{\partial^2 u'}{\partial t'^2} = \mu \nabla^2 u' + \frac{NRT}{\pi a^2 L + 2\pi \int_0^a u'(r',t') r' dr'}. \tag{6.38}$$

Repeating the same arguments as in the previous subsection, we can write down the equation governing temperature,

$$KNR \frac{dT}{dt'} + \pi h a^2 (T - T_h) = -\frac{2\pi NRT}{\pi a^2 L + 2\pi \int_0^a u'(r',t') r' dr'} \int_0^a \frac{\partial u'}{\partial r'} r' dr'. \tag{6.39}$$

Equations (6.38) and (6.39), together with appropriate initial and boundary conditions, represent a closed system for the displacement of the elastic membrane, $u'(r',t')$, and the temperature of the enclosed gas, T. In contrast to the ordinary differential equation model of the previous subsection, this model couples a partial differential equation with an ordinary differential equation. As before, the model is nonlinear, but here an additional complication is that the model is *nonlocal.* That is, the governing equations contain terms that depend on an integral of the solution $u'(r',t')$. This nonlocal behavior is typical of coupled-domain problems. In fact, in the Joule heating model of Chapter 4, the temperature equation contained a nonlocal term. There it arose due to coupling between electrostatic fields and thermal fields; here the nonlocal term arises due to the manner in which the elastic fields are coupled to the thermal fields.

It is convenient to recast our model equations in dimensionless form. As with the mass-spring-piston-cylinder model, we scale the temperature with the heater temperature. We also scale the membrane displacement with cylinder length, the radial variable with cylinder radius, and the time with the natural time scale of the elastic membrane. That is, we define

$$\theta = \frac{T - T_h}{T_h}, \quad u = \frac{u'}{L}, \quad r = \frac{r'}{a}, \quad t = \sqrt{\frac{\mu}{\rho}}\frac{t'}{a}. \tag{6.40}$$

This yields the dimensionless system,

$$\frac{\partial^2 u}{\partial t^2} = \nabla^2 u + \frac{\alpha(1+\theta)}{1 + 2\int_0^1 u(r,t) r dr} \tag{6.41}$$

$$\frac{d\theta}{dt} + \beta\theta = -\frac{2}{K}\frac{(1+\theta)}{1 + 2\int_0^1 u(r,t) r dr}\int_0^1 \frac{\partial u}{\partial t} r dr. \tag{6.42}$$

The parameters α and β are defined as

$$\alpha = \frac{NRT_h}{\pi\mu L^2}, \quad \beta = \frac{\pi h a^3}{KNR}\sqrt{\frac{\rho}{\mu}}. \tag{6.43}$$

As with the mass-spring model the parameter α can be re-expressed as

$$\alpha = \frac{p_h a^2}{\mu L}, \tag{6.44}$$

where p_h is the pressure of N moles of an ideal gas at temperature T_h in a cylinder of radius a and height L. Hence as with the mass-spring model, α is a ratio of a reference pressure force to a reference elastic force in the system. The parameter α again provides a measure of the gas's ability to stretch the membrane. The interpretation of β is also the same as in the mass-spring system; β is a ratio of the rate at which the membrane oscillates to the rate at which thermal energy is exchanged between the gas and the heater.

Our next step is the same as it was for the mass-spring model. Namely, we begin by seeking steady-state solutions. Setting time derivatives to zero in equations (6.41) and (6.42), we immediately see that $\theta = 0$ in the steady state. This is the same result as in the mass-spring-piston-cylinder system and implies that the gas temperature equilibrates with the heater temperature in the steady state. On the other hand, in contrast with the mass-spring system, the steady-state displacement satisfies a differential rather than algebraic equation. In particular the steady deflection $u(r)$ satisfies

$$\nabla^2 u + \frac{\alpha(1+\theta)}{1 + 2\int_0^1 u(r) r dr} = 0. \tag{6.45}$$

Using the expression for the Laplace operator in cylindrical coordinates, this may be rewritten as

$$\frac{d^2u}{dr^2} + \frac{1}{r}\frac{du}{dr} + \frac{\alpha(1+\theta)}{1+2\int_0^1 u(r)rdr} = 0. \tag{6.46}$$

Since this equation is second order, we need to specify two boundary conditions. We assume that the edge of the membrane is held fixed and impose

$$u(1) = 0, \tag{6.47}$$

while we require that the solution is symmetric about the origin and impose

$$\frac{du}{dr}(0) = 0. \tag{6.48}$$

Notice that equation (6.46) is still *nonlocal.* The integral term implying the dependence of the solution at a point on the solution everywhere still appears. Let us make a small change in perspective by rewriting equation (6.46) as two equations,

$$\frac{d^2u}{dr^2} + \frac{1}{r}\frac{du}{dr} + \lambda = 0 \tag{6.49}$$

$$\lambda = \frac{\alpha(1+\theta)}{1+2\int_0^1 u(r)rdr}. \tag{6.50}$$

What kind of an object is λ? A brief inspection reveals that λ is a *constant.* Consequently, we can solve equation (6.49), treating λ as an unknown constant, impose the boundary conditions (6.47) and (6.48), and then compute λ using equation (6.50). Solving equations (6.49), (6.47), and (6.48) we find

$$u(r) = \frac{\lambda}{4}(1-r^2). \tag{6.51}$$

Inserting this into equation (6.50), we find that λ must satisfy

$$\lambda^2 + 8\lambda - 8\alpha = 0. \tag{6.52}$$

The reader should note the similarity between the algebraic equation satisfied by λ and the algebraic equation satisfied by the steady-state solutions to the mass-spring-piston-cylinder model. The analogy is even stronger when we realize that the maximum displacement of the elastic membrane in this model occurs at $r = 0$ and is given by

$$u(0) = \frac{\lambda}{4}. \tag{6.53}$$

That is, the maximum displacement is completely determined by λ and hence by the solution to an algebraic equation. Solving for λ we find

$$\lambda_1 = -4 + 4\sqrt{1+\frac{\alpha}{2}} \tag{6.54}$$

$$\lambda_2 = -4 - 4\sqrt{1 + \frac{\alpha}{2}}. \tag{6.55}$$

As with the mass-spring model, one of these solutions is unphysical. In particular, λ_2 represents a displacement of the membrane through the heater and is an unphysical solution. On the other hand, λ_1 is a physical solution and can be expanded under the assumption that α is small. Doing so we find that the maximum displacement of the elastic membrane is approximately

$$u(0) = \frac{\lambda}{4} \approx \frac{\alpha}{4} = \frac{NRT_h}{4\pi\mu L^2} \tag{6.56}$$

or in terms of dimensional quantities,

$$u'(0) \approx \frac{NRT_h}{4\pi\mu L}. \tag{6.57}$$

Note that this maximum displacement is essentially the same as the displacement of the entire piston head computed from the mass-spring model. Of course, the membrane model captures the deformation of the membrane as well. Understanding the difference between these two results is key in understanding how to optimize devices such as the thermopneumatic pump. Determining the linear stability of this steady-state solution appears as an exercise at the end of the chapter.

6.3.3 A Plate-Based Model

Many further extensions of the basic membrane model are possible. One direction for investigation is to replace the elastic membrane in Figure 6.7 with a thin elastic plate. Using the plate theory presented in Chapter 2 and following the development in the previous subsection, it is easy to see that the governing equation for plate's displacement is

$$\rho\frac{\partial^2 u'}{\partial t'^2} = \mu\nabla^2 u' - D\nabla^4 u' + \frac{NRT}{\pi a^2 L + 2\pi\int_0^a u'(r',t')r'dr'}. \tag{6.58}$$

Recall that D is the *flexural rigidity* of the plate. Note that this reduces to the membrane theory when $D = 0$. The equation governing the temperature of this system remains unchanged from the membrane model and hence T satisfies,

$$KNR\frac{dT}{dt'} + \pi h a^2(T - T_h) = -\frac{2\pi NRT}{\pi a^2 L + 2\pi\int_0^a u(r',t')r'dr'}\int_0^a \frac{\partial u'}{\partial r'}r'dr'. \tag{6.59}$$

In computing the volume we have again assumed symmetry of solutions, i.e., we have assumed that u' depends only on the radial coordinate r' and time. Equations (6.58) and (6.59), together with the appropriate initial and boundary conditions, represent a closed system for the displacement of the elastic plate in our model thermopneumatic actuator. The analysis of this model closely parallels the analysis of the membrane model. The various steps appear as exercises at the end of the chapter.

6.4 The Thermoelastic Rod Revisited

In Chapter 2 the displacement of a one-dimensional thermoelastic rod subjected to an arbitrary axially varying temperature field was computed (see Figure 2.9). Taking the axial direction as the x' direction and denoting the displacement by $u'(x', t')$ we found

$$u'(x', t') = \alpha \frac{3\lambda + 2\mu}{\lambda + 2\mu} \int_0^{x'} T(z, t')dz - \alpha \frac{3\lambda + 2\mu}{\lambda + 2\mu} T_0 x'. \tag{6.60}$$

For comparison with other actuators, it is useful to compute the steady displacement for a constant temperature, T. We find

$$u'(x') = \alpha \frac{3\lambda + 2\mu}{\lambda + 2\mu} x'(T - T_0). \tag{6.61}$$

Under the simplifying assumption that there is no lateral contraction of the rod, i.e., Poisson's ratio may be set equal to zero, this simplifies to

$$u'(x') = \alpha x'(T - T_0). \tag{6.62}$$

At the end of the rod, i.e., at the point $x' = L$, the displacement is then

$$u'(L) = \alpha L(T - T_0), \tag{6.63}$$

and hence the new length of the rod is given by

$$L_{\text{new}} = L(1 + \alpha(T - T_0)). \tag{6.64}$$

Notice that this length varies linearly with the temperature increase, $T - T_0$. Also notice it is independent of the shape or size of the cross-section of the rod. The magnitude of the displacement of such an actuator is governed only by the coefficient of thermal expansion, α, and the increase in temperature.

6.5 Modeling Thermoelastic V-Beam Actuators

Thermally induced buckling of an elastic rod can be a vice or a virtue. In the case of an actuator, it's a virtue; in the case of a MEMS/NEMS structure where the rod is supposed to remain straight, it's a vice. Here we consider the structure in Figure 6.8 and imagine that this is part of an actuator design. We imagine that the beam is heated, perhaps by a resistive heater modeled in Chapter 4. As with thermopneumatic systems and the thermoelastic rod,

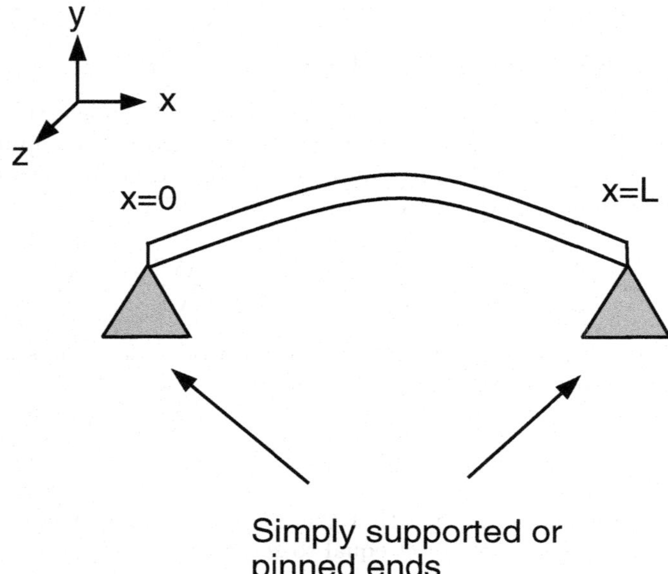

FIGURE 6.8: The basic thermoelastic buckling actuator structure.

we would like to compute the deflection of our actuator as a function of rod temperature. As in previous sections of this chapter, we take the temperature field as a known quantity. The ambitious reader can easily couple the system studied here to the model of Joule heating in Chapter 4.

In Chapter 2 we introduced the calculus of variations as a tool for deriving the beam equation of elasticity. Recall that in the steady-state, the deflection of an elastic beam satisfies

$$\frac{d^2}{dx^2}\left(EI\frac{d^2v}{dx^2}\right) + P\frac{d^2v}{dx^2} = F. \tag{6.65}$$

The function $v(x)$ represents the steady-state deflection of the elastic rod in Figure 6.8 in the x-y plane. There is assumed to be no motion in the y-z plane. We have modified our notation from Chapter 2 slightly. Instead of the second term in equation (6.65) being negative and containing a term proportional to the tension in the beam, we have taken it to be positive and will refer to P as the pressure. This change has been made to avoid confusing the T of tension with the T of temperature. Also, we have introduced a term on the right-hand side of equation (6.65). This term represents an applied load. Here the load will be a consequence of thermal stresses. In general, the load F can be written as

$$F = p(x) - \frac{d^2 M_{Tz}}{dx^2}. \tag{6.66}$$

The function $p(x)$ is the applied load *not* due to thermal stresses. For example it might be gravitational loading, although in MEMS such a term would be negligible. The term M_{Tz} is computed from the thermal field through

$$M_{Tz} = \int_A \alpha ET y dA \tag{6.67}$$

where the integral is an area integral over the cross-section of the rod. The assumption that the rod does not move in the $y - z$ plane requires that the analogous term M_{Ty} defined as

$$M_{Ty} = \int_A \alpha ET z dA, \tag{6.68}$$

be assumed to be identically zero. Now, we will assume that $p(x) = 0$ and that T is a constant. Notice that this implies $M_{Tz} = 0$ as well. We will also assume that the flexural rigidity, EI, is constant. Then, our beam equation becomes

$$\frac{d^4 v}{dx^4} + k^2 \frac{d^2 v}{dx^2} = 0, \tag{6.69}$$

where

$$k^2 = \frac{P}{EI}. \tag{6.70}$$

We assume that the ends of our beam are simply supported and impose the boundary conditions,

$$v(0) = v(L) = \frac{d^2 v}{dx^2}(0) = \frac{d^2 v}{dx^2}(L) = 0, \tag{6.71}$$

as discussed in Chapter 2. Equation (6.69), together with the boundary conditions (6.71), constitutes an *eigenvalue problem* for the beam deflection $v(x)$. Notice that $v(x) = 0$ is always a solution to this problem. No matter the imposed temperature the straight *undeflected* beam is always an equilibrium solution. The question is whether there are other solutions to equations (6.69)–(6.71) and if so when those solutions occur. It is easy to write down the general solution to equation (6.69); we write as

$$v(x) = a_0 + a_1 x + a_2 \cos(kx) + a_3 \sin(kx). \tag{6.72}$$

Imposing the conditions at $x = 0$ immediately implies that $a_2 = 0$ and $a_0 = 0$; hence our solution simplifies to

$$v(x) = a_1 x + a_3 \sin(kx). \tag{6.73}$$

Imposing the conditions at $x = L$, we see that either $a_1 = a_3 = 0$ and we have the $v(x) = 0$ solution already uncovered or $a_1 = 0$ and

$$\sin(kL) = 0. \tag{6.74}$$

That is, the function,

$$v(x) = a_3 \sin(kx), \tag{6.75}$$

with arbitrary constant a_3 solves our boundary value problem provided k satisfies equation (6.74). Since the zeros of the sin function occur at $n\pi$, $n = 1, 2, 3, \dots$, equation (6.74) implies

$$k = \frac{n\pi}{L}. \tag{6.76}$$

These values of k are called the *eigenvalues* for our problem. From equation (6.70) we can relate the eigenvalues to the pressure P in the rod. In particular,

$$P = \frac{n^2\pi^2 EI}{L}. \tag{6.77}$$

The lowest value of P for which we have an eigenvalue is when $n = 1$ or

$$P_c = \frac{\pi^2 EI}{L}. \tag{6.78}$$

This critical value, P_c, is called the *Euler Load* for the structure and when P reaches P_c, we say that *buckling* has occurred. But, we are attempting to model a thermal V-beam actuator. What does P have to do with a heated beam? In the V-beam actuator the pressure or stress in the beam is due to the temperature increase in the beam. In Chapter 2 we introduced the Duhamel-Neumann relation of thermoelasticity. This related the stress in a thermoelastic material to strains and temperature. Assuming the structure of Figure 6.8 is *undeflected* and that the ends of the rod do not move, we can use the Duhamel-Neumann relation to compute P. Setting the Poisson ratio to zero we find

$$P = \alpha AE(T - T_0). \tag{6.79}$$

In turn we can compute the critical temperature at which buckling occurs,

$$T_c = T_0 + \frac{\pi^2 I}{\alpha AL^2}. \tag{6.80}$$

Now, in addition to knowing the critical temperature and pressure at which buckling occurs, we would also like to know the magnitude to the deflection of our actuator as a function of the temperature above the critical temperature. The purely linear theory that we have discussed cannot tell us the amplitude of the deflection. This is a point worth elaborating. If the reader returns to Chapter 2, where the beam equation was derived, they will note that both a linearized form of the stretching energy and bending energy were assumed. If we really want to compute the buckling amplitude, we should return to our beam equation derivation, proceed without linearizing, and derive the full nonlinear equations governing an elastic beam. Such an effort would reveal that the linear beam theory discussed here is really a linear stability theory for the $v = 0$ solution of the full nonlinear equations. Our determination of the critical buckling temperature or pressure really reveals that the $v = 0$ solution becomes linearly unstable at this critical point. Only an analysis of the nonlinear equations or a *nonlinear stability theory* can yield the proper amplitude for the buckled state. Nevertheless, we can make an approximate argument and compute a close approximation to the buckling amplitude. If we revisit our thermoelastic rod computation from earlier in this chapter, we can compute the length a free rod would assume at the critical buckling temperature. We find

$$L_c = L(1 + \alpha(T_c - T_0)). \tag{6.81}$$

Also, we can compute the length a free rod would assume at an arbitrary temperature. We denote this L_n and find

$$L_n = L(1 + \alpha(T - T_0)). \tag{6.82}$$

Now, since our actual rod is constrained up to the buckling temperature, we assume any change in length of the rod beyond the change at the buckling temperature to translate into motion perpendicular to the rod axis. That is, we expect our rod to now have total length,

$$L + L_n - L_c. \tag{6.83}$$

On the other hand, we can compute the length of our rod from $v(x)$.

$$\int_0^L \sqrt{1 + \left(\frac{dv}{dx}\right)^2}\, dx. \tag{6.84}$$

Linearizing this expression, inserting equation (6.75) into (6.84), and equating (6.83) with (6.84) we find

$$a_3 = \frac{2L}{\pi}\sqrt{\alpha(T - T_c)}. \tag{6.85}$$

Hence the approximate buckled shape assumed by our V-beam actuator is

$$v(x) = \frac{2L}{\pi}\sqrt{\alpha(T - T_c)}\sin\left(\frac{\pi x}{L}\right), \tag{6.86}$$

with maximum displacement occurring at $x = L/2$ and having value

$$v(L/2) = \frac{2L}{\pi}\sqrt{\alpha(T - T_c)}. \tag{6.87}$$

We observe that the maximum deflection for the V-beam actuator is proportional to the square root of the deviation of the temperature from buckling temperature. Also, we observe that the maximum deflection is independent of the cross-section of the beam.

6.6 Modeling Thermal Bimorph Actuators

As mentioned earlier the thermal bimorph actuator has been used with great effect in MEMS systems. The basic operating principle is the creation of a thermal gradient between the two arms of the bimorph structure. Recall that the two arms are composed of the same material. In this section we examine a simplified model of a thermal bimorph. Rather than studying a two-arm structure where the arms are at different temperatures, we study the single beam structure shown in Figure 6.9 and impose a thermal gradient perpendicular to the beam axis. In particular, we mimic the two-arm structure

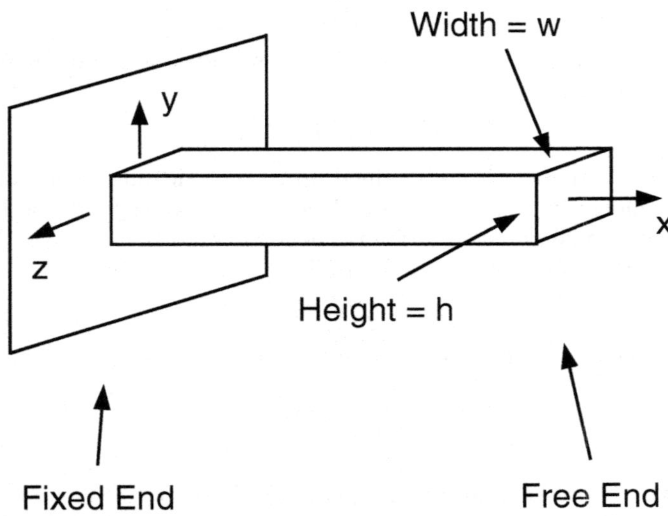

FIGURE 6.9: The beam model of the thermal bimorph structure.

by imposing the temperature profile,

$$T(y) = \frac{T_h - T_c}{h} y + \frac{T_c + T_h}{2}. \tag{6.88}$$

This says the bottom of the beam is at temperature T_c or the cold temperature, while the top of the beam is at temperature T_h or the hot temperature. Hence the top of the beam mimics the small arm of the bimorph structure and the bottom of the beam the large arm.

As in the previous section, we will use the equation for the steady-state deflection of an elastic beam. In particular, denoting the deflection in x-y

plane by $v(x)$ we have

$$\frac{d^2}{dx^2}\left(EI\frac{d^2v}{dx^2}\right) + P\frac{d^2v}{dx^2} = F. \tag{6.89}$$

We assume no motion of the beam in the y-z plane. The notation introduced in the previous section, P and F, is assumed here. The applied load can again be written as

$$F = p(x) - \frac{d^2 M_{Tz}}{dx^2}. \tag{6.90}$$

We assume that the load $p(x)$, i.e. the contribution from effects other than thermal stress, is identically zero. The term M_{Tz} is computed by

$$M_{Tz} = \int_A \alpha ETydA, \tag{6.91}$$

where the integral is an area integral over the cross-section of the beam. Using our assumed form for temperature and assuming that α and E are constant we find

$$M_{Tz} = \frac{\alpha Ewh^2}{12}(T_h - T_c). \tag{6.92}$$

Note that w is the beam width (see Figure 6.9). The assumption that there is no motion in the $y-z$ plane requires us to assume that the analogous quantity M_{Ty} defined as

$$M_{Ty} = \int_A \alpha ETzdz \tag{6.93}$$

be identically zero. This follows from our assumed form for the temperature field. The expression for M_{Tz}, equation (6.92), implies that

$$\frac{d^2 M_{Tz}}{dx^2} = 0, \tag{6.94}$$

and hence our beam equation reduces to

$$EI\frac{d^4v}{dx^4} + P\frac{d^2v}{dx^2} = 0. \tag{6.95}$$

We further assume that $P = 0$ and hence simplify to

$$\frac{d^4v}{dx^4} = 0. \tag{6.96}$$

We assume the left end of the beam is held fixed or anchored and impose the boundary conditions,

$$v(0) = \frac{dv}{dx}(0) = 0. \tag{6.97}$$

The right end of the beam is assumed to be free or unsupported. The appropriate boundary conditions require a modification from those presented in

Chapter 2 to account for the moment due to thermal stresses. In the presence of thermal stresses, the boundary conditions at a free end of a beam are

$$EI\frac{d^2v}{dx^2}(L) = -M_{Tz} \tag{6.98}$$

and

$$EI\frac{d^3v}{dx^3}(L) + P\frac{dv}{dx} = -\frac{dM_{Tz}}{dx}. \tag{6.99}$$

In our case, using equation (6.93), these reduce to

$$EI\frac{d^2v}{dx^2}(L) = -\frac{\alpha Ewh^2}{12}(T_h - T_c) \tag{6.100}$$

and

$$\frac{d^3v}{dx^3}(L) = 0. \tag{6.101}$$

It is easy to integrate and apply the boundary conditions to find

$$v(x) = -\frac{wh^2\alpha}{24I}(T_h - T_c)x^2. \tag{6.102}$$

Since I is defined as

$$I = \int_A y^2 dA, \tag{6.103}$$

this simplifies to

$$v(x) = -\frac{\alpha(T_h - T_c)}{2}\frac{x^2}{h}. \tag{6.104}$$

We note that the maximum deflection occurs at $x = L$ and is given by

$$v(L) = -\frac{\alpha(T_h - T_c)}{2}\frac{L^2}{h}. \tag{6.105}$$

A brief comparison between the bimorph and V-beam actuators is in order. In particular we compare equations (6.87) and (6.105). Earlier we claimed that the bimorph structure was more efficient than the V-beam structure. In comparing equations (6.87) and (6.105), we see that the V-beam maximum displacement increases with the square root of the temperature gradient in the system, $T - T_c$, while the bimorph maximum displacement increases linearly with the imposed temperature gradient, $T_h - T_c$. The bimorph also exhibits a geometric dependence other than the beam length. That is, the maximum displacement depends on h, the beam height. Loosely speaking, this may be identified with the ratio of arm widths in a bimorph structure. Hence a bimorph design can be geometrically optimized.

6.7 Modeling Bimetallic Thermal Actuators

The bimetallic actuator structure is not unique to MEMS and NEMS, and a good discussion of the behavior of such structures can be found in one of any number of texts on elasticity and thermoelasticity. We do not repeat the analysis here, rather we simply give a key result so that the reader may easily compute estimates for deflections of bimetallic MEMS structures. Consider the bimetallic structure shown in Figure 6.4. If we assume the Young's moduli are identical, that each beam is of equal height $h/2$, and that the beam is unconstrained everywhere, then the maximum displacement of the structure occurs at the center and has magnitude $\frac{L^2}{8\rho}$. Here ρ is the radius of curvature of the structure and is given by

$$\frac{1}{\rho} = \frac{3}{2h}(\alpha_2 - \alpha_1)(T - T_0). \tag{6.106}$$

Hence the maximum displacement of the bimetallic structure in terms of the temperature is

$$\frac{3}{16hL^2}(\alpha_2 - \alpha_1)(T - T_0). \tag{6.107}$$

We note that this is linear in the temperature increase of the system above the ambient temperature. Further, this displacement depends on h, the beam height. Hence, geometry plays a role in optimizing displacement for this structure.

6.8 Chapter Highlights

- MEMS and NEMS devices make use of or are affected by thermal-elastic effects in a variety of ways.
- Thermoelastic damping affects the Q of resonant sensors. Thermal stresses can lead to fracture or destruction of a microsystem.
- Thermopneumatic actuators are indirect thermal-elastic systems. Heating of a gas causes a pressure increase which in turn exerts a force on the elastic component of a system.
- Direct thermal-elastic systems take many forms. Thermoelastic rods, V-beam actuators, thermal bimorphs, bimetallic thermal actuators, and shape memory actuators all rely on direct heating of the elastic component.

- Modeling of thermopneumatic systems involves coupling the equations of heat transfer to the equations of elasticity through an equation of state representing the behavior of a gas.

- Thermoelastic rods, V-beam actuators, thermal bimorphs, and bimetallic structures all show different responses to applied thermal gradients. The effectiveness of these structures can be computed and compared via modeling.

6.9 Exercises

Section 6.3

1. Consider the model for a damped linear oscillator:

$$\frac{d^2u}{dx^2} + 2\frac{du}{dx} + u = 0.$$

Rewrite this as a first-order system. Find a Lyapunov function for this system and show that $u = 0$ is a stable equilibrium solution.

2. Use the Lyapunov function, $V(\mathbf{y}) = y_1^2 + y_2^2 + y_3^2$, to show that the origin is a stable steady-state solution of the first-order system:

$$\frac{dy_1}{dt} = -y_2 - y_1 y_2^2 + y_3^2 - y_1^3$$

$$\frac{dy_2}{dt} = y_1 + y_3^3 - y_2^3$$

$$\frac{dy_3}{dt} = -y_1 y_3 - y_3 y_1^2 - y_2 y_3^2 - y_3^5.$$

3. Investigate solutions of the fully nonlinear mass-spring-piston-cylinder model of thermopneumatic actuation by solving the governing equations numerically. As initial conditions take $\theta(0) = 0$, $u(0) = 0$, $u'(0) = 0$. How does the approach to equilibrium depend on β?

4. The mass-spring-piston-cylinder model in this chapter assumed that the resistive heater was at fixed temperature T_h. Improve this aspect of the model by coupling the thermomechanical model presented here with the Joule heating model of Chapter 4.

5. Investigate the stability of the steady-state solution to the mass-spring-piston-cylinder model by using the linear stability approach of Chapter 4.

6. In the mass-spring-piston-cylinder model, change the resistive heater temperature from a fixed constant to the periodic function, $T_h \sin^2(\omega t')$. This might represent a model of pumping action. Rescale the system to recast in dimensionless form. You should encounter a new dimensionless parameter that contains ω. Give a physical interpretation of this parameter. Numerically investigate solutions to this pumping model. Can you say anything about the maximum pumping speed for such a system?

7. The membrane model developed in Section 6.3 has an advantage over the mass-spring model in that it captures the effect of geometry on the maximum displacement achievable. To further examine this effect, redo the analysis of the membrane model, replacing the cylindrical piston with a piston of square cross-section and the circular membrane with a square membrane. Formulate the steady-state problem for this geometry. Solve. Compare with a circular membrane of the same area.

8. Perform a stability analysis of the steady-state solutions of the membrane model. Use the linear stability ideas introduced in Chapter 4. In particular, seek solutions of the form,

$$\theta(t) \sim \epsilon A e^{-\gamma t}$$

$$u(r,t) = \frac{\lambda}{4}(1-r^2) + \epsilon v(r) e^{-\gamma t},$$

expand for $\epsilon \ll 1$, and solve the resulting eigenvalue problem to determine γ.

9. Nondimensionalize the plate model of thermopneumatic actuation, equations (6.58) and (6.59). You should find one more dimensionless parameter arising than in the membrane model. Give a physical interpretation of this new dimensionless parameter.

10. Find all steady-state solutions of the plate model of thermopneumatic actuation in dimensionless form. Assume fixed edge boundary conditions. What is the maximum displacement of the plate? How does this compare to the pure membrane theory?

11. Perform a linear stability analysis of the physical steady-state solutions found in the exercise 10.

12. At the start of the chapter we discussed the "microbellows" actuator developed at Caltech. Develop a first model of the microbellows by using the plate-based model and introducing a boundary condition that models the edges of the plate being attached to a spring with spring constant k. Find the steady-state displacement of this system. How does it compare with the steady-state displacement of the plate with fixed-edge boundary conditions?

Section 6.4

13. Return to the one-dimensional elastic rod discussed in Chapter 2 and revisited in Section 6.4. Compute the displacement in the rod with the same elastic boundary conditions but assuming the temperature at the left end is T_1 and at the right end T_2. Find the steady-state displacement of the point $x = L$. How does this vary with T_1 and T_2?

Section 6.5

14. Use the steady-state lumped model of Joule heating discussed in Chapter 4 to plot an applied voltage versus deflection curve for the V-beam actuator of Section 6.5.

15. Suppose the fabrication process for your V-beam actuator with simply supported ends leaves the beam with a residual tensile stress of magnitude P_r so that when heated $P = P_r + \alpha AE(T - T_0)$. How does this change the critical temperature of buckling?

16. Assuming your V-beam actuator has length 1200μm, CTE of $2.5 \times 10^{-6}\,\mathrm{K}^{-1}$, and that room temperature is $T_0 = 300\,\mathrm{K}$, compute the maximum deflection at $T = 500\,\mathrm{K}$, $T = 1000\,\mathrm{K}$, and $T = 1500\,\mathrm{K}$.

17. For many MEMS materials the CTE is a linearly increasing function of temperature. Suppose the CTE of your V-beam actuator doubles over the temperature range studied in problem 16. How does this affect the maximum deflection at $T = 1500\,\mathrm{K}$?

18. Redo the analysis of the V-beam actuator in Section 6.5, assuming the ends of the beam are fixed rather than simply supported. Compute the critical buckling load and the approximate amplitude of the buckled structure. How does it compare to the simply supported case?

19. Redo the analysis of the V-beam actuator in Section 6.5, assuming one end of the beam is fixed and the other simply supported. Compute the critical buckling load and the approximate amplitude of the buckled structure. How does it compare to the simply supported case?

Section 6.6

20. Assume the bimorph model discussed in Section 6.6 has length 1200μm, CTE of $2.5 \times 10^{-6}\,\mathrm{K}^{-1}$, and height 2μm. Plot the maximum deflection as a function of the thermal gradient, $T_h - T_c$. How large of a thermal gradient is needed to cause a 1μm deflection?

21. Suppose the fabrication process for the bimorph model of this section left a residual tensile stress in the beam, $-P$. Solve for the deflection of the beam. How does this residual stress affect the maximum deflection?

22. Suppose in the model of this section, a parabolic temperature profile were imposed. Assume $T = T_c$ at $z = -h/2$, $T = T_h$ at $y = h/2$ and that $dT/dy = 0$ at $y = -h/2$. Redo the analysis of this section using this temperature profile. How does the maximum deflection compare with the case of the linear temperature profile?

23. Suppose in the model of this section, a discontinuous temperature profile were imposed. Assume $T = T_c$ for $y \in [-h/2, a]$ and $T = T_h$ for $y \in [a, h/2]$. Redo the analysis of this section using this temperature profile. How does the maximum deflection compare with the case of the linear temperature profile? How does the maximum deflection depend on a?

24. In the beam model of this section, assume that the left end is simply supported rather than fixed. The appropriate boundary conditions now become

$$v(0) = 0, \quad EI\frac{d^2v}{dx^2}(0) = -M_{Tz}.$$

Using these boundary conditions solve for the deflection of the maximum deflection of the beam.

25. In the beam model of this section, assume that the cross-section of the beam remains rectangular, but varies linearly in height from h_1 to h_2 as x varies from 0 to L. For each fixed x assume a linear temperature profile in the y direction. Redo the analysis of this section using this temperature profile. Compute the maximum deflection and discuss how it varies with h_1 and h_2.

Section 6.7

26. Assume a MEMS bimetallic structure has length 1200μm, that $\alpha_1 = 2.5 \times 10^{-6}\,\mathrm{K}^{-1}$ and $\alpha_2 = 5 \times 10^{-6}\,\mathrm{K}^{-1}$. Assume the beam has height $h = 10\mu$m. Estimate the maximum deflection of the structure, assuming a temperature increase above room temperature of 100 K, 500 K, 1000 K.

6.10 Related Reading

The text on thermoelasticity is now available in a Dover addition:

B.A. Boley and J.H. Weiner, *Theory of Thermal Stresses*, New York: Dover, 1988.

The texts by Timoshenko are out of print but it's worth hunting for used copies.

S. Timoshenko, *Theory of Plates and Shells*, New York: McGraw Hill 1959.

S. Timoshenko, *Theory of Structures*, New York: McGraw Hill, 1965.

S. Timoshenko and J. Gere, *Theory of Elastic Stability*, New York: McGraw Hill, 1961.

A very nice introduction to thermodynamics is the small Dover book by Van Ness.

H.C. Van Ness, *Understanding Thermodynamics*, New York: Dover, 1983.

Enrico Fermi's book on thermodynamics is more advanced but also very readable.

E. Fermi, *Thermodynamics*, New York: Dover, 1937.

You won't learn much math from the book by Von Baeyer, but it's a great read and a worthwhile history of thermodynamics.

H.C. Von Baeyer, *Warmth Disperses and Time Passes: A History of Heat*, Modern Library, 1999.

A good overview of Lyapunov functions appears in the text by Brauer and Nohel.

F. Brauer and J.A. Nohel, *The Qualitative Theory of Ordinary Differential Equations: An Introduction*, New York: Dover, 1969.

6.11 Notes

1. The reader may wonder why we use the term *thermal-elastic* as opposed to *thermoelastic.* The term *thermoelastic* usually refers to the direct creation of stress in a material due to thermal gradients. We use the term *thermal-elastic* to capture this and effects such as the thermopneumatic effect.

2. A thermal gradient in the direction perpendicular to the axis of a beam might be achieved by doping the beam to vary electrical conductivity in this direction. To our knowledge this idea has not been investigated in practice.

3. At one time Edmund Scientific sold a "jumping quarter" based on the bimetallic structure—JAP.

4. The model studied in this section may be extended in many directions. One interesting extension is to couple together many mass-spring-piston-cylinder systems and to take a *continuum limit.* Doing so one obtains the equations of linear thermoelasticity. The interested reader may find this analysis in [157].

5. This assumption may easily be removed. This adds a constant to the equation of motion

6. See Chapter 2 and Chapter 3 for a discussion of the mechanisms of heat transfer in microsystems.

7. The mean value theorem for integrals is stated in Appendix A.

Chapter 7

Modeling Electrostatic-Elastic Systems

Other springs, more surprises.

Edward Abbey

7.1 Introduction

The two hundredth anniversary of Coulomb's law is already ten years past. Yet, everyday micro- and nanoelectromechanical systems use the Coulomb force to grab, pump, bend, spin, and even slide. The theory of such *electrostatically actuated* devices, begun with Coulomb's famous inverse square law and essential to the continued development of MEMS and NEMS, is the subject of this chapter.

Experimental work in this area dates to 1967 and the work of Nathanson et. al. [143]. In their seminal paper, Nathanson and his coworkers describe the manufacture, experimentation with, and modeling of, a millimeter-sized resonant gate transistor. This early MEMS device utilized both electrical and mechanical components on the same substrate resulting in improved efficiency, lowered cost, and reduced system size. Nathanson and his coworkers even introduced a simple lumped mass-spring model of electrostatic actuation. In an interesting parallel development, the prolific British scientist, G.I. Taylor, investigated electrostatic actuation at about the same time as Nathanson [188]. While Taylor was concerned with electrostatic deflection of soap films rather than the development of MEMS devices, his work spawned a small body of literature with relevance to both MEMS and NEMS.

In this chapter we will investigate the work of both Nathanson and Taylor, as we build a theory of electrostatically actuated devices. We begin in Section 7.2 with an overview of eight electrostatically actuated MEMS and NEMS devices. A striking fact concerning these eight devices is that they are all limited by an instability. This instability, named the "pull-in" instability by Nathanson, occurs when voltage applied to the system exceeds a critical value. Beyond this critical value there is no longer a steady-state configuration of the device where mechanical members remain separate. That is, the separate elements of the device have "pulled in" or collapsed onto one another. An example is the model device shown in Figure 7.6. Pull-in corresponds to the two plates colliding as the applied voltage is increased past some critical value. The voltage at which pull-in occurs is referred to as the "pull-in voltage," while the maximum deflection achieved before the onset of the instability is referred to as the "pull-in distance." As a rule of thumb, the pull-in distance is about one-third the size of the zero voltage gap. Because the pull-in distance is generally so small, this instability is a key limiting factor in the design of almost all electrostatically actuated MEMS devices. Developing an understanding of this instability is a principle goal of this chapter.

In Section 7.3 we study the mass-spring model of electrostatic actuation. This model was introduced into the literature by Nathanson, Newell, Wickstrom, and Davis in 1967 [143]. In their influential study of a resonant gate transistor, Nathanson et. al. introduced and analyzed the mass-spring model as a way of gaining insight into device operation. The clever abstraction of the mass-spring system from the actual beam-plate device design removed the complication of geometry from the analysis and put the focus upon the balance of electrostatic and mechanical forces. This allowed Nathanson and his coworkers to predict and offer a first explanation of the ubiquitous pull-in voltage instability. Since Nathanson's original article the mass-spring model has appeared in the microsystems literature numerous times. In [176], Shi et. al. used a mass-spring model of an electrostatically actuated microsystem to model the behavior of a microtweezer. Other authors have used mass-spring models to understand hysteresis effects in MEMS [65], the dynamics of devices [177], the design of strain gauges [13] and the behavior of electrostatic actuators [32]. Still others have embedded the basic mass-spring model in more complex systems in an effort to understand proposed schemes for the control of the pull-in instability [27, 173, 174]. As new devices are developed, one expects the mass-spring model to be reincarnated as a useful first model of many microsystems. In Section 7.3 we compute the pull-in voltage and pull-in distance predicted by the mass-spring model. We show that the mathematical explanation of the pull-in instability is the same as the explanation of thermal runaway in Chapter 4; the bifurcation diagram contains a fold.

In Section 7.4 we construct a mathematical model of a general electrostatically actuated structure. In particular we consider an elastic diaphragm suspended above a rigid plate. Unlike the plates in the mass-spring model, the diaphragm is allowed to bend and deform. This basic structure is at the heart of the design of numerous devices; micropumps, microvalves, microswitches, electrostatic actuators, the grating light valve, etc., all make use of this arrangement. We carefully formulate both the electrostatic and elastic problems. We show precisely how this is a coupled-domain problem. In the MEMS and NEMS literature, a standard way to simplify this problem is to use an approximate solution for the electric field surrounding the structure. Given the field, the problem reduces to a nonlinear equation for the elastic deformation alone. By scaling arguments and elementary perturbation theory, we show that this approximation is valid in the limit where the aspect ratio of the device is small. Roughly speaking, the lateral dimensions of the system must be large compared to the distance between components. Curiously enough, the first study of electrostatic actuation that made use of this approximation was not a study of MEMS or NEMS. Rather, G.I. Taylor and R.C. Ackerberg [1, 188], used the same approximation in their study of electrostatically deflected soap films. Taylor's interest in this problem was motivated by cloud electrification, a subject many orders of magnitude away from microsystems in both voltage and size! Tangentially, Taylor's elegant experiment provides a simple macroscopic vehicle for studying electrostatic actuation.[1]

In Section 7.5 we specialize the general theory of Section 7.4 to the case when the diaphragm is an elastic membrane. This is a reasonable approximation to Taylor's soap film experiments and to many MEMS devices. We focus on steady-state behavior and the characterization of the pull-in instability. Borrowing tools from the mathematical theory of combustion, we present results concerning solutions to the steady-state equation for arbitrarily shaped membranes. Specifically, we show that the pull-in instability is always present in such systems. We derive a first estimate of the pull-in voltage and explain how this estimate varies with the shape of the membrane. Next, we focus on two specific shapes: the one-dimensional strip and the disk. In the strip case we solve the nonlinear equation governing the deflection exactly. We show that the bifurcation diagram for the system is the same as for the mass-spring model; it contains a single fold. We also discuss the stability of these solutions and show that the strip model has the same behavior as the mass-spring system. We use a Green's function to turn the nonlinear ordinary differential equation governing the deflection into a nonlinear integral equation. We use the integral equation to investigate iterative numerical methods similar to those explored in Chapter 4. In the disk case an exact solution to the nonlinear ordinary differential equation governing the deflection is not possible. However, we introduce *symmetry methods* that allow solutions to be completely characterized. A surprising result is uncovered. For the disk, the bifurcation diagram contains not one, but infinitely many folds! The immediate implication of this is that there are values of the applied voltage for which infinitely many steady-state solutions exist. We discuss the implications of this and indicate directions for future research.

In Section 7.6 we briefly discuss the general theory of Section 7.4 for the case where the diaphragm is an elastic plate or an elastic beam. We present the governing equations for steady-state deflections of these systems. The analysis of these models is partially developed in the exercises.

Finally, in Section 7.7 we apply our theory of electrostatic actuation to the study of capacitive control. The claim is that by adding a series capacitance to the circuit containing an electrostatically actuated device, the effect of the pull-in instability may be reduced. We investigate this claim in the context of a mass-spring model and in the context of a membrane based model. We show that the mass-spring analysis leads to the conclusion that the capacitive control scheme is completely stabilizing. By analyzing the membrane model we see that this conclusion is an artifact of ignoring deformations of the deflected elastic diaphragm. The membrane analysis reveals that the scheme is partially stabilizing at the cost of increased actuation voltages.

7.2 Devices Using Electrostatic Actuation

In this section we take a brief look at eight devices that use electrostatic forces for their operation. Our selections are representative of the variety of ways and variety of systems in which electrostatic forces are used. Many more examples appear in the bibliography; the reader is directed to the literature for other examples of electrostatic-elastic systems.

7.2.1 Grating Light Valve

One of the most celebrated electrostatic-elastic systems is the grating light valve or GLV invented by researchers at Stanford University.[2] A commercial display system based on the grating light valve is being developed by Silicon Light Machines of Sunnyvale, California. The system consists of an array of pixels. Each pixel is a MEMS based electrostatically controlled diffraction grating. Each diffraction grating consists of an array of thin elastic ribbons

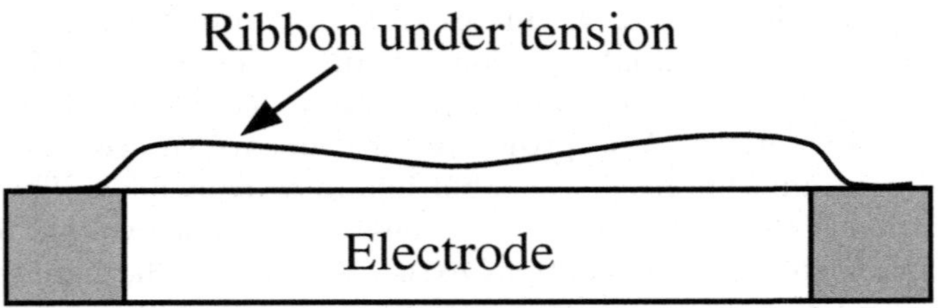

FIGURE 7.1: A single ribbon in the GLV.

held under tension. A single ribbon is shown in Figure 7.1. A given ribbon is deflected by the application of a potential difference between the ribbon and ground electrode. When no ribbons are deflected, the pixel presents a solid mirror-like surface. When alternate ribbons are deflected, as shown in Figure

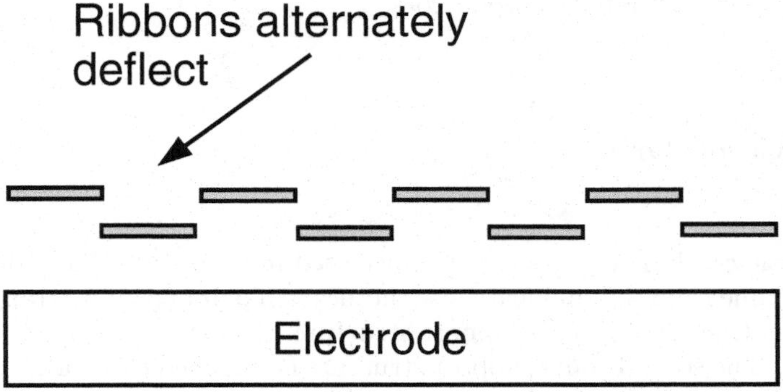

FIGURE 7.2: A side view of the GLV showing diffraction grating effect.

7.2, the pixel becomes a diffraction grating.

We make several observations concerning the GLV. First, the use of an applied potential difference to provide a motive force is typical in electrostatically driven MEMS and NEMS. Next, the design is limited by the pull-in instability. The degree to which the ribbons can be deflected and hence the properties of the diffraction grating are directly controlled by the pull-in effect. It is not desirable to allow the ribbons to make contact with the ground electrode. The phenomenon of stiction[3] may prevent a return to the original configuration even when the applied voltages are removed. Finally, the GLV is an example of a device consisting of elastic membranes held under tension. The geometry is essentially one-dimensional and the ratio of gap size to lateral size is small.

7.2.2 Micromirrors

The torsion micromirror was discussed in Chapter 5 as an example of a simple elastic structure often found in MEMS. Recall that in this system the goal is to change the angle of a rigid mirror in order to control the location of a reflected beam of light. The mirror can then be used as an optical switch for fiber networks or as a component of an adaptive optical system. The basic micromirror structure is shown in Figure 5.1. When electrostatic forces are used to actuate the system, the range of operation is limited by the pull-in instability. The device is actuated by applying a potential difference between the mirror and one of the ground electrodes. This causes a torque on the system, which is countered by the effect of a torsion spring. For this system, the pull-in instability occurs once rotation through a critical angle has taken place. That is, the range of angular motion is limited by the pull-in instability. References [22, 43, 84, 85] provide further discussion of electrostatically actuated micromirrors.

7.2.3 Comb Drive

The basic comb drive structure is illustrated in Figure 7.3. The comb drive takes its name from the similarity in structure to a pair of combs arranged with interwoven tines. One comb support structure is anchored in place and does not move. The second comb support structure is attached to a spring or folded beam and is free to move. A potential difference is applied between the two sets of combs, resulting in an electrostatic force pointing in the direction of the fingers. The comb drive does not suffer from the pull-in instability in the usual sense. That is, its displacement is not limited by the pull-in phenomenon. However the pull-in phenomenon does create a lateral "side-to-side" instability in the comb drive. If the comb structure is disturbed laterally causing a pair of tines to become too close, the pull-in phenomenon will complete the disturbance and cause the tines to collide. Sufficient lateral stiffness must be designed into the comb drive structure to prevent this difficulty. The reader is directed to the literature for a host of uses of the comb drive structure. The electrostatic comb drive enjoys widespread use in MEMS technology.

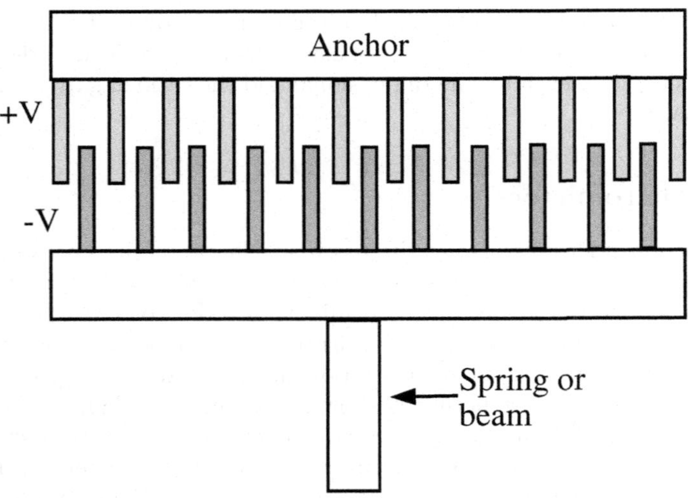

FIGURE 7.3: The basic comb drive structure.

7.2.4 Micropumps

The MEMS micropump has appeared several times throughout this text. This is a testament to the importance of this particular device. In Chapter 6 we examined a micropump driven using thermopneumatic forces. In Chapter 8 we will examine a magnetically driven micropump. Here, we discuss the electrostatically actuated micropump. The basic pump structure is illustrated in Figure 7.4. The pump is operated by applying a potential difference between a deformable diaphragm and a fixed counterelectrode. Electrostatic forces cause the diaphragm to deflect toward the counterelectrode. The change in volume in the chamber below the diaphragm causes fluid to be drawn in through the inlet check valve.[4] The voltage is then removed, the diaphragm returns to its un-deflected state and pushes fluid out through the outlet check valve. This operation is repeated resulting in a pumping action. The design of the electrostatic micropump is directly limited by the pull-in instability. The volume stroke of the pump, which one would like to maximize, is small due to the limited range of stable displacements of the diaphragm. It is undesirable to run the pump in "pull-in" mode where the diaphragm smashes into the counterelectrode. Stiction may prevent re-release of the diaphragm and repeated impacts can damage the diaphragm.

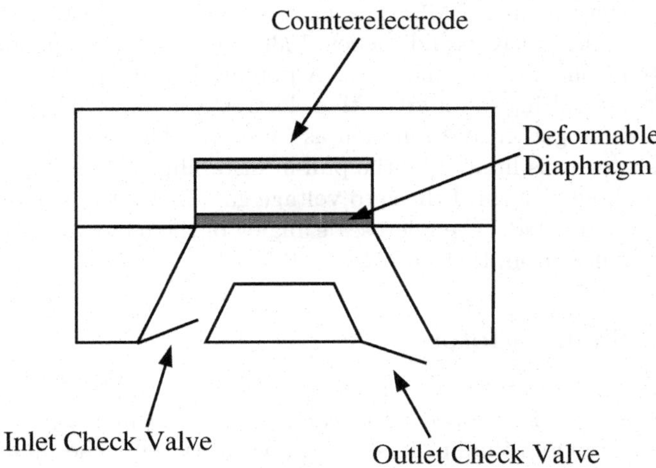

FIGURE 7.4: The basic electrostatic micropump.

7.2.5 Microswitches

The electrostatically actuated microswitch is an example of a system that is often run in "pull-in" mode. The structure of the microswitch varies widely according as the intended application. One simple design is to imitate the structure of a single GLV ribbon, Figure 7.1. By applying a potential difference between the ribbon and substrate, the ribbon can be pulled down. In this way an electrical contact between the ribbon and a secondary electrode on the substrate can be made, resulting in a microswitch.

7.2.6 Microvalves

In addition to micropumps, microvalves are necessary to regulate microscale flow. The electrostatically actuated microvalve provides a second example of a system often run in "pull-in" mode. Like the microswitch a wide variety of microvalve designs are possible. At their core all microvalve designs consist of a deformable plate that can be moved to open or close a valve. In electrostatically actuated microvalves, the plate is moved by applying a potential difference between the plate and a ground electrode. If operated in pull-in mode, a voltage beyond the pull-in voltage is applied causing the plate to collapse onto the substrate, closing the valve.

7.2.7 Micro- and Nanotweezers

Micro- and nanotweezers were discussed in Chapter 5 as examples of common elastic structures in MEMS. The basic structure is illustrated in Figure 5.4. In the design of Kim and Lieber [103], the arms of the nanotweezers are constructed from carbon nanotubes. A potential difference is applied between the arms, resulting in an attractive electrostatic force. The tips of the arms move closer together and function as tweezers. The design of the electrostatic nanotweezer is limited by the pull-in instability. The range of stable displacement, about one third the zero voltage gap, limits the size of objects that may be manipulated. The microscale analog of the nanotweezer, i.e., the microtweezer, suffers from the same design difficulty.

7.2.8 The Shuffle Motor

A final novel example of the use of electrostatic forces in MEMS is provided by the shuffle motor. Developed by researchers at the University of Twente (in the Netherlands) and Phillips Research Laboratory [187], the shuffle motor is a microscale linear electrostatic stepper motor capable of producing large forces and small motions. The basic principle of operation is illustrated in Figure 7.5. An elastic plate is suspended between two "clamps." The clamps are free to slide on the substrate. A potential difference between either clamp and an electrode embedded in the substrate may be applied. When this is

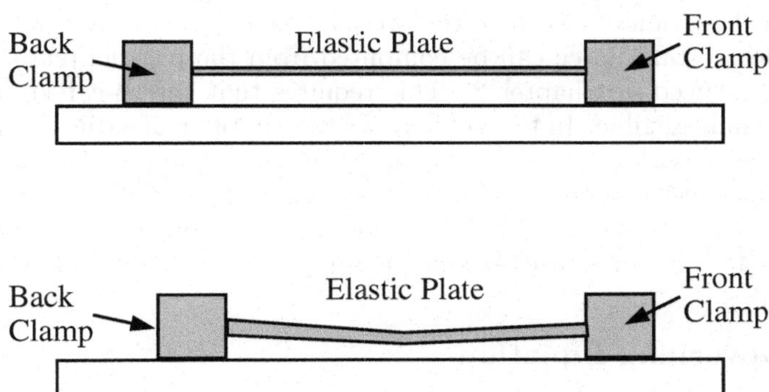

FIGURE 7.5: The electrostatic shuffle motor.

done, the clamp is effectively locked in place. If the front clamp is locked and a potential difference is then applied between the elastic plate and the substrate electrode, the subsequent deflection of the plate drags the rear clamp forward. The back clamp may then be locked, the front clamp released, and the voltage removed from the plate. This forces the front clamp forward resulting in a return to the original configuration but shifted to the right. Repeating the process allows the motor to "walk" in either direction. Again because of the phenomenon of stiction, it is not desirable to have the elastic plate make contact with the substrate.[5] Hence the shuffle motor's steps are limited by the pull-in phenomenon.

7.3 The Mass-Spring Model

In this section, we examine the mass-spring model of electrostatically actuated MEMS devices in detail. Note that the foundation of this model rests upon several approximations. The first, alluded to above, concerns geometry. In particular, the mass-spring model neglects geometric effects such as the bending of beams or the deflections of membranes. Rather, parallel components of the system are assumed to remain parallel throughout all deflections. This implies that the mechanical state of the system may be specified by a single dependent variable, u, the deflection, which itself is only a function of time. Of course, this idealization is rarely realized in practice. However, as in Nathanson's original study, one imagines that the spring constant, k, serves as a "lumped" parameter approximately characterizing the mechanical properties of the system. The second assumption is also geometrical in nature, but perhaps less obviously so. This assumption concerns the electric field and hence the calculation of electrostatic forces in the system. Here, we'll assume that the electrostatic force can be computed from the parallel plate approximation introduced in Chapter 2. This requires that the so-called "fringing fields" remain negligible. In this section, we will simply make this assumption, but in the next section, we will examine it in detail and see that this really requires the aspect ratio of the device to be small. That is, the characteristic length of the deflected component should be large compared to the size of the deflections. It is in this sense that this assumption is geometrical in nature.

7.3.1 Governing Equation

We begin by analyzing the balance of forces for the system pictured in Figure 7.6. The governing equation for our mass-spring system follows directly

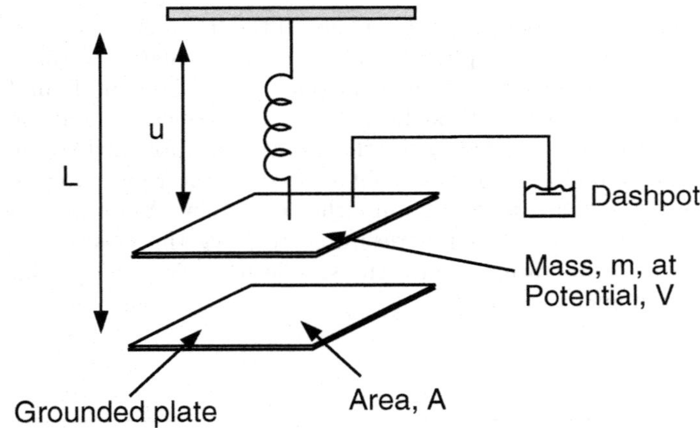

FIGURE 7.6: Sketch of the electrostatically actuated damped mass-spring system.

from Newton's second law. That is,

$$m\frac{d^2u}{dt'^2} = \sum \text{forces}. \tag{7.1}$$

Here, u is the displacement of the top plate from the top wall and m is the top plate's mass. We assume that the bottom plate is held in place. The forces acting on our system are the spring force, F_s, a damping force represented by the dashpot in Figure 7.6, F_d, and the electrostatic force, F_e, due to the applied voltage difference between the plates. We assume that the spring is a linear spring and follows Hooke's law,

$$F_s = -k(u - l), \tag{7.2}$$

where l is the rest length of the spring and k is the spring constant. We assume that damping is linearly proportional to the velocity, that is,

$$F_d = -a\frac{du}{dt'}. \tag{7.3}$$

To compute the electrostatic force we need to recall that the energy, U, contained in an electric field is given by

$$U = \frac{\epsilon_0}{2}\int |\mathbf{E}|^2, \tag{7.4}$$

where ϵ_0 is the permittivity of free space and the integral is a volume integral over all space. From Chapter 2 we know that the magnitude of the E field between the plates may be approximated by

$$E = \frac{Q}{\epsilon_0 A}, \tag{7.5}$$

where Q is the total charge on a plate and A is the plate area. Hence,

$$U = \frac{Q^2(u - L)}{2\epsilon_0 A}. \tag{7.6}$$

We would like an expression in terms of voltage rather than charge. From Chapter 2 we know that the charge can be expressed in terms of voltage and capacitance as

$$Q = CV = \frac{\epsilon_0 AV}{(u - L)} \tag{7.7}$$

and hence,

$$U = \frac{\epsilon_0 V^2 A}{2(u - L)}. \tag{7.8}$$

Now, to compute the force, we simply compute the change in this energy for an infinitesimal change in the gap between the plates. That is,

$$F_e = -\frac{\partial U}{\partial(u - L)} = \frac{1}{2}\frac{\epsilon_0 AV^2}{(L - u)^2}. \tag{7.9}$$

Here, we generalize the electrostatic force, F_e, by allowing the applied voltage to vary in time. We assume the variation is time harmonic and has the form $V\cos(\omega t')$. Under the assumption that the field between the plates can still be computed using electrostatics, the only change to equation (7.9) is the addition of the cosine term,

$$F_e = \frac{1}{2}\frac{\epsilon_0 AV^2}{(L - u)^2}\cos^2(\omega t'). \tag{7.10}$$

Inserting equations (7.2), (7.3) and (7.10) into equation (7.1) yields

$$m\frac{d^2u}{dt'^2} + a\frac{du}{dt'} + k(u-l) = \frac{1}{2}\frac{\epsilon_0 AV^2}{(L-u)^2}\cos^2(\omega t'). \tag{7.11}$$

Equation (7.11) is our first model of damped electrostatically forced MEMS devices. Before attempting to analyze equation (7.11), it is useful to rescale the equation and introduce dimensionless variables. Since the proper choice of scaling depends upon the parameter range under consideration and hence, upon the type of system one wishes to study, we consider several possible choices of time scale. With regard to the length scale, we introduce

$$v = \frac{u-l}{L-l}. \tag{7.12}$$

That is, we scale the displacement of our upper plate with the distance $L-l$. Note that this is a characteristic displacement. In fact, it is the maximum distance that the upper plate can deflect from its zero voltage configuration. Hence, this choice of scale allows us to assume that v is $O(1)$.

7.3.2 A Time Scale for Systems Dominated by Inertia

If the system we wish to model is dominated by inertia, that is, if we expect inertial effects to be more important than damping effects, then we should scale with the natural frequency of oscillation of the mass-spring system. We define the dimensionless time,

$$t = \sqrt{\frac{k}{m}}\,t'. \tag{7.13}$$

Introducing equations (7.12) and (7.13) into equation (7.11) yields

$$\frac{d^2v}{dt^2} + \alpha\frac{dv}{dt} + v = \frac{\lambda}{(1-v)^2}\cos(\Omega_1 t), \tag{7.14}$$

where

$$\alpha = \frac{a}{\sqrt{mk}}, \quad \Omega_1 = \omega\sqrt{\frac{k}{m}}, \quad \lambda = \frac{1}{2}\frac{\epsilon_0 AV^2}{k(L-l)^3}. \tag{7.15}$$

The dimensionless parameter α may be interpreted as a damping coefficient that measures the relative strength of the viscous damping force as compared to the spring force. In Section 5.3 we noted that α is the reciprocal of the quality factor or Q for the system. Notice that α multiplies the damping term, $\dot{v}$. In the inertially dominated regime, we expect a high Q and hence α to be small, and we see that we have scaled correctly as α occurs in front of $\dot{v}$. The parameter Ω_1 measures the frequency of forcing relative to the natural frequency of oscillation of the system. The parameter λ is a key parameter in electrostatic MEMS and warrants further consideration. If we rewrite λ as

$$\lambda = \frac{1}{2}\frac{\epsilon_0 AV^2}{(L-l)^2} \times \frac{1}{k(L-l)}, \tag{7.16}$$

then we see that

$$\lambda = \frac{\text{reference electrostatic force}}{\text{reference spring force}}. \tag{7.17}$$

That is, λ measures the relative strengths of elastic and electrostatic forces in our device. As it is proportional to V^2, it also serves as the "tuning" parameter for our system. Thinking experimentally, λ is the parameter that will be varied as the applied voltage is varied in an experiment. The fact that λ contains the spring constant k, the lengths, l and L, the area of the plates, A, and the permittivity of free space, ϵ_0, says that the results of our analysis may be conveniently summarized in terms of λ. If we understand the behavior of our equations as a function of λ, then we understand the behavior of *all* such devices for which our simplifying assumptions are valid.

7.3.3 A Time Scale for Viscosity-Dominated Systems

If the system we wish to model is viscosity dominated, that is, if we expect damping effects to be more important than inertial effects, then we should introduce a time scale based on the damping time for the system. We define a dimensionless time scale by

$$t = \frac{k}{a}t'. \tag{7.18}$$

Introducing equations (7.12) and (7.18) into equation (7.11) yields

$$\frac{1}{\alpha^2}\frac{d^2v}{dt^2} + \frac{dv}{dt} + v = \frac{\lambda}{(1-v)^2}\cos(\Omega_2 t), \tag{7.19}$$

where here α and λ are as before and Ω_2 is the ratio of damping and forcing times given by

$$\Omega_2 = \omega\frac{k}{a}. \tag{7.20}$$

When damping effects dominate over inertial effects, i.e., for a small Q device, we expect the parameter α to be large. With the scaling just introduced, a factor of $1/\alpha^2$ appears in front of the inertial term, rendering it small, as expected.

7.3.4 Steady-State Solutions

Many devices, such as microtweezers, micropumps, linear motors, or even actuators are operated in the direct current regime. A switch is thrown, a voltage applied, the device responds and comes to rest in a new configuration.

Mathematically, to capture this scenario, we set the AC frequency, ω, to zero in equation (7.11) and then set all time derivatives to zero to obtain an algebraic equation governing the steady-state behavior of our mass-spring system. In dimensionless variables, regardless of the time scale, this algebraic equation is

$$v = \frac{\lambda}{(1-v)^2}. \tag{7.21}$$

Defining

$$f(v) = \frac{\lambda}{(1-v)^2} - v \tag{7.22}$$

allows us to characterize steady-state solutions for our mass-spring system as real roots of $f(v)$. In Figure 7.7, we plot $f(v)$ for various values of λ. First, notice that f has a singularity at $v = 1$. This is to be expected as $v = 1$ corresponds to our top plate colliding with our bottom plate and the electrostatic force becoming infinite. Since $v > 1$ corresponds to the top plate passing through the bottom plate, any roots of f greater than one are unphysical and not of interest. The top plate is not allowed to move through the bottom plate! On the other hand, roots of f less than one correspond to physically relevant solutions. We see in Figure 7.7 that f either has two such roots, precisely one such root or none at all. The transition occurs with

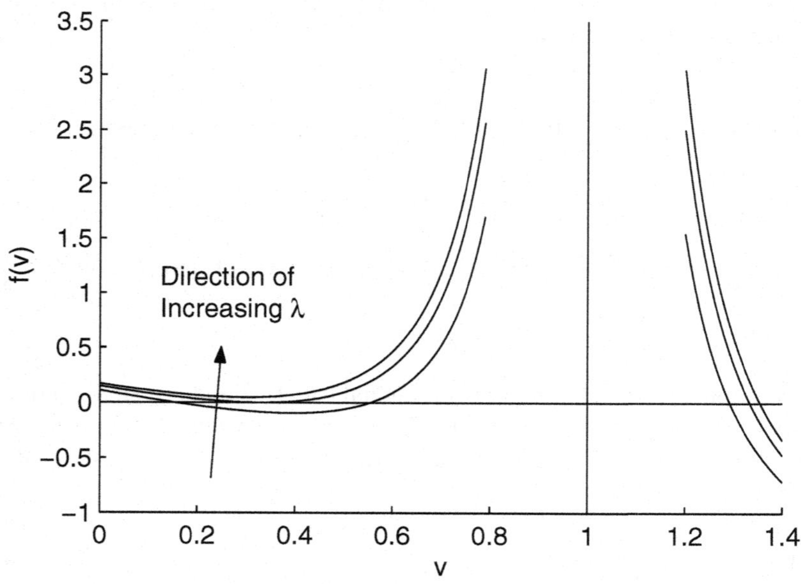

FIGURE 7.7: Plot of $f(v)$. Real roots of f define steady states for the mass-spring system.

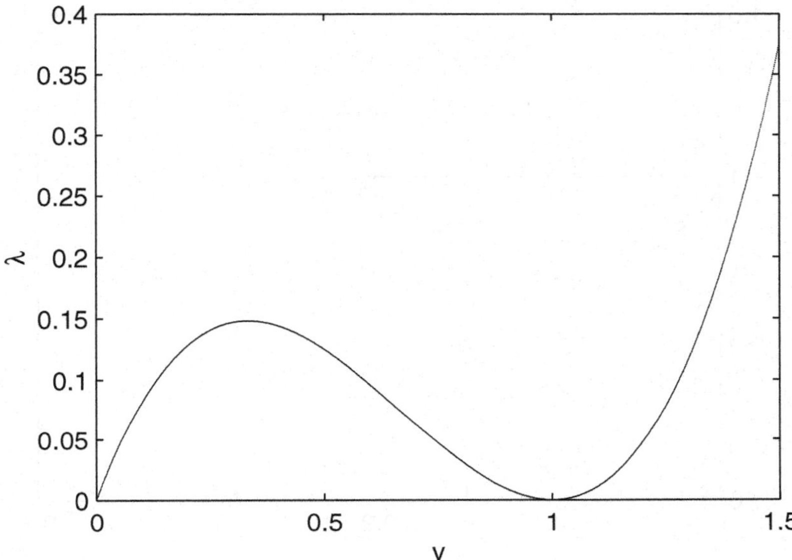

FIGURE 7.8: λ plotted as a function of v.

increasing λ. For small λ, we have two roots. Why? When λ is sufficiently small, the electrostatic force is weak enough so that the linear spring force may exactly balance it, creating a stationary state. As λ is increased, the electrostatic force increases, eventually overwhelming the linear spring force and all steady states disappear. This is the origin of the pull-in voltage instability! The complete disappearance of all physically possible steady-state solutions implies that the top plate must have collapsed or "pulled into" the bottom plate.

Another characterization of the pull-in instability is possible. If we solve equation (7.21) for λ we obtain

$$\lambda = v(1-v)^2. \tag{7.23}$$

For the moment, let's think of λ as being a function of v, defined by equation (7.23) and plot $\lambda(v)$. This is done in Figure 7.8. If we now flip the axes in Figure 7.8 as is done in Figure 7.9, we obtain what is known as the bifurcation diagram for our model. Along the x axis, we pick the value of λ that corresponds to our experiment and then read off the possible steady-state solutions along the v axis. When λ passes the value λ^* in Figure 7.9, no such physical solutions exist. Hence, λ^* is the dimensionless pull-in voltage! Here, we can explicitly compute λ^*. To do so, we simply compute the location and height of the first maxima of the curve in Figure 7.8. We find that the dimensionless pull-in voltage is $\lambda^* = 4/27$. The distance our device has deflected

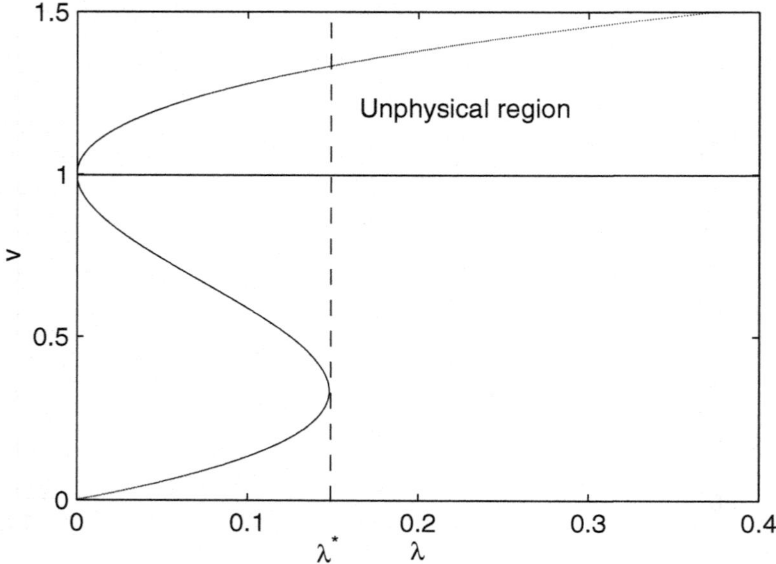

FIGURE 7.9: λ as a function of v with the axes flipped. This is the bifurcation diagram for our system.

at λ^*, denoted by v_p, is also easily computed. We find $v_p = 1/3$. As we shall see in the next section, v_p is the maximum stable deflection, hence v_p is the dimensionless pull-in distance. We can use the definition of λ to derive the dimensional value of the pull-in voltage for our system. Denoting this value by V_p, we find

$$V_p = \sqrt{\frac{8}{27}\frac{k(L-l)^3}{\epsilon_0 A}}. \tag{7.24}$$

Similiarly, we can use our scaling for v to compute the dimensional value of the pull-in distance. Denoting this value by u_p, we find

$$u_p = \frac{1}{3}(L-l). \tag{7.25}$$

Notice that for the mass-spring model, the pull-in distance is precisely one third of the zero voltage gap.

The object pictured in Figure 7.9 is called a *fold*. The curve of solutions, as visualized in the bifurcation diagram folds back upon itself as the bifurcation parameter, λ, is increased. We have already encountered a fold in a bifurcation diagram; in Chapter 4 the same structure arose in our model of Joule heating. Later in this chapter, we will see that the pull-in instability for more complicated devices can again be characterized in terms of a fold in the bifurcation diagram. While the v axis in Figure 7.9 will be replaced by some other measure of the solution, the underlying structure we still be a simple fold.

7.3.5 Stability of Steady States

When there is more than one physically relevant steady-state solution to a given model, as is the case with our system, knowing the stability of these steady states is crucial. First, let's work with our model in the viscosity dominated regime and determine stability under the simplifying assumption that $\alpha \gg 1$. In this case, we ignore the inertial term in equation (7.19) and work with the simpler first order equation,

$$\frac{dv}{dt} + v = \frac{\lambda}{(1-v)^2}, \tag{7.26}$$

where, of course, we have also set $\omega = 0$. It is convenient to rewrite this equation as

$$\frac{dv}{dt} = \frac{\lambda}{(1-v)^2} - v. \tag{7.27}$$

We recognize that the right-hand side is just the function $f(v)$, defined above, and we may rewrite our equation as

$$\frac{dv}{dt} = f(v). \tag{7.28}$$

Now, stability may be determined by simply examining Figure 7.7. Consider the case where λ is beyond the pull-in voltage, i.e., λ is so large that no roots of f less than one exist. Then, from Figure 7.7, the right-hand side of equation (7.28) is always positive. So, starting from any physical initial point for v, the displacement of the top plate will continuously increase, colliding with the bottom plate when $v = 1$. This should come as no surprise; we are beyond the pull-in voltage. Now, consider the case where $\lambda < \lambda^*$ so that two roots for f exist in Figure 7.7. Consider the smaller of the two roots. To the left of this root, $f(v)$ is positive, while to the right, provided we are less than the second root, $f(v)$ is negative. Hence, from equation (7.28), if we start to the left or to the right and smaller than the second root, we approach this solution. This means that this solution is stable. On our bifurcation diagram, Figure 7.9, the lowest branch of solutions, i.e. before the first fold, is a stable branch of solutions. Now, consider the larger of two roots. It is easy to see that this root is unstable. If we start to the left or to the right, the right-hand side of equation (7.28) is either negative or positive, pushing us away from this solution. On the bifurcation diagram, these roots correspond to the second branch and hence we conclude that this branch is a branch of unstable solutions. The uppermost branch is unphysical and we don't concern ourselves with its stability.

Notice that for this simplified first-order model, equation (7.28), we have also determined the global *dynamics* of solutions. If we pick an initial value for v and a value for λ and consider the motion on our bifurcation diagram, we see that for $\lambda < \lambda^*$, solutions monotonically approach the lower stable branch provided they start below the middle branch. If they start above the unstable middle branch, the device is pulled in monotonically. If $\lambda > \lambda^*$, the device is also pulled in monotonically.

We would also like to know the stability of solutions for the full second-order model. For second-order ordinary differential equations, global stability results are not as easily obtained as they are in the first-order case. Here, we rely upon the local method of linear stability theory outlined in Chapter 5. We work with equation (7.19); the reader may easily repeat our analysis for the systems scaled on damping time. Let us denote any of the steady-state solutions, i.e., roots of $f(v)$, by v^*. Now, we seek a solution to equation (7.19) in the form,

$$v(t) = v^* + \epsilon e^{-\mu t}, \tag{7.29}$$

where $\epsilon \ll 1$ and μ is an eigenvalue parameter to be determined. Notice that the real part of μ positive (negative) corresponds to linear stability (instability). Inserting (7.29) into equation (7.19), we obtain

$$\epsilon \frac{\mu^2}{\alpha^2} e^{-\mu t} - \epsilon \mu e^{-\mu t} = f(v^* + \epsilon e^{-\mu t}). \tag{7.30}$$

Now, we expand f in a Taylor series about $\epsilon = 0$, ignore terms of order ϵ^2 and obtain the characteristic equation for μ,

$$\frac{\mu^2}{\alpha^2} - \mu = f'(v^*), \tag{7.31}$$

with roots,

$$\mu = \frac{\alpha^2}{2} \pm \frac{\alpha^2}{2} \sqrt{1 + \frac{4f'(v^*)}{\alpha^2}}. \tag{7.32}$$

It is easy to see that the real part of μ is positive if and only if $f'(v^*) < 0$. This implies that once again we may read off stability from Figure 7.7. The function f passes through the smallest root with negative slope and hence this solution is stable, while the opposite is true for the other physical steady-state solution.

7.3.6 Iterative Numerical Methods

In Chapter 4 we explored an iterative numerical scheme designed to solve the steady-state, coupled electrostatic-thermal problem of Joule heating. A similar approach is often used to solve the steady-state electrostatic-elastic problem of electrostatic actuation. A flowchart outlining the basic iterative algorithm used to simulate electrostatic actuation is presented in Figure 7.10.

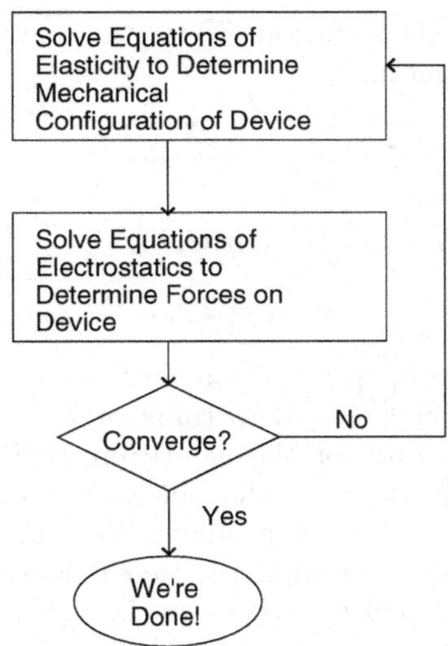

FIGURE 7.10: Flowchart of the basic iterative method for numerical electrostatic-elastic solvers.

This iterative scheme is at the core of many algorithms and codes used to compute the steady state configuration of an electrostatically actuated microsystem [3, 62, 177]. The idea behind the scheme is quite natural. We begin with some guess as to the elastic configuration of the device, holding this configuration fixed, we compute the electrostatic field and hence the electrostatic forces on the system, then using the computed forces, solve the elastic problem to find the updated configuration of the device. We iterate until things no longer change, i.e., until convergence. In addition to simplicity, this scheme has the advantage that it allows the researcher to simply couple together well-developed and well-understood elastic and electrostatic solvers to create an electrostatic-elastic solver.

We can apply this scheme to the solution of the steady-state problem for our mass-spring model. Here, the configuration of the device is given by the deflection, which we denote v_n. Solving for the electrostatic force simply means computing the value of

$$\frac{\lambda}{(1 - v_n)^2}, \tag{7.33}$$

and solving for the new elastic configuration, v_{n+1}, simply means balancing the spring force with the electrostatic force, i.e.,

$$v_{n+1} = \frac{\lambda}{(1 - v_n)^2}. \tag{7.34}$$

It is useful to introduce the notation, T, for the operator on the right-hand side of equation 7.34. That is,

$$Tv_n = \frac{\lambda}{(1-v_n)^2}. \tag{7.35}$$

Then, our iterative scheme may be written

$$v_{n+1} = Tv_n. \tag{7.36}$$

In an abstract sense, the operator T stands for the inversion of the elastic problem. The reader is encouraged to think of T in this way and envision equation (7.36) as shorthand for the algorithm in Figure (7.10). That is, T should be viewed as the operator that allows us to invert the equations of elasticity and hence solve the elastic problem. The numerically minded reader may think of T as the code, perhaps a finite element code, which solves the elastic problem. We will revisit this notion in the next section when for a membrane based theory, we see that T is the integral operator that inverts the Laplacian.

In Chapter 4 we introduced the contraction mapping theorem and explained its utility in the analysis of iterative schemes. We can apply the contraction mapping theorem to the study of our iterative scheme, equation (7.36), with T defined by equation (7.35). To investigate the convergence of our numerical scheme, we must investigate the contractivity of T on a suitably defined Banach space. In this case, the Banach space is simply a closed interval on the real line. Let u, v be real numbers, let $\lambda \geq 0$ and consider

$$||Tu - Tv|| = |Tu - Tv| = \lambda \left| \frac{1}{(1-u)^2} - \frac{1}{(1-v)^2} \right|. \tag{7.37}$$

Clearly, the right-hand side of this expression cannot be bounded if the interval (Banach space), S, contains the number one. Consider the interval $S = [0, 1-\epsilon]$ where $1 > \epsilon > 0$. Over S, we can bound the right-hand side of equation (7.37).

$$|Tu - Tv| = \lambda |u - v| \left| \frac{2-(u+v)}{(1-u)^2(1-v)^2} \right| \leq \frac{\lambda(4+2\epsilon)}{\epsilon^4} |u - v|. \tag{7.38}$$

Hence, we see that T is a contraction provided

$$\lambda < \frac{\epsilon^4}{4+2\epsilon}. \tag{7.39}$$

That T with λ satisfying this condition maps S into itself is easy to see and hence, by the contraction mapping theorem, our iterative scheme converges when λ satisfies this restriction.

While it is nice to have proven convergence of our numerical method for a range of λ the implications of our analysis for numerical difficulties are of primary interest. First, we notice that the contractivity of the mapping T depends on λ. In particular, as λ is increased T becomes "less contractive." This is not surprising; after all, we know that the method must fail once λ exceeds λ^*. That is, for λ beyond λ^*, there is no solution to the problem in S and hence our method should not converge. What is perhaps surprising is that the rate of convergence of the iterative method will depend on λ. As λ approaches λ^* and the mapping becomes less contractive, the number of iterations to convergence will tend to infinity. This is illustrated in Figure 7.11 where, starting from the initial guess $v_0 = 0$, we plot the number of iterations to convergence as a function of λ. The implications of this are severe.

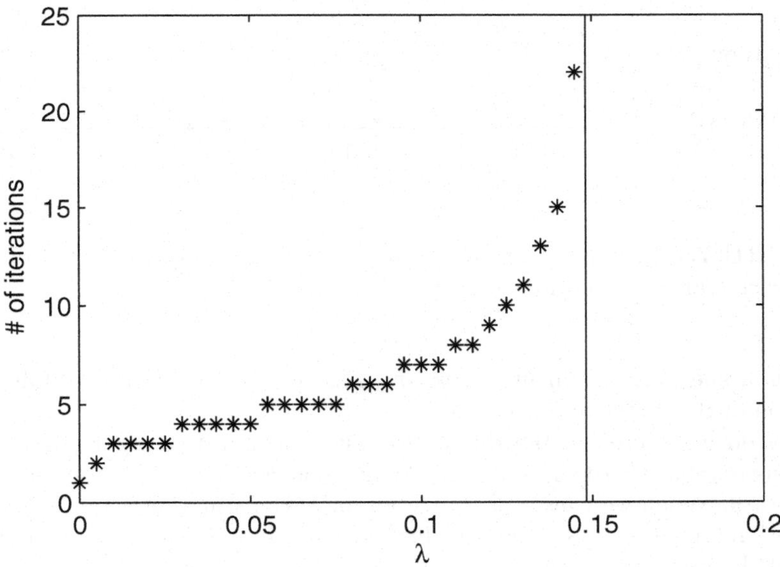

FIGURE 7.11: Iterations to convergence as a function of λ.

While the simple iterative algorithm will work well at low voltages, it will become more and more expensive as voltages are increased. If we wish to numerically attempt to compute the pull-in voltage, the iterative algorithm will after considerable work, yield only a crude approximation. Finally, the iterative scheme is only returning the lower branch of solutions. That is, we are only computing the lowest branch of solutions on the bifurcation diagram, Figure 7.9. The results of such a computation using the iterative method are shown in Figure 7.12. Here, one may argue, that the middle branch is unstable and the upper branch unphysical and hence they are not of interest.

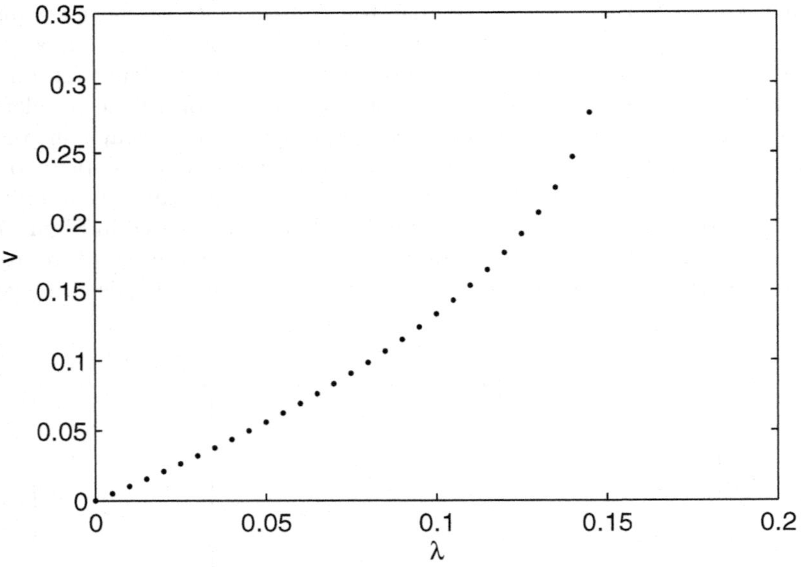

FIGURE 7.12: Output of the iterative method. Note we have captured the lower branch of solutions only.

This is a specious argument. *A priori*, one does not know the stability or location of the upper branches.

How do we overcome the difficulties with the iterative algorithm? Our analysis points toward the solution. Again consider the bifurcation diagram, Figure 7.9. When we view v as a function of λ, the situation is quite complicated. The function $v(\lambda)$ is multivalued and has infinite slope at two points. One of these points is the pull-in voltage, λ^*. The bifurcation diagram contains a fold. On the other hand, when we view λ as a function of v, Figure 7.8, the situation is quite tame. The function $\lambda(v)$ is single valued and differentiable everywhere. The key to resolving the difficulties with the iterative scheme lies in this change in perspective. Here, for our mass-spring model, we can ask to find $\lambda(v)$ rather than $v(\lambda)$. Earlier we wrote

$$\lambda = v(1 - v)^2 \tag{7.40}$$

and simply plotted $\lambda(v)$ rather than solving for $v(\lambda)$. The same idea can be used to rescue our numerical method. While for the mass-spring system, this is trivial, for more complicated models choosing the correct perspective is not. Much research has focused upon the efficient numerical computation of bifurcation diagrams which contain a fold. As noted in Chapter 4, the work of H.B. Keller [100], and the method of pseudo-arc-length continuation is central to resolving these difficulties. The interested reader is referred to the section on related reading at the end of the chapter.

7.4 Modeling General Electrostatic-Elastic Systems

The biggest deficiencies of the mass-spring model of electrostatic actuation are the inability to capture real geometry and the inability to capture real elastic effects.[6] In this section we take another step in developing a theory of

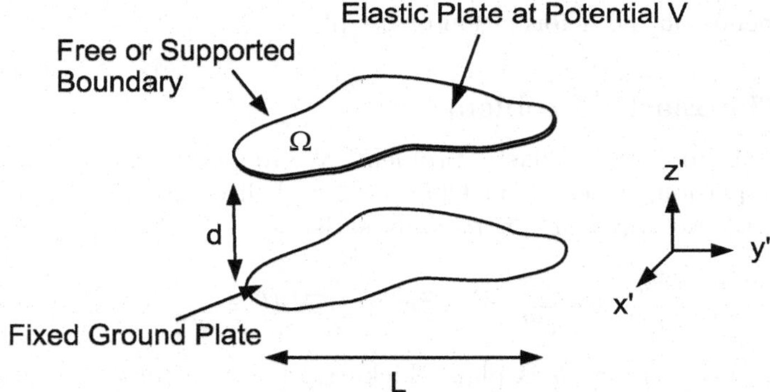

FIGURE 7.13: The general electrostatic-elastic system.

electrostatic-elastic systems by constructing a model of the structure shown in Figure 7.13. This canonical structure is at the core of many of the devices discussed at the start of the chapter. Modeling this structure forces us to confront the limitations of the mass-spring model. The geometric limitation is relaxed by allowing the flexible plate in Figure 7.13 to deform. In turn, this distorts the electric field. Rather than computing the field for a simple parallel plate structure, we must compute the electric field between a ground plate and a body undergoing arbitrary deformations. Realistic elastic effects may be incorporated into our model by using membrane, beam, or plate theory to model the deformations of the elastic plate.

7.4.1 The Electrostatic Problem

We begin by formulating the equations governing the electrostatic potential for the system shown in Figure 7.13. We assume that the elastic plate is held at potential V, while the fixed ground plate is held at zero potential. The electrostatic potential, ϕ, satisfies the Laplace equation,

$$\nabla^2 \phi = 0, \tag{7.41}$$

everywhere in the region between the elastic plate and ground plate and in the region surrounding the device. The boundary conditions are

$$\phi = V \text{ on elastic plate} \tag{7.42}$$

and

$$\phi = 0 \text{ on ground plate.} \tag{7.43}$$

Note that the electrostatic problem is coupled to the elastic problem through the boundary conditions. In particular the boundary condition, equation (7.42), depends on the deformation of the plate.

7.4.2 The Elastic Problem

Next, we formulate the elastic problem. We model the elastic plate using the plate equation presented in Chapter 2 and discussed in Chapter 5. In particular, the deflection w' of the plate satisfies

$$\rho h \frac{\partial^2 w'}{\partial t'^2} + a \frac{\partial w'}{\partial t'} - \mu \nabla_\perp^2 w' + D \nabla_\perp^4 w' = -\frac{\epsilon_0}{2} |\nabla \phi|^2. \tag{7.44}$$

Recall that ρ is density, h is plate thickness, μ is the tension in the plate, D is the flexural rigidity, and ϵ_0 is the permittivity of free space. We are using the notation $\nabla_\perp$ to differentiate the operator in equation (7.44) from the Laplace operator in equation (7.41). In equation (7.44) the differentiation is only with respect to x' and y', while in equation (7.41) differentiation is with respect to x', y', and z'. We have modified our standard plate equation in two ways. First, we have added a damping term. The parameter a is a damping constant. We have assumed the damping term is proportional to velocity. This is the same assumption we made in our mass-spring model. Second, we have included a source term that captures the force on the plate due to the electric field. Notice that this force is proportional to the norm squared of the gradient of the potential and couples the solution of the elastic problem to the solution of the electrostatic problem. A derivation of this source term may be found in [93]. We will not formulate boundary or initial conditions for equation (7.44) here, rather we will pose such conditions as the need arises.

7.4.3 Scaling

It is useful to rescale our system, equations (7.41)–(7.44), and rewrite in dimensionless form. We scale the electrostatic potential with the applied voltage, time with a damping timescale of the system, the x' and y' variables with a characteristic length of the device, and z' and w' with the size of the gap between the ground plate and elastic plate for zero applied voltage. We define

$$w = \frac{w'}{d}, \quad \psi = \frac{\phi}{V}, \quad x = \frac{x'}{L}, \quad y = \frac{y'}{L}, \quad z = \frac{z'}{d}, \quad t = \frac{\mu t'}{aL^2}. \tag{7.45}$$

and substitute these into equations (7.41)–(7.44) to find

$$\epsilon^2 \nabla_\perp^2 \psi + \frac{\partial^2 \psi}{\partial z^2} = 0 \tag{7.46}$$

$$\psi = 1 \text{ on elastic plate} \tag{7.47}$$

$$\psi = 0 \text{ on ground plate} \tag{7.48}$$

$$\frac{1}{\alpha^2}\frac{\partial^2 w}{\partial t^2} + \frac{\partial w}{\partial t} - \nabla_\perp^2 w + \delta \nabla_\perp^4 w = -\lambda \left[\epsilon^2 |\nabla_\perp \psi|^2 + \left(\frac{\partial \psi}{\partial z} \right)^2 \right]. \tag{7.49}$$

The parameter α is the inverse of the quality factor for the system. It is defined by

$$\alpha = \frac{aL}{\sqrt{\rho h \mu}}. \tag{7.50}$$

The parameter δ measures the relative importance of tension and rigidity and is defined,

$$\delta = \frac{D}{L^2 \mu}. \tag{7.51}$$

The parameter ϵ is the aspect ratio of the system and is defined,

$$\epsilon = \frac{d}{L}. \tag{7.52}$$

The parameter λ has the same interpretation as λ in the mass-spring model. It is a ratio of a reference electrostatic force to a reference elastic force and is defined,

$$\lambda = \frac{\epsilon_0 V^2 L^2}{2d^3 \mu}. \tag{7.53}$$

Notice that it is proportional to the square of the applied voltage and serves as a "tuning" parameter for our system. Understanding the behavior of a given device is equivalent to characterizing solutions for all positive λ.

7.4.4 The Small-Aspect Ratio Limit

Our model of electrostatic actuation, equations (7.46)–(7.49), is a system of nonlinear coupled partial differential equations. Even for very simple geometries, we have little hope of finding an exact solution. However, we can simplify the system by examining a restricted parameter regime. In particular, we consider the small-aspect ratio limit. That is, the limit where $\epsilon \ll 1$. Physically, this means that the lateral dimensions of the device in Figure 7.13 are large compared to the size of the gap between the elastic plate and ground plate. For many MEMS systems this is an excellent approximation. For example, the GLV discussed at the start of the chapter has an aspect ratio $\epsilon = 0.0065$, while the electrostatic shuffle motor has an aspect ratio $\epsilon = 0.01$. We exploit the small-aspect ratio by sending ϵ to zero in equation (7.46). This reduces the electrostatic problem to

$$\frac{\partial^2 \psi}{\partial z^2} = 0, \tag{7.54}$$

which we may solve to find the approximate potential,

$$\psi \approx Az + B. \tag{7.55}$$

We are primarily concerned with the field between the plates and hence apply the boundary conditions on ψ by requiring that

$$\psi(x, y, w, t) = 1 \tag{7.56}$$

and

$$\psi(x, y, 0, t) = 0. \tag{7.57}$$

This yields

$$\psi \approx \frac{z}{w}. \tag{7.58}$$

In engineering parlance, this approximation is equivalent to ignoring fringing fields. The potential is computed as if the plates were locally parallel. Now, we send ϵ to zero in equation (7.49) and use this approximate potential to find

$$\frac{1}{\alpha^2}\frac{\partial^2 w}{\partial t^2} + \frac{\partial w}{\partial t} - \nabla_\perp^2 w + \delta \nabla_\perp^4 w = -\frac{\lambda}{w^2}. \tag{7.59}$$

Equation (7.59) is a reduced model for the elastic behavior of our system. This equation is *uncoupled* from the potential equation and may be solved independently. Keep in mind however, that equation (7.59) is still nonlinear. Except in very simple situations, we do not expect to find exact solutions.

7.5 Electrostatic-Elastic Systems—Membrane Theory

In this section we further simplify equation (7.59) by assuming that our elastic plate has no rigidity. That is, we take $\delta = 0$ in equation (7.59) and only consider the effects of tension. This approximation is reasonable for a restricted class of MEMS devices. The GLV and certain MEMS micropumps are examples of this class. However, we remind the reader that this approximation is certainly not valid for *all* MEMS devices. Further, in this section we focus on steady-state deflections and set the time derivatives in equation (7.59) to zero. In particular, we study

$$\nabla^2 w = \frac{\lambda}{w^2} \text{ in } \Omega, \tag{7.60}$$

where Ω is the domain pictured in Figure 7.13. We assume our membrane is held fixed on its boundary. With the scalings introduced in the previous section this implies that we must require

$$w = 1 \text{ on } \partial\Omega. \tag{7.61}$$

We have dropped the $\perp$-notation. The reader should keep in mind that the Laplacian in this section takes derivatives in the x and y directions only.

7.5.1 General Results

Equation (7.60) is a *semilinear elliptic equation.* Similar problems arise in the mathematical modeling of combustion [8], microwave heating [109, 110, 111, 112, 113], and chemical reactor theory [34, 35, 36, 37]. In fact, the steady-state version of our reduced-order Joule heating model from Chapter 4 was a semilinear elliptic problem. Mathematical methods developed to analyze such problems are easily adapted to the study of equation (7.60). Here we prove a relevant theorem using standard techniques from the theory of elliptic equations. Before stating the theorem we make the boundary conditions, equation (7.61), homogenous. To accomplish this, we let $w = 1 + u$ and study

$$-\nabla^2 u = -\frac{\lambda}{(1+u)^2} \tag{7.62}$$

with

$$u = 0 \text{ on } \partial\Omega. \tag{7.63}$$

The key result of this section is

THEOREM 7.1
Let Ω be a bounded domain in $\Re^2$ with smooth boundary $\partial\Omega$. Consider equation (7.62) with Dirichlet boundary condition equation (7.63). Then, there exists a λ^* such that no solution u exists for any $\lambda > \lambda^*$.

Before proving Theorem 7.1 some physical interpretation is in order. In the mass-spring system the pull-in instability was characterized in terms of a fold in the bifurcation diagram. In the bifurcation diagram for the mass-spring model, Figure 7.9, we see that when λ exceeds the critical value λ^*, there is no longer a steady-state solution. We say that the system has pulled in and interpret λ^* as the pull-in voltage. Theorem 7.1 says that the pull-in phenomenon occurs for a large class of electrostatically actuated systems. Unlike our mass-spring analysis, the theorem does not tell us what the bifurcation diagram looks like. Rather, it simply tells us that at some point no steady-state solutions exist. Note that the assumptions behind the theorem restrict us to claiming that the pull-in instability is present in devices where the deflected component may be modeled as an elastic membrane held fixed on its boundary. Further the assumption of small aspect ratio is required. Now let's prove Theorem 7.1.

PROOF Let κ_1 be the lowest eigenvalue of

$$-\nabla^2 u = \alpha u \text{ on } \Omega \tag{7.64}$$

$$u = 0 \text{ on } \partial\Omega \tag{7.65}$$

with v_1 the associated eigenfunction. It is well known that κ_1 is *simple.*[7] It is also well known that v_1 may be chosen strictly positive in Ω. Now, rewrite equation (7.62) as

$$-\nabla^2 u - \kappa_1 u = -\frac{\lambda}{(1+u)^2} - \kappa_1 u. \tag{7.66}$$

An equation of the form (7.66) will have a solution if and only if its right-hand side is orthogonal to all solutions of the homogeneous equation. In particular, here the solvability condition is

$$\int_\Omega \left(\frac{\lambda}{(1+u)^2} + \kappa_1 u \right) v_1 = 0. \tag{7.67}$$

Since v_1 is strictly positive, the term in parentheses must either be identically zero, or it must change sign. That it is not zero is clear. Hence, we are led to consider $\lambda(1+u)^{-2} + \kappa_1 u$. If this expression is to change sign, at some u, we must have $\lambda(1+u)^{-2} = -\kappa_1 u$. A simple plot of each side of this expression as a function of u reveals that as λ is increased beyond some value λ^*, the two curves no longer intersect and hence no solution exists for $\lambda > \lambda^*$. □

This proof of Theorem 7.1 is a straightforward adaptation of similar theorems from the theory of elliptic equations. The reader seeking more details on this argument is referred to [150, 184]. An important corollary of Theorem 7.1 provides an upper bound on the pull-in voltage

COROLLARY 7.1
The pull-in voltage for equations (7.62) and (7.63), λ^*, satisfies $\lambda^* \leq \frac{4}{27}\kappa_1$.

This corollary follows immediately from the proof of Theorem 7.1. In the proof the existence of λ^* relied only on the fact that for some value of λ the graph of $f(u) = \lambda(1+u)^{-2}$ and $g(u) = -\kappa_1 u$ did not intersect. It is easy to see that the smallest value of λ for which this is true is precisely $\frac{4}{27}\kappa_1$. That this provides an *upper bound* on λ^* is simply a reflection of the fact that nothing in the proof says that solutions can't fail to exist earlier. Notice that this upper bound depends only on the lowest eigenvalue of the negative Laplace operator over Ω. This eigenvalue is a function of the shape of Ω! So, Corollary 7.1 gives us a way of studying how the pull-in voltage changes as the shape of the membrane changes.

Now, in addition to the nonexistence result, Theorem 7.1, and the upper bound on the pull-in voltage, Corollary 7.1, we can prove an existence result and obtain a lower bound on the pull-in voltage. To do so, we rely upon the well-known method of upper and lower solutions. Here, we sketch the main elements of the theory. For more details the reader is referred to [150, 184]. We begin with

DEFINITION 7.1 *The function $\bar{u} \in C^2(\bar{\Omega})$ is called an upper solution if*

$$-\nabla^2 \bar{u} \geq -\frac{\lambda}{(1+\bar{u})^2} \quad \text{in } \Omega \tag{7.68}$$

$$\bar{u} \geq 0 \quad \text{on } \partial\Omega. \tag{7.69}$$

The function $\underline{u} \in C^2(\bar{\Omega})$ is called a lower solution if the opposite inequalities are satisfied.

An upper solution is easy to find as is shown in the following lemma.

LEMMA 7.1
Any positive constant, $C > 0$, is an upper solution for all $\lambda \geq 0$.

PROOF Obvious. ☐

Now, to construct a lower solution, we first define the domain Ω' as a bounded domain with smooth boundary, which contains Ω as a proper subset. Next, consider the eigenvalue problem,

$$-\nabla^2 u = \mu u \text{ on } \Omega' \tag{7.70}$$

$$u = 0 \text{ on } \partial\Omega', \tag{7.71}$$

on this enlarged domain and let μ_1 be the lowest eigenvalue with w_1 the associated eigenfunction. Notice that since Ω is a proper subset of Ω', w_1 may be chosen strictly positive on Ω. Now, we attempt to construct a lower solution of the form Aw_1 where A is a scalar. We need $Aw_1 \leq 0$ on $\partial\Omega$, which means we must require $A < 0$. Next, we need

$$-\nabla^2(Aw_1) \leq -\frac{\lambda}{(1+Aw_1)^2} \tag{7.72}$$

to hold in Ω. But, since the Laplacian is a linear operator and w_1 an eigenvector, we may rewrite this requirement as

$$-\mu_1 Aw_1 - \frac{\lambda}{(1+Aw_1)^2} \geq 0. \tag{7.73}$$

There are two difficulties with satisfying the inequality (7.73). First, we must ensure that $1+Aw_1 > 0$. Suppose w_1 is normalized so that its maximum over Ω is one and let m be its minimum over Ω. We can satisfy $1+Aw_1 > 0$ by satisfying $1+A > 0$ or equivalently, $A > -1$. So, thus far, we must choose A such that $0 > A > -1$. Now, we still must choose A so that equation (7.73) is satisfied. But, it is easy to see by graphical analysis that this inequality may be satisfied for a range of λ by choosing A such that $-1 < A < (3m)^{-1}$ provided $m > 1/3$. Now, it is clear that the domain Ω' must be chosen so that the minimum of the first eigenfunction, w_1, is greater than $1/3$ over Ω. Finally, an easy calculation shows that this choice of Ω' and A leaves the inequality (7.73) satisfied for $\lambda \leq \frac{4}{27}\mu_1$. To summarize, we have shown

LEMMA 7.2
There exists a constant A such that the function Aw_1 is a lower solution for all $\lambda \leq \frac{4}{27}\mu_1$.

Now, using Lemmas 7.1 and 7.2 and the well-known theorem that a solution exists between an ordered pair of upper and lower solutions, we obtain the existence result:

THEOREM 7.2
There exists a solution to (7.62)-(7.63) for all $\lambda \leq \frac{4}{27}\mu_1$.

An immediate corollary is the lower bound on the pull-in voltage,

COROLLARY 7.2
The pull-in voltage for equations (7.62)-(7.63), λ^*, satisfies $\lambda^* \geq \frac{4}{27}\mu_1$.

This lower bound on the pull-in voltage also allows us to study how the pull-in voltage changes as the shape of the membrane changes. Here, an estimate of the lower bound can be made for *any* shape domain. To do so, we simply must enclose the membrane domain; Ω is a circle sufficiently large so that the first eigenfunction of the negative Laplacian over the circle is everywhere greater than 1/3 on Ω. Since the eigenfunctions of the Laplacian on a circle are easily computed by hand, this computation is not difficult. See the exercises.

7.5.2 One-Dimensional Membranes

There are several simple geometries for which more detailed results on equations (7.62) and (7.63) are possible. One such case is the one-dimensional membrane or strip [10]. In this case equations (7.62) and (7.63) reduce to

$$\frac{d^2u}{dx^2} = \frac{\lambda}{(1+u)^2} \tag{7.74}$$

and

$$u(-1/2) = u(1/2) = 0. \tag{7.75}$$

Equation (7.74) can be solved exactly. It is worth carrying out one step in the process. If we multiply equation (7.74) by du/dx we can rewrite as

$$\frac{d}{dx}\left[\frac{1}{2}\left(\frac{du}{dx}\right)^2 + \frac{\lambda}{1+u}\right] = 0, \tag{7.76}$$

which can be integrated one time to obtain

$$E = \frac{1}{2}\left(\frac{du}{dx}\right)^2 + \frac{\lambda}{1+u}. \tag{7.77}$$

We have denoted the constant of integration by E because this expression is in fact an *energy* for our system. The first term in equation (7.77) represents the elastic energy while the second term represents the energy in the electric field. Equation (7.77) can be solved for du/dx and integrated again to yield the implicit formula for $u(x)$,

$$\sqrt{\frac{(u+1)(u+1-\lambda/E)}{2E}} + \frac{\lambda}{E\sqrt{2E}}\tanh^{-1}\sqrt{\frac{u+1-\beta/E}{u+1}} = x. \tag{7.78}$$

Here we have made use of the fact that solutions are symmetric about the point $x = 0$ in order to eliminate the second constant of integration. Applying the boundary condition at $x = 1/2$ yields the following equation for E:

$$\sqrt{\frac{1-\lambda/E}{2E}} + \frac{\lambda}{E\sqrt{2E}}\tanh^{-1}\sqrt{1-\lambda/E} = \frac{1}{2}. \tag{7.79}$$

Equation (7.79) can be solved numerically to yield E as a function of λ. The result of such a computation is shown in Figure 7.14. Notice that for $\lambda \in [0, \lambda^*]$, there are two values of E for each λ. Beyond λ^* no solutions to equation (7.79) exist. The two solutions found when $\lambda < \lambda^*$ correspond to two different solutions for $u(x)$. In Figure 7.15 we plot $u(x)$ for various

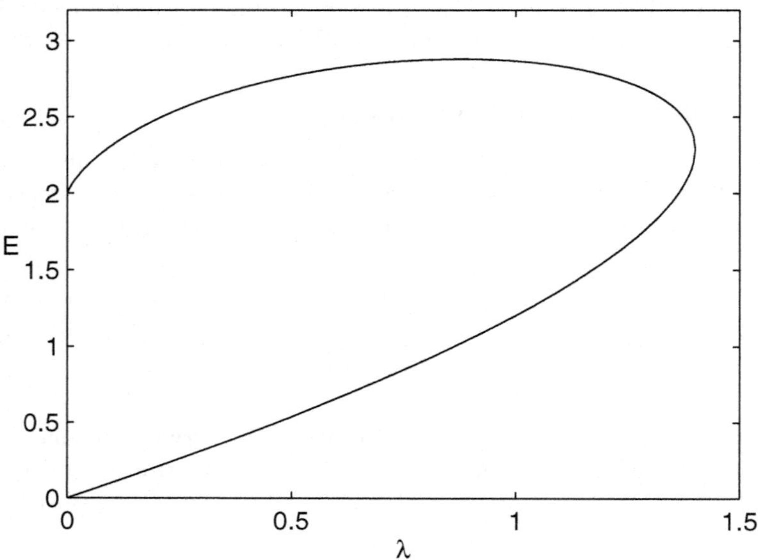

FIGURE 7.14: The energy for the 1-D membrane.

values of λ. Notice that for all of these solutions the maximum deflection occurs at $x = 0$. Hence, we can use the information relating E and λ to construct a bifurcation diagram for equations (7.74)–(7.77). We do so in Figure 7.16. Here the "y" axis is not the solution itself, but rather a measure of the solution. In particular, it is the magnitude of the maximum of the given solution. Once again, we find a bifurcation diagram with a fold. The pull-in instability for this 1-D system, which we knew was present from the general theory, has the same mathematical explanation as did the pull-in instability for the mass-spring model.

An alternative look at equations (7.74) and (7.75) is worth pursuing. Recall that in Section 5.4 we solved a *Green's function* problem to determine the deflection of a 1-D membrane subjected to a concentrated load. In Section 5.4 we then used the Green's function solution to determine the deflection of a 1-D membrane subjected to an *arbitrary* load. While we cannot use the Green's function to solve equations (7.74) and (7.75) exactly, we can use the Green's function of Section 5.4 to turn our nonlinear ordinary differential equation into a nonlinear integral equation. The procedure is identical to that of Section 5.4, where the integral representation of the solution to the arbitrary loading problem was constructed. That is, we multiply equation (7.74) by the Green's function $g(x, x_0)$, multiply equation (5.56) by $u(x)$, subtract the two

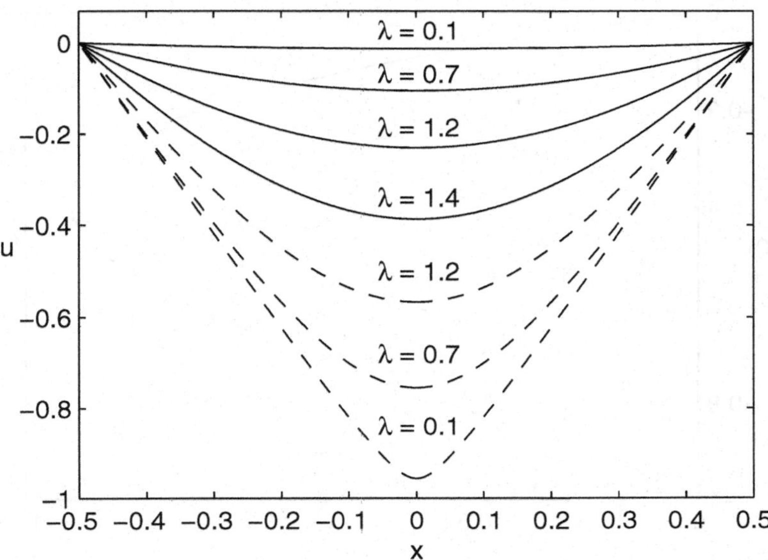

FIGURE 7.15: The displacement of the 1-D membrane for various values of the applied voltage.

equations and integrate over the domain. This yields

$$\int_{-1/2}^{1/2}\left[g(x,x_0)\frac{d^2u}{dx^2} - u(x)\frac{d^2g}{dx^2} - \frac{\lambda g(x,x_0)}{(1+u(x))^2} - u(x)\delta(x-x_0)\right]dx = 0. \tag{7.80}$$

Integrating the first two terms in this expression by parts one time and applying the boundary conditions, we find that all contributions from these terms vanish. Recalling that the delta function picks out the value of a function at a point we are left with

$$u(x_0) = -\lambda\int_{-1/2}^{1/2}\frac{g(x,x_0)}{(1+u(x))^2}dx. \tag{7.81}$$

Noting that g is a symmetric function of x and x_0 we may rename the variables and write

$$u(x) = -\lambda\int_{-1/2}^{1/2}\frac{g(x,x_0)}{(1+u(x_0))^2}dx_0. \tag{7.82}$$

Equation (7.82) is equivalent to equations (7.74) and (7.75). Rather than a nonlinear differential equation, equation (7.82) presents us with a nonlinear integral equation for the deflection $u(x)$. The utility of equation (7.82) becomes apparent when we consider applying the iterative numerical method

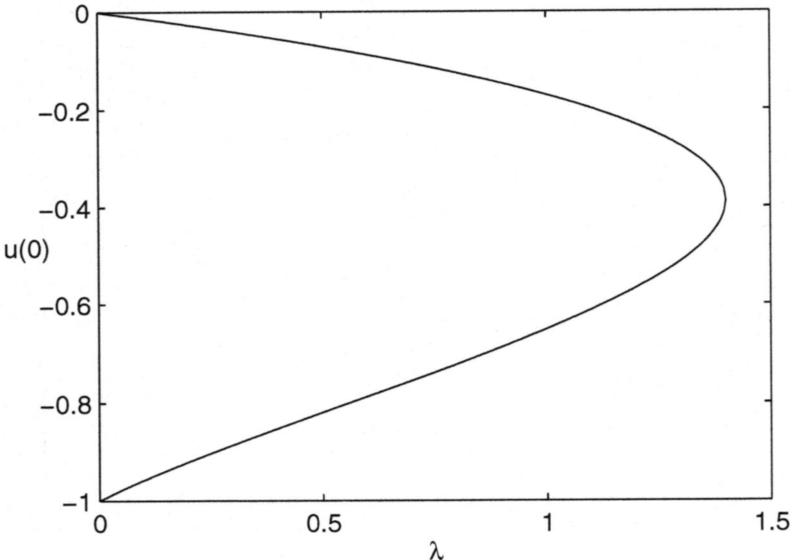

FIGURE 7.16: The bifurcation diagram for the 1-D membrane.

of Figure 7.10 to the strip model. If we start with some initial guess for the elastic deflection, say $u_0(x)$, our next step is to compute the electrostatic force on the device. For the strip, this is just the right-hand side of equation (7.74), i.e.,

$$\frac{\lambda}{(1+u_0(x))^2}. \tag{7.83}$$

Next, we need to solve the elastic problem using this expression for the force. But, this is precisely what the integral equation, (7.82), does. That is, equation (7.82) inverts or solves the elastic problem. Hence, the next iterate, $u_1(x)$ is given by

$$u_1(x) = -\lambda \int_{-1/2}^{1/2} \frac{g(x,x_0)}{(1+u_0(x_0))^2} dx_0. \tag{7.84}$$

In general, we can express our iterative scheme as

$$u_{n+1}(x) = -\lambda \int_{-1/2}^{1/2} \frac{g(x,x_0)}{(1+u_n(x_0))^2} dx_0. \tag{7.85}$$

If we name the operator on the right-hand side as we did in the mass-spring theory, that is, let

$$Tu(x) = -\lambda \int_{-1/2}^{1/2} \frac{g(x,x_0)}{(1+u(x_0))^2} dx_0, \tag{7.86}$$

our iterative scheme becomes

$$u_{n+1}(x) = Tu_n(x). \tag{7.87}$$

Formally, this is precisely the same as the iterative scheme for our mass-spring model. The operator T is once again the operator that inverts or solves the elastic problem. In this case, this means inverting the one-dimensional Laplace operator, an operation we carried out using the Green's function. The contraction mapping theorem may be applied to equation (7.87) in the same manner as it was applied to the mass-spring iterative scheme. The reader is directed to the exercises for this analysis. Unsurprisingly, the result is the same as for the mass-spring model. Increasing the value of λ reduces the "contractivity" of T and slows convergence of the iterative scheme. When started with zero deflection the iterative scheme yields only the lowest branch of solutions on the bifurcation diagram. Numerical computation of the pull-in voltage via the iterative scheme is slow and expensive.

7.5.3 Circular Membranes

A second geometry for which a detailed analysis of equations (7.62) and (7.63) is possible is the circular disk. The analysis of the disk geometry and comparison with the previous subsection will yield insight into the effect of geometry on solutions to equations (7.62) and (7.63). For the disk geometry we assume all solutions of equations (7.62) and (7.63) depend only on r, i.e., the radial coordinate.[8] Retaining only the radial part of the Laplace operator our model reduces to

$$\frac{d^2w}{dr^2} + \frac{1}{r}\frac{dw}{dr} = \frac{\lambda}{w^2}, \tag{7.88}$$

with boundary condition

$$w(1) = 1. \tag{7.89}$$

Notice that this is a second-order nonlinear ordinary differential equation but that we have imposed only one boundary condition. As a second condition we impose

$$\frac{dw}{dr}(0) = 0. \tag{7.90}$$

We may think of this condition as requiring that solutions be symmetric about the point $r = 0$. The analysis of the disk model, equations (7.88)–(7.90), is not as straightforward as the analysis of the strip model. We cannot integrate equations (7.88)–(7.90) directly. However, after examining equation (7.88) for awhile one notices that this equation contains a symmetry. In particular, if we make the change of variables,

$$w^* = e^{\epsilon}w, \quad r^* = e^{\frac{3\epsilon}{2}}r, \tag{7.91}$$

in equation (7.88) we get the same equation back. That is, equation (7.88) is *invariant* under this change of variables. Further notice that the change of variables, equation (7.91) contains a free parameter, ϵ. Hence our governing equation is invariant under a one-parameter family of transformations. This family is called a *stretching group* or a *Lie group*. The significance of this invariance is that it implies that our boundary value problem may be converted to an initial value problem. In some sense we can "trade in" the invariance for a simplification of our problem. Since the analysis of initial value problems is substantially easier than that of boundary value problems, this is a good trade. This technique of mapping solutions of a boundary value problem to solutions of an initial value problem is well known in fluid dynamics. Sometimes called exact shooting, this method was first introduced by Toepfer [198], in 1912, in connection with the Blasius problem.[9] In the 1960s and 1970s this method was extended to a wide class of problems primarily through the work of Klamkin and Na. The interested reader is referred to the related reading, in particular the text by Na. Here, we proceed with the study of equations (7.88)–(7.90) by noting that $w(r)$ is a solution if and only if

$$w(r) = \alpha y(\gamma r), \tag{7.92}$$

where

$$\alpha = \frac{1}{y(\gamma)} \tag{7.93}$$

$$\frac{\lambda}{\gamma^2 \alpha^3} = 1, \tag{7.94}$$

and y satisfies the initial value problem,

$$\frac{d^2 y}{dr^2} + \frac{1}{r}\frac{dy}{dr} = \frac{1}{y^2} \tag{7.95}$$

$$y(0) = 1, \quad \frac{dy}{dr}(0) = 0. \tag{7.96}$$

With these identifications, the bifurcation diagram for our problem, equations (7.88)–(7.90), is parameterized in terms of γ and y. That is, to understand solutions, we wish to plot λ versus some measure of the solution. We define this measure as $||w||_\infty = 1 - w(0)$. But from equations (7.92)–(7.94) we have $\lambda = \gamma^2/y(\gamma)^3$ and $||w||_\infty = 1 - 1/y(\gamma)$. It is easy to numerically integrate our initial value problem and use the result to compute the complete bifurcation diagram for our problem. The result of such a computation is shown in Figure 7.17. We note that as $r \to \infty$, $y(r)$ and hence λ and w change very little. This makes it difficult to compute the bifurcation diagram as it approaches the barrier $||w||_\infty = 1$. Hence, it is useful to study the equations for $y(r)$ analytically. To proceed, we change variables in equation (7.95) by setting $\eta = \log(r)$ and $y(r) = r^{2/3}v(\eta)$. This yields the autonomous equation,

$$\frac{d^2 v}{d\eta^2} + \frac{4}{3}\frac{dv}{d\eta} + \frac{4}{9}v - \frac{1}{v^2} = 0. \tag{7.97}$$

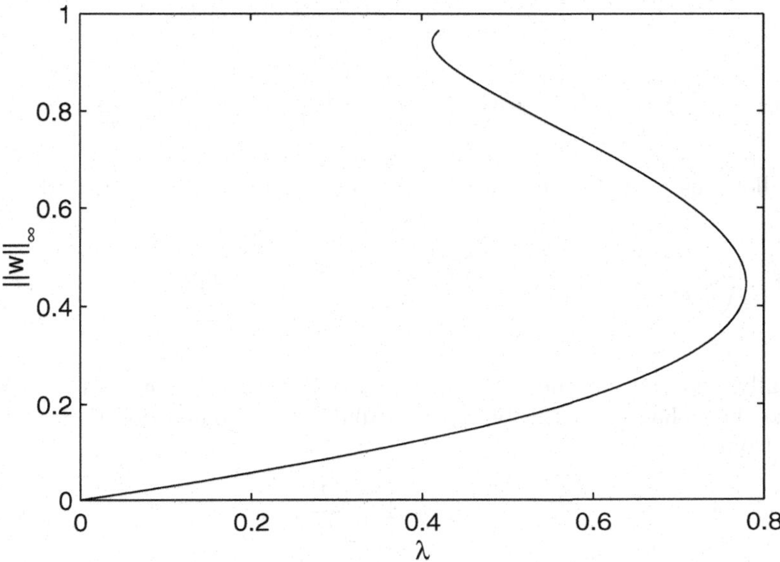

FIGURE 7.17: The bifurcation diagram for the disk-shaped elastic membrane.

We may rewrite as a first-order system:

$$\frac{dv}{d\eta} = h \tag{7.98}$$

$$\frac{dh}{d\eta} = -\frac{4}{3}h - \frac{4}{9}v + \frac{1}{v^2}. \tag{7.99}$$

We observe that this system has a single critical point located at $v = (\frac{9}{4})^{1/3}, h = 0$. Linearizing about this critical point we find that the linear system has eigenvalues,

$$\mu = -\frac{2}{3} \pm \frac{2\sqrt{2}i}{3}, \tag{7.100}$$

and hence this point is a stable spiral. In fact, using the Lyapunov function[10]

$$V(v,h) = \frac{h^2}{2} + \frac{2}{9}v^2 + \frac{1}{v} - \left(\frac{4}{9}\right)^{1/3} - \frac{2}{9}\left(\frac{9}{4}\right)^{2/3}, \tag{7.101}$$

it is easy to see that our critical point is the global attractor for the system. We note that we may restrict our attention to the physical region of phase space, $v > 0$. The large η asymptotics for $v(\eta)$ are now simple to obtain, we find

$$v(\eta) \sim \left(\frac{9}{4}\right)^{1/3} + Ae^{-2\eta/3}\cos\left(\frac{2\sqrt{2}}{3}\eta + B\right) + o\left(e^{-2\eta/3}\right) \quad \text{as } \eta \to \infty. \tag{7.102}$$

This allows us to deduce the large r asymptotics for $y(r)$ and we find

$$y(r) \sim \left(\frac{9}{4}\right)^{1/3} r^{2/3} + A\cos\left(\frac{2\sqrt{2}}{3}\log(r) + B\right) + o(1) \quad \text{as } r \to \infty. \tag{7.103}$$

Finally, we may use the behavior we have obtained for $y(r)$ as $r \to \infty$ to deduce the behavior of the bifurcation diagram for equations (7.88)-(7.89). First, from

$$||w||_\infty = 1 - \frac{1}{y(\gamma)} \tag{7.104}$$

and equation (7.103), we see that $||w||_\infty \to 1$ as we monotonically let $\gamma \to \infty$, while from

$$\lambda = \frac{\gamma^2}{y(\gamma)^3} \tag{7.105}$$

and (7.103), we see that $\lambda \to 4/9$ as $\gamma \to \infty$. We also observe that the curve of solutions must oscillate infinitely many times as it heads to the point $\lambda = 4/9$, $||w||_\infty = 1$. A close-up view of this part of the bifurcation diagram is plotted in Figure 7.18.

Our analysis of the circular membrane has revealed similarities and differences between the circular membrane and the 1-D strip membrane of the previous section. The origin of the pull-in instability is the same. The bifurcation diagram for the circular membrane folds back upon itself so that beyond some critical value of λ no solutions to the steady-state system exist. The structure of the bifurcation diagram is quite different. Rather than folding back upon itself a single time, the bifurcation diagram folds infinitely many times. This implies that there is a value of λ for which infinitely many steady-state solutions exist. In particular, when $\lambda = 4/9$, there are infinitely many steady-state solutions. Moreover, given any positive integer n, there is a range of λ for which precisely n solutions exist. This surprising result illustrates the dramatic effect of geometry on solutions to the basic model. Stability of solutions is easy to investigate. It can be shown that the lowest branch of solutions is stable, while all other branches are linearly unstable. The disk analysis is partially present in the work of Taylor and Ackerberg [1, 188], and was made rigorous in [155].

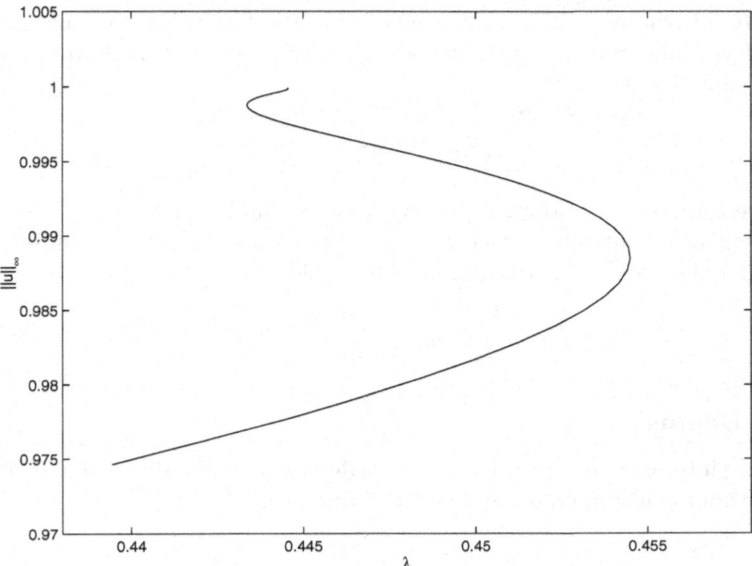

FIGURE 7.18: Close-up of the bifurcation diagram for the disk-shaped elastic membrane.

7.6 Electrostatic-Elastic Systems—Beam and Plate Theory

Many further extensions of the membrane theory of the previous section are possible. One direction for investigation is to relax the zero flexural rigidity assumption and study the small aspect ratio, steady-state, plate-based model. That is,

$$-\nabla^2 w + \delta \nabla^4 w = -\frac{\lambda}{w^2}. \tag{7.106}$$

Equation (7.106) contains both the effects of tension and flexural rigidity. In a system where tension is negligible and rigidity dominates, we would like to ignore the ∇^2 terms. In order to do so we divide by δ in equation (7.106) and note that

$$\frac{\lambda}{\delta} = \frac{\epsilon_0 V^2 L^4}{2d^3 D}. \tag{7.107}$$

We give this parameter a name, i.e.,

$$\beta = \frac{\lambda}{\delta} = \frac{\epsilon_0 V^2 L^4}{2d^3 D}. \tag{7.108}$$

Note that β is the ratio of a reference electrostatic force to a reference elastic force. Here, however, the reference elastic force is based on rigidity rather than tension. So, we have

$$-\frac{1}{\delta}\nabla^2 w + \nabla^4 w = -\frac{\beta}{w^2}. \tag{7.109}$$

To ignore tension we send δ to infinity. Bear in mind that to be able to take β to be our new control parameter, we must send δ to infinity by taking T to zero rather than taking D to infinity. This yields

$$\nabla^4 w = -\frac{\beta}{w^2}. \tag{7.110}$$

7.6.1 Beams

If we are interested in the electrostatic deflection of a one-dimensional elastic beam without tension, equation (7.110) reduces to

$$\frac{d^4 w}{dx^4} = -\frac{\beta}{w^2}. \tag{7.111}$$

The appropriate boundary conditions depend on the support at the endpoints. For example, in the case of a cantilevered beam, we would impose

$$w(0) = 1, \quad \frac{dw}{dx}(0) = \frac{d^2 w}{dx^2}(1) = \frac{d^3 w}{dx^3}(1) = 0. \tag{7.112}$$

In the case of other end supports, the appropriate boundary conditions may be chosen following the discussion in Chapter 2. Analysis of the beam model is not as straightforward as that of the 1-D membrane considered earlier. Some progress is possible. In particular, the Green's function for the cantilevered beam with concentrated load derived in Chapter 5 may be used to recast our beam model as a nonlinear integral equation. The procedure is the same as for the 1-D membrane. The reader is directed to the exercises for this analysis.

7.6.2 Circular Plates

If we are interested in the electrostatic deflection of a circular plate without tension and assume radial symmetry, equation (7.110) reduces to

$$\left(\frac{d^2}{dr^2} + \frac{1}{r}\frac{d}{dr}\right)\left(\frac{d^2 w}{dr^2} + \frac{1}{r}\frac{dw}{dr}\right) = -\frac{\lambda}{w^2}. \tag{7.113}$$

The appropriate boundary conditions depend on the support at the edges. For example, in the case of a fixed edge, we would impose

$$w(1) = 1, \quad \frac{dw}{dr}(1) = 0. \tag{7.114}$$

In the case of other edge supports, the appropriate boundary conditions may be chosen following the discussion in Chapter 2. As with the beam, analysis of the circular plate model is difficult. We draw the reader's attention to the interesting open question of the bifurcation diagram for this system. In particular, does the bifurcation diagram for the circular plate contain infinitely many folds? Does it resemble the case of a circular elastic membrane? If not, how and why does it differ?

7.7 Analysis of Capacitive Control Schemes

As has been stressed throughout this chapter, the pull-in instability is a ubiquitous feature of electrostatically actuated systems. Predictably, much research has focused on extending the stable operation of electrostatically actuated systems beyond the pull-in distance. In this section we apply our models of electrostatic actuation to the study of one type of control scheme, the capacitive control scheme. We study the scheme in the context of the mass-spring model and in the context of the membrane model of electrostatic actuation. The contrast between these two provides another example of the effect geometry may have on the behavior of an electrostatically actuated system.

7.7.1 The Basic Circuit

The basic capacitive control scheme was first proposed by Seeger and Crary in [173, 174]. In this scheme, control was achieved by the addition of a series capacitance to the circuit containing the MEMS device. The basic circuit is sketched in Figure 7.19. Seeger and Crary analyzed this scheme via a mass-spring type model. As we shall see, in this analysis, the nonlinearity due to the electrostatic force is mitigated by the additional capacitance in the circuit. This leads to the result that the scheme can stabilize the device over the entire range of travel! A subsequent study by Chan and Dutton [27] indicated that this was a spurious result due to the neglect of deformation of the device. Chan and Dutton included deformation of the device in a modified mass-spring model in an *ad hoc* manner and found that while the stable operating range was extended it was not extended to the entire range of travel. In [153, 154] a membrane-based model of the capacitive control scheme was analyzed confirming the *ad hoc* results of Chan and Dutton.

The circuit shown in Figure 7.19 exerts a stabilizing influence on the device because it acts as a voltage divider. A straightforward application of Kirchoff's laws to the circuit reveals that the voltage drop, V, across the device is related

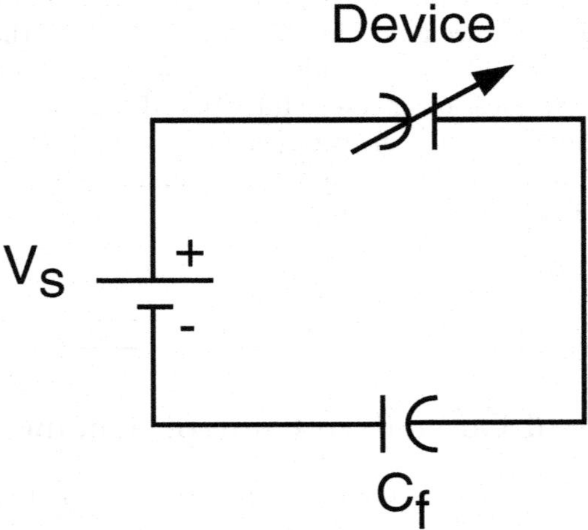

FIGURE 7.19: The basic capacitive control circuit.

to the source voltage, V_s, via

$$V = \frac{V_s}{1 + C/C_f}. \tag{7.115}$$

Here, C is the capacitance of the device while C_f is the capacitance of the fixed series capacitor. The capacitance C of a given device is a function of the deflection of the system. This capacitance will increase as the gap between elements of the system decreases. In turn, this causes the voltage drop across the device, V, to decrease, reducing the electrostatic force and stabilizing the device.

7.7.2 A Mass-Spring Model

To analyze the effectiveness of the capacitive control scheme, we embed our electrostatically actuated mass-spring system, Figure 7.6, in the control circuit, Figure 7.19. Recalling that the capacitance of our parallel plate capacitor is given by

$$C = \frac{\epsilon_0 A}{u - L}, \tag{7.116}$$

we see that the voltage drop across our mass-spring system is

$$V = \frac{V_s}{1 + \frac{\epsilon_0 A}{C_f}\left(\frac{1}{u - L}\right)}. \tag{7.117}$$

Hence, assuming the source is DC, the deflection of the mass-spring system satisfies

$$m\frac{d^2u}{dt'^2} + a\frac{du}{dt'} + k(u-l) = \frac{1}{2}\frac{\epsilon_0 A V_s^2}{(L-u)^2}\left[1 + \frac{\epsilon_0 A}{C_f}\left(\frac{1}{u-L}\right)\right]^{-2}. \tag{7.118}$$

Scaling on the inertial timescale of Section 7.3, the dimensionless version of our controlled mass-spring system is

$$\frac{d^2v}{dt^2} + \alpha\frac{dv}{dt} + v = \frac{\lambda}{(1-\chi-v)^2}. \tag{7.119}$$

Recall that α is the reciprocal of the quality factor or Q of the system. The parameter λ is the same as in Section 7.3 where the voltage in λ is the source voltage. The parameter χ is defined by

$$\chi = \frac{\epsilon_0 A}{C_f(L-l)} \tag{7.120}$$

and is a ratio of capacitances. In particular it the ratio of the capacitance of the undeflected device to the capacitance of the fixed series capacitor. Notice that the location of the singularity in the electrostatic force term is shifted. The degree of shifting is controlled by χ. This shift is the mathematical manifestation of the circuit's controlling influence.

If we look at steady-state deflections of the controlled mass-spring system, the value of adding the series capacitance becomes clear. Steady-state deflections satisfy

$$v = \frac{\lambda}{(1-\chi-v)^2}. \tag{7.121}$$

Solving for λ yields

$$\lambda = v(1-\chi-v)^2. \tag{7.122}$$

In Figure 7.20 we plot λ as a function of v for $\chi = 0.25$. Notice that this differs from the $\chi = 0$ plot, Figure 7.9, in two ways. First, the location of minima, previously at $v = 1$, has shifted to the left. This has the effect of bringing part of the previously unphysical branch of solutions into the physical regime. Second, the height of the maximum has decreased. The entire curve is becoming "straightened out." In Figure 7.21 we plot λ as a function of v for $\chi = 1$. We see that the straightening out process is complete. The multiple branch structure has disappeared, leaving a single branch of solutions. If we flip the axes and sketch the bifurcation diagram as is done in Figure 7.22, the implications of this straightening become clear. For any desired displacement v between 0 and 1, there is a single value of the source voltage, λ, which will yield the desired displacement. An easy stability analysis reveals that this single branch of solutions is linearly stable. Hence, the mass-spring analysis of the capacitive control scheme implies that for $\chi \geq 1$ the pull-in instability has been completely removed!

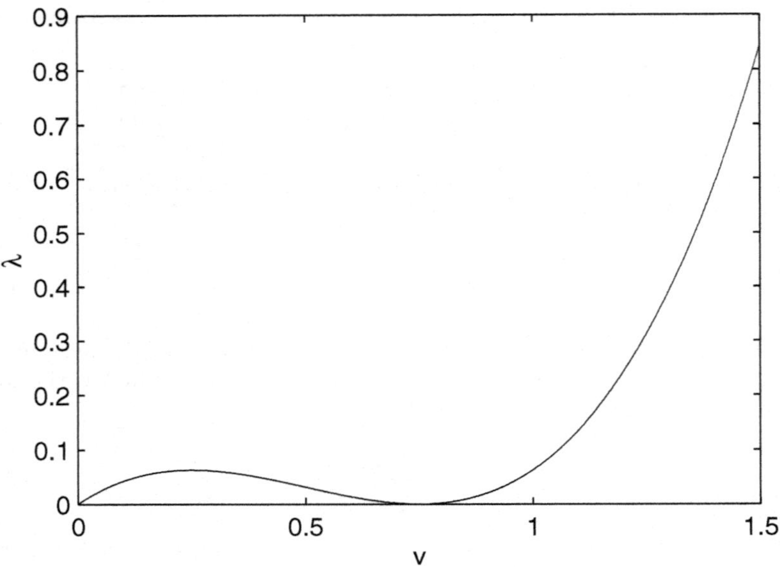

FIGURE 7.20: The parameter λ as a function of v for $\chi = 0.25$.

7.7.3 Control of a General Structure

As with most things that appear too good to be true, the mass-spring result predicting total removal of the pull-in instability for $\chi \geq 1$ collapses when closely examined. As we shall see in this section, when realistic deformations of the structure are accounted for, the pull-in instability reappears. The key difference between the mass-spring system of Figure 7.6 and the general device structure of Figure 7.13 lies in the computation of the capacitance. In the mass-spring model, the capacitance of the parallel plate structure is a function of the size of the gap between the plates. As we have already seen in Chapter 5, for the general device of Figure 7.13, the capacitance is an integral over the domain of the deflection of the elastic component. That is, in general, the capacitance of the structure in Figure 7.13 is given by

$$C = \frac{\epsilon_0}{V} \int_\Omega \nabla\phi \cdot \mathbf{n} \;\; dx' dy'. \tag{7.123}$$

In the small aspect ratio limit, the gradient of ϕ reduces to a derivative in the z' direction only and hence,

$$C \approx \frac{\epsilon_0}{V} \int_\Omega \frac{\partial \phi}{\partial z'} dx' dy'. \tag{7.124}$$

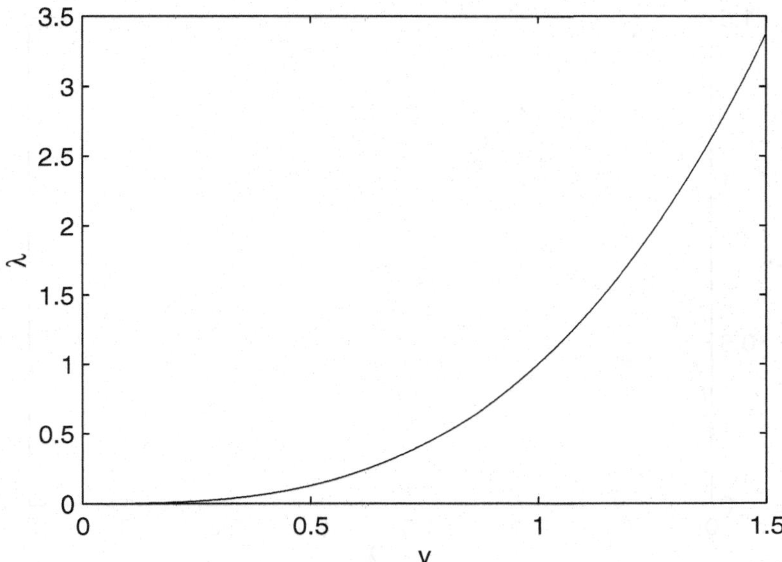

FIGURE 7.21: The parameter λ as a function of v for $\chi = 1$.

Using the approximate potential of Section 7.4 and scaling as in Section 7.4 we find

$$C \approx \frac{\epsilon_0 L^2}{d} \int_\Omega \frac{dxdy}{w(x,y)}. \tag{7.125}$$

The implies that the voltage drop, V, across the device of Figure 7.13 when embedded in the control circuit of Figure 7.19 is given by

$$V = \frac{V_s}{1 + \chi \displaystyle\int_\Omega \frac{dxdy}{w(x,y)}}. \tag{7.126}$$

Here

$$\chi = \frac{\epsilon_0 L^2}{C_f d} \tag{7.127}$$

and is again a ratio of capacitances. Consequently our reduced model for the elastic deflection of the system, equation (7.59), becomes

$$\frac{1}{\alpha}\frac{\partial^2 w}{\partial t^2} + \frac{\partial w}{\partial t} - \nabla_\perp^2 w + \delta \nabla_\perp^4 w = -\frac{\lambda}{w^2 \left(1 + \chi \displaystyle\int_\Omega \frac{dxdy}{w(x,y)}\right)^2}. \tag{7.128}$$

In addition to being nonlinear equation (7.128) is also *nonlocal.* We saw a similar situation arise in Chapter 4 during our study of Joule heating. In

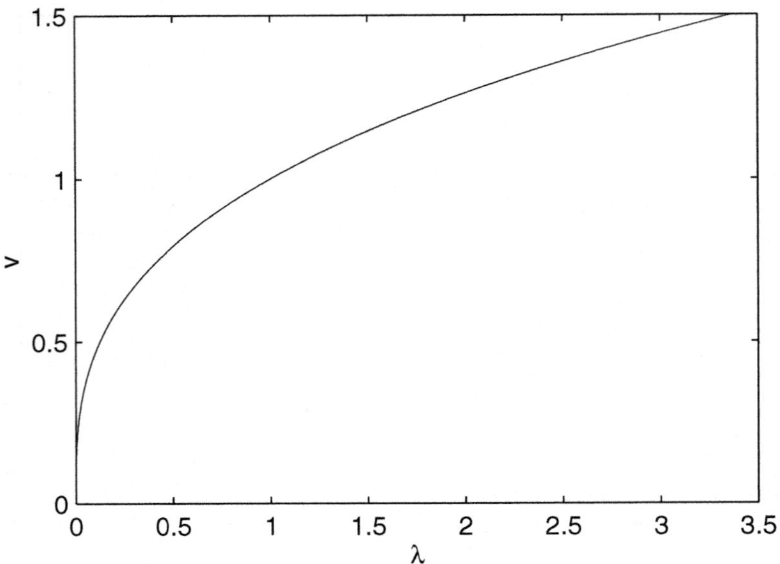

FIGURE 7.22: The bifurcation diagram for the mass-spring model for the value $\chi = 1$.

some sense the nonlocal term is an artifact of the model reduction and coupled domain structure of the problem. Here, the nonlocal term captures the effect of the control circuit. If we look at steady-state solutions and work in the $\delta \to 0$ limit, equation (7.128) reduces to

$$\nabla^2 w = \frac{\lambda}{w^2 \left(1 + \chi \int_\Omega \frac{dxdy}{w(x,y)}\right)^2} \text{ in } \Omega. \tag{7.129}$$

We have dropped the $\perp$ symbol; bear in mind that here the Laplace operator only acts in the x and y directions. Also recall that sending δ to zero means we are treating the diaphragm as an elastic membrane. Assuming the edges of the membrane are held fixed leads to the boundary condition,

$$w = 1 \text{ on } \partial\Omega. \tag{7.130}$$

Equation (7.129), while apparently different from our earlier model of electrostatically deflected membranes is actually virtually identical. We observe that a solution $w(x, y)$ of equations (7.129) and (7.130) is a solution of the local problem,

$$\nabla^2 w = \frac{\beta}{w^2} \text{ in } \Omega \tag{7.131}$$

$$w = 1 \quad \text{on} \quad \partial\Omega \tag{7.132}$$

for

$$\beta = \frac{\lambda}{\left(1 + \chi \int_{\Omega} \frac{dxdy}{w(x,y)}\right)^2}. \tag{7.133}$$

Similarly, a solution of the local problem, equations (7.131) and (7.132), is a solution of the nonlocal problem, equations (7.129) and (7.130) provided

$$\lambda = \beta \left(1 + \chi \int_{\Omega} \frac{dxdy}{w(x,y)}\right)^2. \tag{7.134}$$

This implies that the solution sets of the local and nonlocal problems are identical. Hence results concerning the local problem may be translated into results concerning the nonlocal problem through the β-λ relationship. Fortunately, we already have many results concerning the local problem. As an example, in the case where Ω is a circular disk, we know how to sketch the bifurcation diagram. Using these solutions we can sketch the bifurcation diagram for the controlled disk for any value of χ by simply using the β-λ relationship. An example computation is shown in Figure 7.23. Notice that

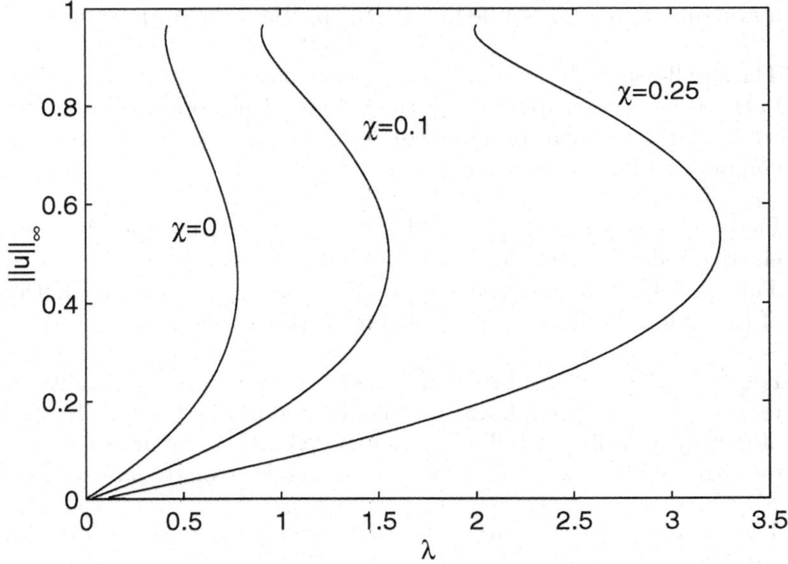

FIGURE 7.23: The bifurcation diagram for the membrane disk model for various values of χ.

for any value of χ, the bifurcation diagram still folds. That is, the pull-in instability is always present. On the other hand, the nose of the first fold has moved up and to the right. This implies that the pull-in distance has increased. The control scheme exerts a partially stabilizing influence. The fact that the nose moves to the right implies that higher and higher voltages are necessary in order to achieve a desired displacement.

7.8 Chapter Highlights

- Electrostatic actuation is the most widely used form of actuation for MEMS and NEMS devices.

- The mass-spring parallel plate system serves as a useful first model of electrostatic actuation. The mass-spring model is rich enough to capture the pull-in phenomenon and provides a first estimate of the pull-in voltage and pull-in distance. The pull-in phenomenon is a result of a fold in the bifurcation diagram.

- Realistic deformations of electrostatically actuated systems can be modeled using membrane, beam, or plate theory. The general model of such a system consists of coupled nonlinear partial differential equations.

- The small-aspect ratio approximation is a useful simplifying assumption in the study of electrostatic-elastic systems. It is appropriate for devices with a lateral dimension large compared to the spacing between elastic components.

- In the small-aspect ratio zero flexural rigidity limit, the electrostatic-elastic model reduces to a single nonlinear partial differential equation. This model is rich enough to capture the pull-in phenomenon. Estimates of the pull-in voltage are possible for arbitrarily shaped membranes.

- For simple geometries the small-aspect ratio zero flexural rigidity model reduces to a single nonlinear ordinary differential equation. In the case of a strip or a disk, a full analysis is possible. The bifurcation diagram for either system contains a fold, yet depends strongly on geometry.

- The models of electrostatic actuation developed in this chapter can be applied to the study of a wide variety of systems. One example is the study of capacitive control schemes. The contrast between the mass-spring model and the membrane model provides another example of the effect of geometry on behavior.

7.9 Exercises

Section 7.3

1. Rewrite equation (7.19) as a first-order system and set Ω_2 to zero. Identify the critical points for this system and plot them in the phase plane. Perform a linear stability analysis about each critical point and sketch the local behavior of solutions on your phase plane.

Section 7.5

2. Compute the upper bound on the pull-in voltage stated in Corollary 7.1 for a circular elastic membrane. Repeat for a square membrane. (Hint: You simply need to compute the lowest eigenvalue of the negative Laplacian for both domains.)

3. Compute the upper bound on the pull-in voltage stated in Corollary 7.1 for a rectangular elastic membrane. Assume one side length has been scaled to one and the other side length is a dimensionless ratio, say β. How does your bound vary with β? Does this agree with your intuition as to how the pull-in voltage should vary with β?

4. Compute the lower bound on the pull-in voltage stated in Corollary 7.2 for a circular elastic membrane. Compare this lower bound with the upper bound from problem 2. How "tight" are your bounds on λ^*?

5. Consider your bounds from Problems 2 and 4. If your circular elastic membrane has radius 100μm, is situated 2μm above the ground plate, and is held under tension $T = 150$MPa, translate your bounds on λ^* into bounds on the applied voltage V.

6. Consider the domain shown in Figure 7.24. Take this as the shape of an electrostatically actuated elastic membrane. Using Corollary 7.2 compute a lower bound for the pull-in voltage for this system.

7. Consider the iterative scheme for the 1-D membrane, equation (7.87). Taking $u_0(x) = 0$ as your initial guess, use the iterative scheme to compute $u_1(x)$ and $u_2(x)$ by hand.

8. Consider the iterative scheme for the 1-D membrane, equation (7.87). Apply the contraction mapping theorem to this scheme to prove that the method converges to a unique solution in a suitably chosen Banach space. Show that the contractivity of T depends on λ.

9. The symmetry method used to study the disk geometry may be used to study the strip geometry as well. Consider the strip model with the

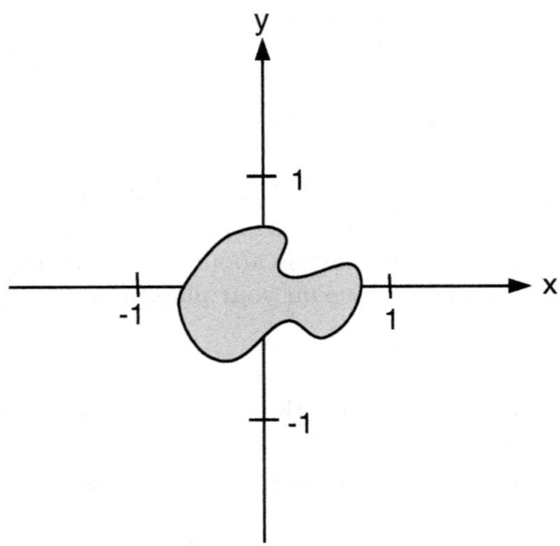

FIGURE 7.24: The domain for problem 6.

additional assumption that solutions are symmetric about $x = 0$. That is, consider the system,

$$\frac{d^2u}{dx^2} = \frac{\lambda}{(1+u)^2}$$

$$\frac{du}{dx}(0) = 0, \quad u(1/2) = 0.$$

By setting $u(x) = -1 + aw(bx)$, show that all solutions to this boundary value problem may be generated from the related initial value problem for $w(x)$,

$$\frac{d^2w}{dx^2} = \frac{1}{w^2}$$

$$w(0) = 1, \quad \frac{dw}{dx}(0) = 0.$$

Solve this initial value problem numerically and use the result to sketch the bifurcation diagram for the strip geometry.

10. Perform a linear stability analysis for the 1-D membrane model in the viscosity dominated regime. That is, take α to infinity and study stability using

$$\frac{\partial w}{\partial t} - \frac{\partial^2 w}{\partial x^2} = -\frac{\lambda}{w^2}.$$

You will need to solve the resulting eigenvalue problem numerically. You should find that the lower branch of solutions is linearly stable, while the upper branch is linearly unstable. Compare with the mass-spring model.

Section 7.6

11. Recast the model of an electrostatically deflected cantilevered beam as a nonlinear integral equation by using the Green's function of Chapter 5. Using the integral equation, write down an iterative scheme that may be used to find steady-state deflection. Starting with a zero initial guess, carry out the first few iterations by hand.

Section 7.7

12. An alternative to the capacitive control scheme was proposed by Seeger and Crary [174]. In place of the fixed capacitor, it was suggested that one use a *varactor*, i.e., a voltage variable capacitor. Assume that the varactor capacitance satisfies

$$C_f = C_m \left(\frac{1}{1 - V_f/V_0} \right)^{1/2},$$

where V_0 and C_m are fixed known parameters and V_f is the voltage drop across the varactor. Repeat the steady-state mass-spring analysis for this case. How do your results differ from the case of fixed series capacitance?

13. In addition to a fixed series capacitance Chan and Dutton analyzed the case of a circuit with a fixed series capacitance and a fixed parallel "parasitic" capacitance. Repeat the steady-state mass-spring analysis for this case. How do your results differ from the case of the fixed series capacitance only?

14. Consider the case where the controlled membrane is an elastic strip allowed to deflect along the x coordinate only. Use your knowledge of solutions to the $\chi = 0$ problem to sketch the bifurcation diagram for $\chi = 0.1, 0.5$. What do you observe?

7.10 Related Reading

The book by Na is a nice introduction to symmetry methods for boundary value problems.

T.Y. Na, *Computational Methods in Engineering: Boundary Value Problems*, Academic Press, 1979.

A more mathematical introduction to symmetry methods and Lie groups is the book by Bluman and Kumei.

G.W. Bluman and S. Kumei, *Symmetries and Differential Equations*, Springer Verlag, 1989.

An excellent introduction to numerical methods for bifurcation problems is the book by H.B. Keller.

H.B. Keller, *Lectures on Numerical Methods in Bifurcation Problems*, Springer Verlag, 1988.

7.11 Notes

1. Taylor's setup can be built for a few dollars and a very small investment of time. The only possible stumbling block for the basement experimenter is in finding a high voltage source. A Van de Graaff generator works fine. JAP has had undergraduate students replicate the Taylor experiments. If you are interested, email JAP or visit our website for more information.

2. GLV and grating light valve are registered trade marks of Silicon Light Machines.

3. The term *stiction* is used to describe many things by the MEMS and NEMS community. Generally speaking stiction refers to the unwanted adhesion that occurs between surfaces at the microscale. The adhesion may be due to electrostatic, capillary, or Van Der Waals forces.

4. A check valve is a valve that only opens in one direction.

5. Of course, the clamps also suffer from the problem of stiction. This is a design question. How do you make the clamps sufficiently strong, yet minimize their surface area and hence stiction?

6. These are by no means the only deficiencies of the mass-spring model. The model lacks fluid effects, the effect of squeeze film damping, etc. The work of Wenjing Ye and Jacob White nicely illustrates how such effects may make a large difference to the Q of an electrostatic-elastic system.

7. That is, the eigenfunction corresponding to κ_1 is unique.

8. The symmetry implied by this assumption is by no means obvious. For the disk geometry, that solutions to equations (7.62) and (7.63) are symmetric follows from the Gidas-Ni-Niremberg theorem. For a symmetric geometry with a hole such as an annulus, all bets are off.

9. The Blasius problem concerns steady-state flow past a flat plate.

10. Lyapunov functions are discussed in Chapter 6.

Chapter 8

Modeling Magnetically Actuated Systems

There's another disadvantage to the use of the flashlight: like many other mechanical gadgets it tends to separate a man from the world around him.

Edward Abbey

8.1 Introduction

Unlike thermal and electrostatic actuators, magnetic actuators are a familiar part of the macroscopic world. Small magnetic motors (in particular those found in servos) can be found in almost every electronic system in the world. This is in addition to relays, d'Arsonval meters, loudspeakers, and many other devices.

Magnetic actuation is often compared to electrostatic and thermal actuation via the time honored list of pros and cons. We do not resist the temptation. In the following list, each advantage of the magnetic case over the electrostatic or thermal case is followed by a similar disadvantage.

- No large electric fields are generated that require insulation from neighboring electronics. But strong magnetic fields can interfere with neighboring magnetic storage devices.
- Materials that are handled or processed by the device are not exposed to large electric fields. But they are exposed to large magnetic fields. Fortunately nonmagnetic materials can withstand huge fields without themselves becoming magnetized.[1]
- The active parts of a magnetic actuator may come into contact with each other without causing a short circuit that can melt or fuse them together. Of course, this is not true if the active parts are both current carrying wires.
- Comparable or perhaps larger energy densities can be achieved in magnetic devices [72, 73]. However, this comes at the expense of the large

power consumption and the large currents ($\approx$ 100mA) that are often required.

- Joule heating, which raises the ambient temperature of the environment, is not required for actuation. But coils are often used, which leads to Joule heating, and the large power consumptions will still produce a large amount of waste heat.

- Magnetic devices can operate inside a conducting medium (fluid or gas). However, the magnetic field will act on the free charges inside a conductor, complicating the design of such a device.

- Since many materials, such as silicon, are transparent to magnetic fields, the actuating field can be produced by a mechanism physically separated from the actuator itself. But this means that special shielding must be used if the device is in proximity to other magnetically sensitive components.

This chapter follows the format of the previous chapters. In Section 8.2 we give descriptions of a brief subset of devices and effects that are currently in use in micromagnetics. This is followed in Section 8.3 by a series of lumped models describing a specific class of devices that evolve in the following two sections, 8.4 and 8.5, into two models of a magnetically actuated membrane micropump. Although linear, the final model is too complicated to solve analytically. Approximate solutions are obtained using the finite difference method mentioned in Section 2.8.

8.2 Magnetically Driven Devices

Micromagnetic devices can be roughly broken down into three categories:

- Devices that use soft magnetic cores[2] and are activated by currents in "energizing coils." This type of activation is typical of macroscopic magnetic systems and relies on two body effects: the current in the coils and the induced magnetic moment in the core. Because of the combination of these two effects, these devices do not scale well and remain of a considerable size (e.g., millimeters). Examples include micromotors [2], a micromagnetic bearing actuator [64], linear actuators [12, 72], microrelays [190], and microvalves [167].

- Devices that use permanent magnets. Permanent (hard) magnets require no energy to activate and so are preferable to coils in many applications. However, permanent magnetization is still a body effect, since the total magnetic moment is the sum of the individual atomic or

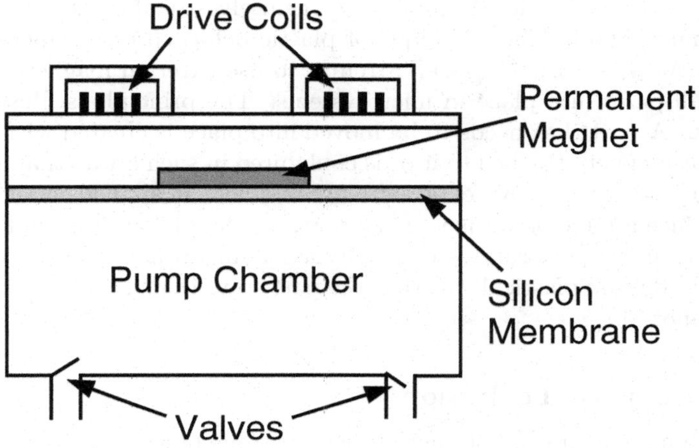

FIGURE 8.1: Sketch of a simple magnetically actuated micropump.

molecular moments. Recent advances in the fabrication of permanent thin film magnets [30] are making this type of device more attractive.

- Devices that utilize other magnetic material properties. This includes magnetostrictive actuators, giant magnetostriction, and coupled magnetic-thermal effects.

8.2.1 Micropumps

Magnetically actuated micropumps and valves have been investigated by a number of researchers, e.g. [102, 168, 219]. The basic configuration is shown in Figure 8.1. A typical device consists of a chamber with inlet and outlet valves, a flexible membrane, a permanent magnet, and a set of drive coils. Either the magnet or the coils may be attached to the membrane. When a current is driven through the coils, the resulting magnetic field creates an attraction or repulsion between the coils and the permanent magnet, which provides the actuating force.

8.2.2 Plastic Deformation Magnetic Assembly

In macroscopic engineering, systems are generally constructed by assembling them out of a supply of parts. The assembly itself consists of humans or robots physically moving the parts into contact with each other in a prescribed order. In microsystems this form of assembly is not effective for a number of reasons, not the least of which is that in order to operate at the necessary level of precision, the robots that perform the assembly must not be much bigger than the device itself, and hence themselves qualify as microsystems (see Trimmer [199]). The technique of plastic deformation magnetic assembly (PDMA) [179, 214, 215] is an attempt to use external magnetic fields to move small parts into place in microsystems. The principle is illustrated in Figure 8.2. A part that needs to be moved into place is created with either a permanent magnet attached to it or is configured in such a way that a current can be run through it. An external magnetic field is applied in such a way as to produce a force or torque on the part, bending it into place. The hinge around which the part rotates (typically a soft metal like gold) is purposely designed to deform plastically so that the part stays in place once the external field is removed.

8.2.3 Magnetic Levitation

The magnetic levitation of microdevices was one of the earliest areas of investigation in microsystems research [158]. Since microsystems can have parts moving at very high speeds,[3] friction is an important effect. However, because of the very low Reynolds numbers encountered, lubrication by conventional fluids such as oils is not practical. One alternative is to use levitation to prevent the parts from coming into contact. As we saw in Section 3.2.4, levitation in micromachines is relatively easy since gravity is a body force. Returning to the actuator discussed in Section 3.3, let's look at the power required to levitate one of the wires above the other. The condition for levitation is

$$\frac{\mu_0 I^2}{2\pi\beta} = \pi\alpha^2 L^3 g\delta, \tag{8.1}$$

where δ is the mass density of the wire (ρ being reserved for the resistivity). The current required for levitation is

$$I = \sqrt{\frac{2\beta\alpha^2\pi^2 L^3 g\delta}{\mu_0}}, \tag{8.2}$$

and the voltage drop and power dissipated in each wire is

$$V = \sqrt{\frac{2\beta L g\delta\rho^2}{\mu_0\alpha^2}}, \quad P = \frac{2\beta\pi L^2 g\delta\rho}{\mu_0}. \tag{8.3}$$

Hence levitation is possible with low voltages and power consumptions. In this case the heat generation rate due to Joule heating of the wires is

$$Q = \frac{2\beta g\delta\rho}{\alpha^2\mu_0 L}, \tag{8.4}$$

and using (3.67) the central temperature rise in each wire is

$$\Delta T = \frac{2\beta\gamma L g\delta\rho}{\alpha^2\mu_0}, \tag{8.5}$$

so that the levitating wire does not even warm up significantly.

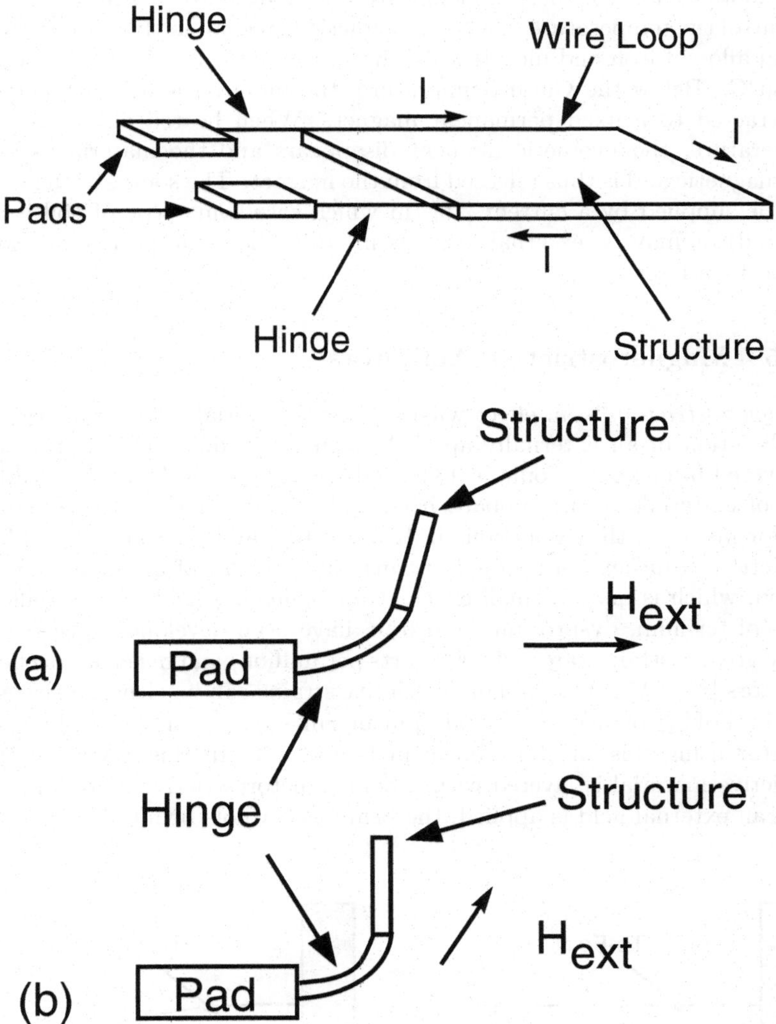

FIGURE 8.2: A basic plastic deformation magnetic assembly process.

8.2.4 Temperature-Controlled Magnetic Actuators

The *Curie temperature* of a ferromagnetic material is the temperature above which it loses its permanent magnetic moment and becomes paramagnetic. Hashimoto et. al. [75, 76] have demonstrated an actuator which uses a heater to control the temperature of a hard magnetic material, in this case Permalloy 50 (an alloy of iron and nickel) which has a relatively low Curie temperature of 530°C. Below the Curie temperature, the material is ferromagnetic and is attracted to a fixed permanent magnet. When heated above the Curie temperature, the magnetic moment disappears and the material is simply paramagnetic and is thus released from the magnet. The source of the heating may be supplied by a current [75] (in which case some type of coil is again required) or may be external, e.g., as in [76], where the permalloy block is heated by a laser.

8.2.5 Magnetostrictive Actuators

Magnetostriction is an effect whereby some materials change in size along the direction of an externally applied magnetic field.[4] The effect was first discovered in nickel by Joule in 1842 and has since been found in a wide variety of materials such as cobalt, iron, and various alloys of these materials (see for instance, the discussions in Kovacs [108] and Madou [128]). Magnetostriction in nickel can produce strains on the order of about 50 parts per million, which is quite a small effect. *Giant magnetostrictive* materials, e.g., alloys of terbium, dysprosium, and iron, have been developed specifically to create greater strains (around 1000 parts per million) and hence more effective actuators [81, 127, 134].[5] Magnetostrictive strains can be either positive (300 ppm in $TbFe_2$) or negative (-400 ppm in $SmFe_2$). A simple cantilever-type actuator using this effect is shown in Figure 8.3. In this device a magnetostrictive material is layered with a nonmagnetorestrictive material so that when an external field is applied, the beam as a whole bends.

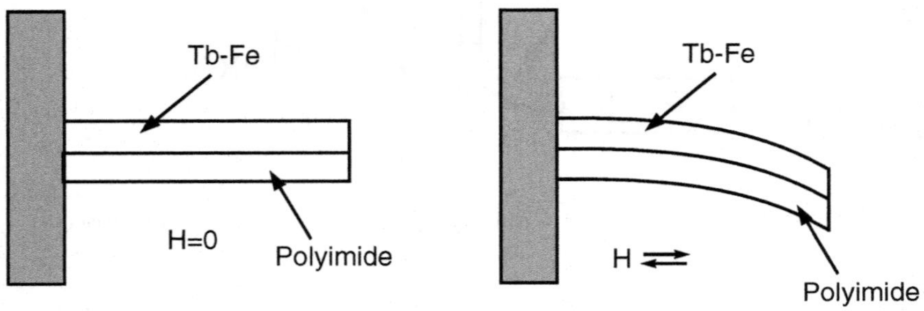

FIGURE 8.3: Sketch of a simple magnetostrictive actuator.

Recently it has been observed that some ferromagnetic shape memory alloys, such as Ni_2MnGa, exhibit large magnetostrictive strains (up to 6% at 0.4 Tesla) in the martensitic phase [137]. These materials are called *ferromagnetic shape memory alloys*, and the strains that can be produced are comparable to or greater than those found in giant magnetostrictive materials. The martensitic phase transformation temperature in these materials is lower than the Curie temperature so the materials remain ferromagnetic in both phases. Note that the actuation time for magnetostriction (which is driven by an external field) is typically significantly smaller than that for the shape memory effect, which is driven by a temperature change.

8.3 Mass-Spring Models

In this section we begin our investigation of the magnetically actuated micropump by examining a series of lumped models using magnetic actuation. These models will give us an idea of the qualitative features we will encounter in magnetic actuators. In subsequent sections we will move on to more realistic and complicated membrane-based models. From these lumped models, we will see that the pull-in phenomenon, which is ubiquitous in electrostatic actuation, is not necessarily present in typical magnetic actuation systems.

8.3.1 Two Parallel Wires

Following [5] consider the mass-spring model for the magnetic actuator sketched in Figure 8.4. The model consists of two parallel conducting wires of length L, separated by a distance d, and carrying parallel currents I. Let one of the wires be fixed in space and let the other be suspended above the first by a spring of constant k. We assume $d \ll L$ and that the radius of the wire is negligible.

Using the small aspect ratio assumption, the magnetic field created by the bottom wire is

$$B = \frac{\mu_0 I}{2\pi y'}. \tag{8.6}$$

Assuming that the wires remain parallel at all times, and that y' is the location of the top wire, the force felt by the top wire is

$$F = ILB = \frac{\mu_0 I^2 L}{2\pi y'}. \tag{8.7}$$

Hence the equation governing the equilibrium position of the top wire is

$$k(d - y') = \frac{\mu_0 I^2 L}{2\pi y'}, \tag{8.8}$$

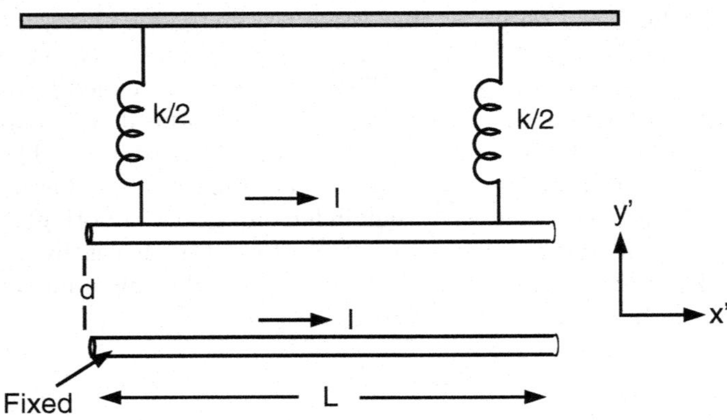

FIGURE 8.4: The simplest mass-spring model for a magnetic actuator.

where $0 < y' \leq d$. Nondimensionalizing by $y = y'/d$ gives

$$1 - y = \frac{\beta}{y}, \quad \beta = \frac{\mu_0 I^2}{2\pi\epsilon^2 kL}, \quad \epsilon = \frac{d}{L}. \tag{8.9}$$

The bifurcation diagram itself consists simply of the parabola,

$$y = \frac{-1 \pm \sqrt{1 - 4\beta}}{2}. \tag{8.10}$$

So in terms of the bifurcation diagram, the situation here is identical to that of the electrostatic mass-spring model described in section 7.3. For each value of the dimensionless parameter β, up to a critical value $\beta^* = 1/4$, there are two equilibrium positions of the top wire. There is thus a critical current $I^* = \epsilon\sqrt{\pi kL/2\mu_0}$ at which the deflection of the top wire is $y' = d/2$. The dynamical stability of these states is also the same as that in the electrostatic model (see exercise 2). The bifurcation diagram is plotted in Figure 8.5.

Despite their qualitative similarity, there are two important differences between this model and the electrostatic mass-spring model. The first is that the parameter β may become negative; this is the case in which the two wires have oppositely directed currents and repel each other. Hence the bifurcation diagram extends to the left of $\beta = 0$ in a physically meaningful way. This is characteristic of magnetic devices that are actuated by a coil: they are bidirectional. The second is that the nonlinearity of the model results from the nonlinear distribution of the magnetic field caused by the lower wire, which is fixed in place. In the electrostatic model the nonlinearity is due to the interaction of the charge on the mass and its position; that is, the position of the mass determines the charge density on it and vice versa. In the magnetic model, the position of the top wire does not influence the magnetic field of

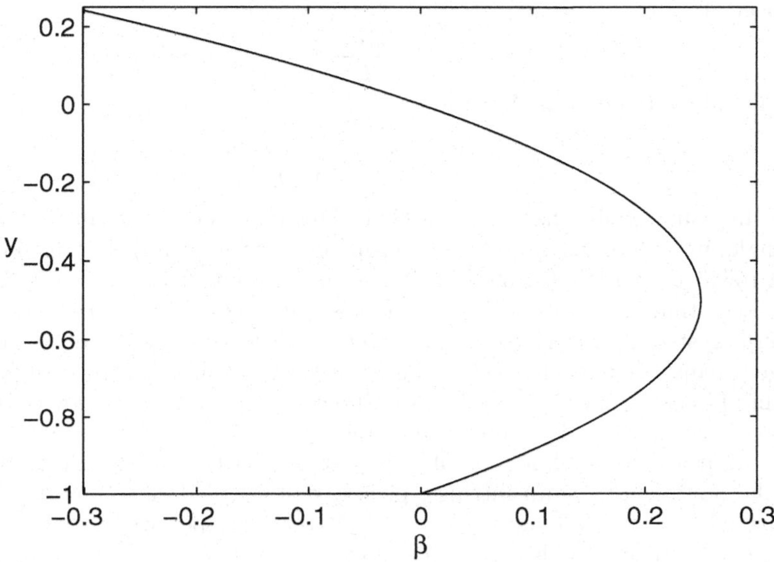

FIGURE 8.5: Bifurcation diagram for the parallel wire model.

the bottom wire and the nonlinearity of the resulting model is due solely to the fact that the top wire is moving through a magnetic field with a nonlinear spatial dependence. This is characteristic of magnetic systems involving fixed actuating coils. If the coils are not fixed in space, but are allowed to deform or move, then the situation becomes similar to that of electrostatics (see Exercise 5). However, even in this case, the two are not completely analogous since the position of the wires has no influence on the current they carry.

8.3.2 A Parallel Plate Model

Now suppose we replace the wires in this model with current sheets. We model the current sheets as if they contained many parallel and very thin wires each carrying a current i. Let the model consist of two such sheets, both square and of side length L, parallel to each other, and separated initially by a distance d, with the usual $d \ll L$. Again, the bottom sheet is fixed and the top sheet is suspended above it by a spring of constant k. Working in the small aspect ratio limit, it is easy to show that the magnetic field produced by the bottom sheet is

$$B = \frac{\mu_0 I}{2L}, \tag{8.11}$$

where I is the total current flowing through the sheet. Note that now B is independent of the separation of the sheets. This leads to a total force on the top sheet of

$$F = \frac{\mu_0 I^2}{2}, \tag{8.12}$$

and an equilibrium equation of the form,

$$y = 1 - \beta, \quad \beta = \frac{\mu_0 I^2}{2k\epsilon L}, \tag{8.13}$$

where the usual scaling has been applied. This case is different in that there is a single, unique solution and no "pull-in" current. Instead, the top sheet is pulled toward the bottom sheet until they make contact at $I = \sqrt{2k\epsilon L/\mu_0}$. This shows that the pull-in phenomenon is not necessarily characteristic of magnetic systems. In electrostatic systems, there will always be a pull-in voltage because there is always a charge density that will overwhelm an elastic restoring force, no matter what the specific configuration of the device is. In magnetic systems this is not the case and the reason for this is again due to the origin of the nonlinearity of the system. That is, in the electrostatic case, the nonlinearity is an intrinsic property of coupled elastic-electrostatic systems, whereas in the magnetic case it is merely a property of a specific type of field configuration. This can be made nonlinear, as in the first example, or not, as in this example.

8.3.3 A Coil-Activated Model

Now let's examine a simple mass-spring model involving a permanent magnet. Following [54] the model consists of a circular wire loop of radius R and carrying current I. Along the line perpendicular to the loop and through its center, we place a point magnetic dipole of strength μ a distance d from the plane of the loop. The dipole is attached to a spring of constant k and we assume that the dipole is restricted to motion along the center line only (Figure 8.6).

Using the Biot-Savart law (2.183),

$$d\mathbf{B} = \frac{\mu_0 I}{4\pi} \frac{d\mathbf{l} \times \mathbf{r}'}{r'^3}, \tag{8.14}$$

the magnetic field along the center line is

$$B_{z'} = \frac{\mu_0 I R^2}{2} \frac{1}{(z'^2 + R^2)^{3/2}}. \tag{8.15}$$

The force on a magnetic dipole $\mathbf{m}$ in the presence of a steady-state external magnetic field $\mathbf{B}$ is (e.g., Jackson [93])[6]

$$\mathbf{F} = (\mathbf{m} \cdot \nabla)\mathbf{B}. \tag{8.16}$$

In this case both $\mathbf{m}$ and $\mathbf{B}$ are directed along the z' axis so that

$$F = \mu \frac{\partial B_{z'}}{\partial z'} = -\frac{3z'}{2} \frac{\mu\mu_0 I R^2}{(z'^2 + R^2)^{5/2}}, \tag{8.17}$$

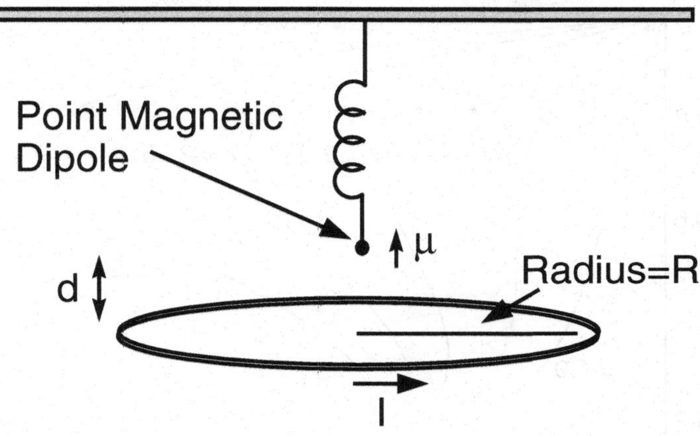

FIGURE 8.6: A mass-spring model for a coil-activated micropump.

where $\mu = |\mathbf{m}|$. The restoring spring force is $F_s = -k(z' - d)$; hence the equilibrium condition is

$$z' - d + \frac{3z'}{2k} \frac{\mu\mu_0 I R^2}{(z'^2 + R^2)^{5/2}} = 0. \tag{8.18}$$

By now it should be second nature to define $z = z'/d$ and $\epsilon = d/R$ and substitute to get

$$1 - z - \frac{\beta z}{(\epsilon^2 z^2 + 1)^{5/2}} = 0, \quad \beta = \frac{3\mu\mu_0 I}{2kR^3}. \tag{8.19}$$

As usual z, ϵ, and β are all dimensionless. It can be shown [54] that (8.19) has only one solution for $\epsilon < \epsilon^* \approx 1.6237976$ and either one or three solutions, depending on the value of β, for $\epsilon > \epsilon^*$. The bifurcation diagram for a few values of ϵ is shown in Figure 8.7. In this case, we have three branches for certain values of ϵ and β, while for other values we have only one branch. The stability of the various branches are explored in exercise 8 and 9.

In this case we have a more complicated bifurcation diagram, consisting of two folds, simply because of the more complicated, nonlinear nature of the field produced by the coil. Notice also that there is no pull-in and that in the limit $\beta \to \infty$, we have a unique solution, $z \approx \beta^{-1}$. This means that no matter how large the current becomes, the dipole is never pulled all the way down to the plane of the coil.

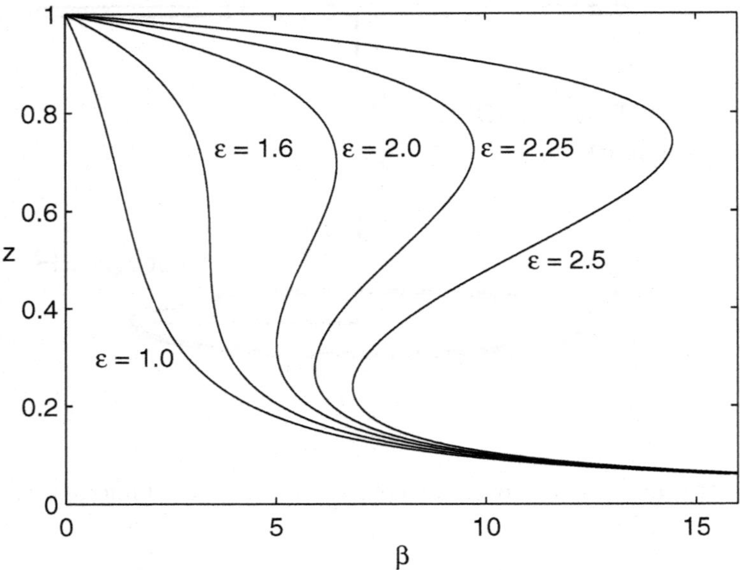

FIGURE 8.7: Bifurcation diagram for the mass-spring coil-activated model.

8.4 A Simple Membrane Micropump Model

Following the previous section, replace the spring with a circular, elastic membrane of the same radius as the activating coil and under tension T (Figure 8.8). In the center of the membrane, place a small circular permanent magnet of radius r_0' and magnetic moment μ. Following Section 5.4, the equation determining the deflection of the membrane is

$$\nabla^2 u' = \frac{P'(u', r')}{T}, \tag{8.20}$$

where $P'(u', r')$ is the pressure on the membrane caused by the permanent magnet.[7] For the moment, assume this is is of the form

$$P' = \begin{cases} P_0' & , r' \leq r_0' \\ 0 & , r' > r_0' \end{cases}, \tag{8.21}$$

where P_0' is a constant. As before, we scale according to $u = u'/d$, $r = r'/R$, and $r_0 = r_0'/R$. If we look for radial solutions only, then (8.20) along with the appropriate boundary conditions on u reduces to

$$\frac{\partial^2 u}{\partial r^2} + \frac{1}{r}\frac{\partial u}{\partial r} = p, \quad u(1) = 0, \quad \frac{\partial u}{\partial r}(0) = 0, \tag{8.22}$$

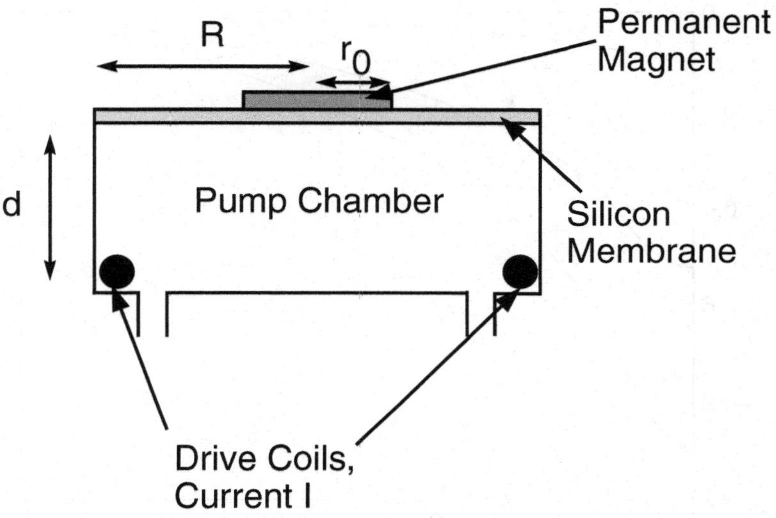

FIGURE 8.8: A membrane model for a coil-activated, permanent magnet actuator.

where

$$P = \begin{cases} p \, , r \leq r_0 \\ 0 \, , r > r_0 \end{cases}, \quad p = \frac{RP_0'}{\epsilon T}. \tag{8.23}$$

We can solve this for the two regions $r \leq r_0$ and $r > r_0$ separately and match the solutions and their derivatives at $r = r_0$ to obtain

$$u(r) = \begin{cases} \dfrac{r^2 p}{4} - \dfrac{r_0^2 p}{4}(1 - 2\log r_0), \; 0 \leq r \leq r_0 \\ \\ \dfrac{r_0^2 p}{2} \log r, \qquad\qquad\qquad r_0 < r \leq 1 \end{cases}. \tag{8.24}$$

In order to use expression (8.17) for the force on the magnet we make the assumption that $r_0 \ll 1$. To model the pressure on the membrane due to the magnet in this limit, we take the force (8.17) and assume it to be applied evenly over the interior of $r \leq r_0$. In this case the constant p becomes dependent on the central deflection $u_0 = u(0)$:

$$u_0 = -\frac{r_0^2 p}{4}(1 - 2\log r_0). \tag{8.25}$$

Using (8.17) and $u' = z' - d$ we see that the pressure caused by the permanent magnet on the membrane is

$$P_0' = \frac{F}{\pi r_0'^2} = \frac{1}{\pi r_0^2} \frac{3\mu\mu_0 I}{2R^4} \frac{\epsilon(1 + u_0)}{(1 + \epsilon^2(1 + u_0)^2)^{5/2}}, \tag{8.26}$$

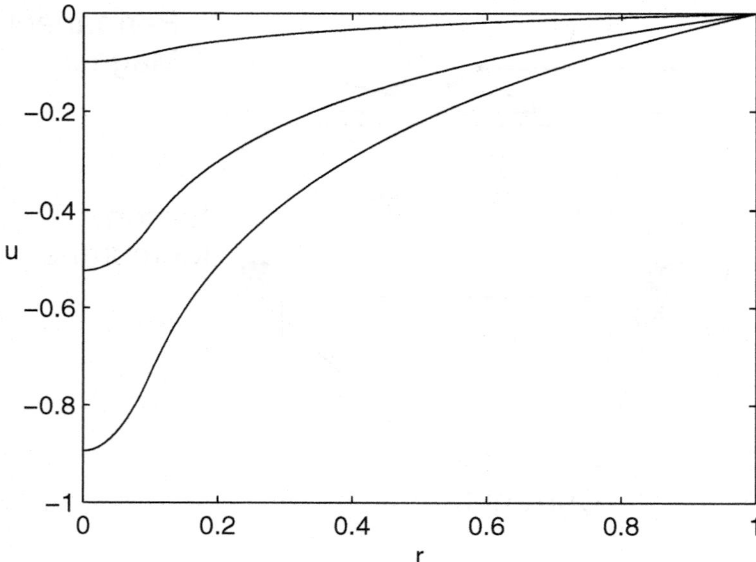

FIGURE 8.9: Solutions for the deflection of the simple membrane micropump model. (In this plot $r_0 = 0.1$, $\epsilon = 2.5$, and $\gamma = 12$.)

which gives

$$p = \frac{1}{\pi r_0^2} \frac{3\mu\mu_0 I}{2TR^3} \frac{(1+u_0)}{(1+\epsilon^2(1+u_0)^2)^{5/2}}. \tag{8.27}$$

Plugging this into (8.25) yields

$$-u_0 = \frac{\gamma(1+u_0)}{(1+\epsilon^2(1+u_0)^2)^{5/2}}, \tag{8.28}$$

where

$$\gamma = \frac{3\mu\mu_0 I(1-2\log r_0)}{8\pi R^3 T}. \tag{8.29}$$

Thus we obtain the same nonlinear algebraic equation for the central deflection as in the mass-spring model. In addition since u_0 determines p and p determines the entire solution, the bifurcation diagram of this model is exactly the same. This makes sense since the membrane here essentially acts as a spring as long as the force on it is restricted to a small region near the center. Thus for $\epsilon > \epsilon^* \approx 1.6$, we will have three equilibrium states of the membrane. These are plotted in Figure 8.9 for $\epsilon = 2.5$ and $\gamma = 12$. Note that the current is now given by γ and r_0, rather than γ alone, as in the mass-spring model.

8.5 A Small-Aspect Ratio Model

For typical micropumps the aspect ratio ϵ is very small, much smaller than the critical value above of $\epsilon^* \approx 1.6$. Based on the analysis above, we would expect such pumps to have unique equilibrium solutions. This makes sense since it is the nonlinearity of the coil field that produces the folds in the bifurcation diagram, so if we linearize the field near the $z = 0$ plane, then we automatically eliminate the folds. Nevertheless, it is instructive to examine such a model since it will allow us to consider the situation in which the size of the magnet is comparable to the radius of the membrane.

8.5.1 Formulation of the Model

We use the same geometry as in the previous section, but this time let $r_0 \approx R$, so that the magnet covers most of the membrane. For the moment we will work in the unscaled variables. For the outer portion $r' > r_0'$ where the pressure vanishes we have the solution,

$$u_{\mathrm{o}}' = -k \log\left(\frac{r'}{R}\right). \tag{8.30}$$

If we denote the inner solution u_{i}', then the boundary condition at $r' = 0$ is the same as before, $u_{\mathrm{i},r'}' = 0$, while the matching conditions for the deflection and its first derivative at $r' = r_0'$ are

$$u_{\mathrm{i}}'(r_0') = -k \log\left(\frac{r_0'}{R}\right), \quad \frac{\partial u_{\mathrm{i}}'}{\partial r'}(r_0') = -\frac{k}{r_0'}. \tag{8.31}$$

Eliminating k yields the conditions on u_{i}',

$$\frac{\partial u_{\mathrm{i}}'}{\partial r'}(0) = 0, \quad \frac{\partial u_{\mathrm{i}}'}{\partial r'}(r_0') - \frac{u_{\mathrm{i}}'(r_0')}{r_0' \log(r_0'/R)} = 0. \tag{8.32}$$

The first of these is called a *Neumann* boundary condition while the second is called a *mixed* or *Robin* boundary condition. Hence the problem is reduced to finding the inner solution on the domain $0 \le r' < r_0'$ with these boundary conditions.

The magnetic field at an arbitrary point around the coil is given by [182]

$$\begin{aligned} B_{r'} &= \frac{\mu_0 I}{2\pi} \frac{z'}{r'\sqrt{(R+r')^2 + z'^2}} \left(E(k) \frac{R^2 + r'^2 + z'^2}{(R-r')^2 + z'^2} - K(k) \right) \\ B_{z'} &= \frac{\mu_0 I}{2\pi} \frac{1}{\sqrt{(R+r')^2 + z'^2}} \left(E(k) \frac{R^2 - r'^2 - z'^2}{(R-r')^2 + z'^2} + K(k) \right), \end{aligned} \tag{8.33}$$

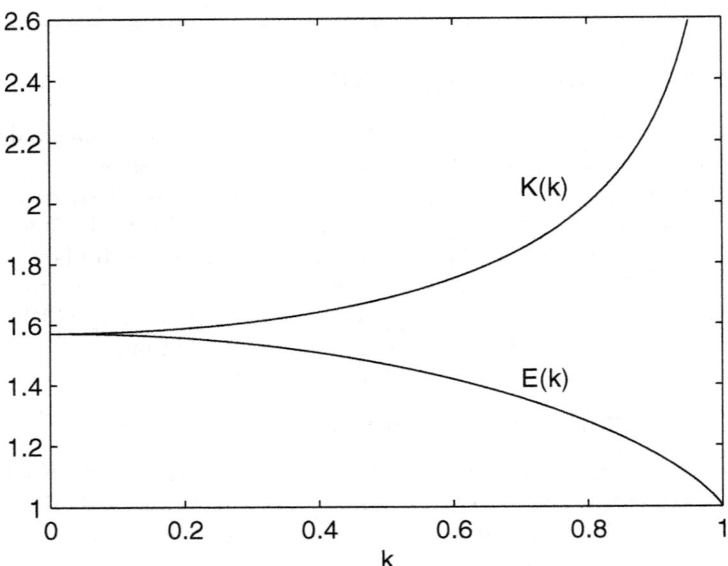

FIGURE 8.10: The complete elliptic integral functions $K(k)$ and $E(k)$.

where

$$k = \sqrt{\frac{4Rr'^2}{(R + r'^2)^2 + z'^2}} \tag{8.34}$$

and K and E are complete elliptic integrals of the first and second kind respectively, and are plotted in Figure 8.10. Note that we have abused the notation for r': it refers to both the radial coordinate covering the membrane and the radial cylindrical coordinate in which (8.33) is written. This is justified later when the small aspect ratio assumption is made. Let $\mathbf{m}$ be the dipole moment per unit area of the magnet. Then the pressure on the membrane is the normal component of (8.16),

$$P'(u', r') = \mathbf{n} \cdot (\mathbf{m} \cdot \nabla)\mathbf{B}. \tag{8.35}$$

Now assume that we are looking for radial solutions only, i.e., $u' = u'(r')$. If the magnet is of negligible thickness then $\mathbf{m}$ will remain perpendicular to the membrane at all times,

$$\mathbf{m} = \mu\mathbf{n}, \tag{8.36}$$

where $\mathbf{n}$ is the upward pointing unit normal to the membrane and $\mu = |\mathbf{m}|$. If the membrane has deflection $u'(r')$ and location $z'(r') = d + u'(r')$ then the unit normal to the membrane is

$$\mathbf{n} = \frac{(-u'_{,r'}, 0, 1)}{\sqrt{1 + (u'_{,r'})^2}}. \tag{8.37}$$

At this point we employ the scaling from the previous section to cast the entire model into dimensionless form. This yields

$$\mathbf{n} = \frac{(-\epsilon u_{,r}, 0, 1)}{\sqrt{1 + \epsilon^2 (u_{,r})^2}}. \tag{8.38}$$

The left-hand side operator becomes

$$\frac{\partial^2 u'}{\partial r'2} + \frac{1}{r'}\frac{\partial u'}{\partial r'} = \frac{\epsilon}{R}\left(\frac{\partial^2 u}{\partial r^2} + \frac{1}{r}\frac{\partial u}{\partial r}\right). \tag{8.39}$$

We write (8.33) as

$$B_r = B_0 \frac{\epsilon z}{r}\frac{F}{H}, \quad B_z = B_0 \frac{G}{H}, \quad B_0 = \frac{\mu_0 I}{2\pi R}, \tag{8.40}$$

where

$$F(r,z) = E(k)\frac{1 + r^2 + \epsilon^2 z^2}{(1-r)^2 + \epsilon^2 z^2} - K(k) \tag{8.41}$$

$$G(r,z) = E(k)\frac{1 - r^2 - \epsilon^2 z^2}{(1-r)^2 + \epsilon^2 z^2} + K(k) \tag{8.42}$$

$$H(r,z) = \sqrt{(1+r)^2 + \epsilon^2 z^2} \tag{8.43}$$

$$k(r,z) = \sqrt{\frac{4r}{(1+r)^2 + \epsilon^2 z^2}}. \tag{8.44}$$

In addition the operator $\mathbf{m} \cdot \nabla$ becomes

$$\mathbf{m} \cdot \nabla = \frac{\mu}{d\sqrt{1 + \epsilon^2 (u_{,r})^2}}\left(-\epsilon^2 u_{,r}\frac{\partial}{\partial r} + \frac{\partial}{\partial z}\right). \tag{8.45}$$

After multiplying through by the factor R/ϵ from (8.39) the pressure (8.35) is

$$P = \frac{\mu}{\epsilon^2 T}\frac{1}{1 + \epsilon^2 (u_{,r})^2}\left(\epsilon^3 (u_{,r})^2 \frac{\partial B_r}{\partial r} - \epsilon u_{,r}\frac{\partial B_r}{\partial z} - \epsilon^2 u_{,r}\frac{\partial B_z}{\partial r} + \frac{\partial B_z}{\partial z}\right). \tag{8.46}$$

Note that the functions F, G, and H are $O(1)$. Using (8.41)–(8.44) we see that

$$\frac{\partial B_r}{\partial r} = O(\epsilon), \quad \frac{\partial B_r}{\partial z} = O(\epsilon), \quad \frac{\partial B_z}{\partial r} = O(1), \quad \frac{\partial B_z}{\partial z} = O(\epsilon^2). \tag{8.47}$$

So the first term in (8.46) is of order ϵ^2, while the last three terms are of order unity. Carrying out the differentiation, using

$$\frac{\partial K}{\partial k} = \frac{E}{k(1-k^2)} - \frac{K}{k}, \quad \frac{\partial E}{\partial k} = \frac{E-K}{k}, \tag{8.48}$$

and eliminating the ϵ^2 terms gives

$$\frac{1}{\epsilon}\frac{\partial B_r}{\partial z} = \frac{1}{r(1+r)}\left(E\frac{1+r^2}{(1-r)^2} - K\right) \tag{8.49}$$

$$\frac{\partial B_z}{\partial r} = \frac{1}{r(1-r^2)}\left(E\frac{1+r+r^2}{1-r} - K\frac{1+r-r^2}{1+r}\right) \tag{8.50}$$

$$\frac{1}{\epsilon^2}\frac{\partial B_z}{\partial z} = -\frac{z}{(1-r)(1+r)^2}\left(E(k)\frac{7+r^2}{(1-r)^2} - K\right). \tag{8.51}$$

Finally, combining this with (8.39) and noting that the force is applied on the membrane at which $z = 1 + u$ yields the dimensionless forcing term,

$$P(u,r) = \beta\hat{P} = \beta\left(f(r)(1+u) + \frac{\partial u}{\partial r}\frac{g}{r}\right), \tag{8.52}$$

where

$$\beta = \frac{\mu\mu_0 I}{2\pi RT} \tag{8.53}$$

and

$$f = \frac{1}{(1-r)(1+r)^2}\left(E(k)\frac{7+r^2}{(1-r)^2} - K(k)\right) \tag{8.54}$$

$$g = \frac{1}{1-r^2}\left(E(k)\frac{2+r+2r^2}{1-r} - K(k)\frac{2+r-2r^2}{1+r}\right). \tag{8.55}$$

Note that β is indeed dimensionless, since in this case μ is a magnetic moment per unit area. Finally, the differential equation governing the deflection of the membrane is

$$\frac{\partial^2 u}{\partial r^2} + \frac{1}{r}\frac{\partial u}{\partial r} = \beta\hat{P} \tag{8.56}$$

and the boundary conditions (8.32) become

$$\frac{\partial u}{\partial r}(0) = 0, \quad \frac{\partial u}{\partial r}(r_0) - \frac{u(r_0)}{r_0\log r_0} = 0. \tag{8.57}$$

This type of equation is known as a *second-order nonhomogeneous equation* with the *separated homogeneous boundary conditions* (8.57). Analysis of equations in this class can be found in many standard textbooks on ordinary differential equations. The interested reader is referred to [14, 16]. At this point we merely point out that if one were to write (8.56) in the standard form,

$$L(u,\beta) - \beta\rho(r)u = f, \tag{8.58}$$

then the linear operator L depends on β, which complicates the situation somewhat.

8.5.2 A Numerical Interlude

Due to the complexity of f and g, equation (8.56) cannot be solved analytically for the deflection as we did in the previous section. Instead we turn to a standard numerical approximation method known as the *finite difference* method. For the moment we will assume that solutions to (8.56) exist and are unique, although this is by no means certain.

As in all numerical methods, the finite difference method relies on a discretization to reduce the degrees of freedom of a set of continuous functions to a finite number. In this case the discretization takes place on the coordinate system that covers the domain of the computation. For this problem the domain is $0 \le r \le r_0$, which is covered by the coordinate r. We choose the uniform discretization,[8]

$$r_i = i\Delta, \quad \Delta = \frac{r_0}{N}, \quad i = 0, 1, \dots, N. \tag{8.59}$$

Of course, any set of intervals may be chosen. The key, as we shall see in a moment, is to choose a discretization for which every $\Delta_i = r_{i+1} - r_i$ vanishes in the limit $N \to \infty$.

We write (8.56) as

$$\frac{\partial^2 u}{\partial r^2} + \frac{1 - \beta g}{r}\frac{\partial u}{\partial r} - \beta f u = \beta f. \tag{8.60}$$

We write the boundary conditions (8.57) as

$$\frac{\partial u}{\partial r}(0) = 0, \quad \frac{\partial u}{\partial r}(r_0) - ku(r_0) = 0, \quad k = \frac{1}{r_0 \log r_0}. \tag{8.61}$$

Because computers can really only do algebra, we seek to replace this differential equation with a set of algebraic equations using the discretization (8.59). To this end, consider the the Taylor expansion of u about r_i,

$$u(r) = u(r_i) + (r - r_i)\frac{\partial u}{\partial r}(r_i) + \frac{(r - r_i)^2}{2}\frac{\partial^2 u}{\partial r^2}(r_i) \tag{8.62}$$

$$+\frac{(r - r_i)^3}{6}\frac{\partial^3 u}{\partial r^3}(r_i) + \frac{(r - r_i)^4}{24}\frac{\partial^4 u}{\partial r^4}(r_i) + O(\Delta^5). \tag{8.63}$$

If we evaluate this at $r = r_{i+1}$ and use $r_{i+1} - r_i = \Delta$ we get

$$u(r_{i+1}) \approx u(r_i) + \Delta\frac{\partial u}{\partial r}(r_i) + \frac{\Delta^2}{2}\frac{\partial^2 u}{\partial r^2}(r_i) + \frac{\Delta^3}{6}\frac{\partial^3 u}{\partial r^3}(r_i) + \frac{\Delta^4}{24}\frac{\partial^4 u}{\partial r^4}(r_i), \tag{8.64}$$

where we have sacrificed equality in order to drop the $O(\Delta^5)$ term. The same expansion in the opposite direction is

$$u(r_{i-1}) \approx u(r_i) - \Delta\frac{\partial u}{\partial r}(r_i) + \frac{\Delta^2}{2}\frac{\partial^2 u}{\partial r^2}(r_i) - \frac{\Delta^3}{6}\frac{\partial^3 u}{\partial r^3}(r_i) + \frac{\Delta^4}{24}\frac{\partial^4 u}{\partial r^4}(r_i). \tag{8.65}$$

Subtracting the first expansion from the second yields

$$\frac{\partial u}{\partial r}(r_i) \approx \frac{u_{i+1} - u_{i-1}}{2\Delta} + \frac{\Delta^2}{3}\frac{\partial^3 u}{\partial r^3}(r_i), \tag{8.66}$$

while adding the two together gives

$$\frac{\partial^2 u}{\partial r^2}(r_i) \approx \frac{u_{i+1} - 2u_i + u_{i-1}}{\Delta^2} + \frac{\Delta^2}{12}\frac{\partial^4 u}{\partial r^4}(r_i). \tag{8.67}$$

Observe two things: the first is that the serendipitous cancellation of terms in the above two expressions was due to the uniform grid spacing Δ. The second is that as long as the function u is well-behaved, the last two terms in each expression will vanish in the limit $\Delta \to 0$. Therefore dropping these terms will result in an error of order Δ^2. These terms (and all those that follow them in the Taylor series) are called the *truncation error.* Hence the approximations,

$$\frac{\partial u}{\partial r}(r_i) \approx \frac{u_{i+1} - u_{i-1}}{2\Delta}, \quad \frac{\partial^2 u}{\partial r^2}(r_i) \approx \frac{u_{i+1} - 2u_i + u_{i-1}}{\Delta^2}, \tag{8.68}$$

are called *second order* because the error of the approximation scales as Δ^2 as long as the higher derivatives of u are finite.[9]

The reader will already be able to guess where we're going: we insert the approximations (8.68) into the differential equation (8.60) at each point of the grid r_i. Since there are N points, we have N algebraic equations in the N unknowns u_i and as the reader will well know, such a system can be expressed by the matrix equation,

$$\mathbf{A}\mathbf{u} = \mathbf{b}, \tag{8.69}$$

where $\mathbf{A}$ is an $N \times N$ matrix and $\mathbf{u}$ and $\mathbf{b}$ are column vectors of length N. Once $\mathbf{A}$ and $\mathbf{b}$ are known, we need only invert the matrix to find the solution $\mathbf{u}$, which is the approximation to the deflection of the membrane.

However, one or two subtleties remain. Substituion of (8.68) into (8.60) gives the *differenced form* of the equation,

$$u_{i+1}\left(1 + \frac{\Delta(1 - \beta g_i)}{2r_i}\right) - u_i(2 + \Delta^2 \beta f_i) + u_{i-1}\left(1 - \frac{\Delta(1 - \beta g_i)}{2r_i}\right) = \Delta^2 \beta f_i, \tag{8.70}$$

where we have multiplied through by Δ^2. This holds for all points in the interior of the domain but not at $r = 0$ or $r = r_0$. The differenced form of the right hand boundary condition (8.61) is

$$\frac{u_{N+1} - u_{N-1}}{2\Delta} - k u_N = 0, \tag{8.71}$$

where we have postulated the existence of a fictitious grid point located at $r = (N+1)\Delta$, where u would have the value u_{N+1}. This gives

$$u_{N+1} = 2k\Delta u_N + u_{N-1} \tag{8.72}$$

and now we may use (8.70) to get the difference equation at $r = r_0$,

$$u_N \left(2k\Delta - 2 - \Delta^2 \beta f_N + \frac{\Delta^2 k(1 - \beta g_N)}{r_N} \right) + 2u_{N-1} = \Delta^2 \beta f_N. \qquad (8.73)$$

This leaves us with only the equation at $r = 0$. First notice that the coefficient function in front of the first derivative term is singular at $r = 0$ so that we cannot simply apply (8.70) there. Instead we use a standard trick, which is to apply L'Hospital's rule to this term,

$$\lim_{r \to 0} \frac{1 + \beta g}{r} \frac{\partial u}{\partial r} = (1 + \beta g_0) \frac{\partial^2 u}{\partial r^2}(0) \qquad (8.74)$$

so that at $r = 0$ we use

$$(2 + \beta g) \frac{\partial^2 u}{\partial r^2} - \beta f u = \beta f \qquad (8.75)$$

instead of (8.60). As at the right-hand endpoint, we denote the value of u at the fictitious grid point $r = -\Delta$ by u_{-1} and use the differenced form of the boundary condition (8.61) to obtain

$$\frac{u_1 - u_{-1}}{2\Delta} = 0 \qquad (8.76)$$

to get $u_{-1} = u_1$. Hence the differenced form of (8.75) is

$$2(2 + \beta g_0)(u_1 - u_0) - \Delta^2 \beta f_0 u_0 = \Delta^2 \beta f_0. \qquad (8.77)$$

If we order the equations in the same way as the grid points, then the rows of **A** can be read off from each equation. For instance, the first row is given by

$$A_{11} = -2(2 + \beta g_0) - \Delta^2 \beta f_0 \qquad (8.78)$$
$$A_{12} = 2(2 + \beta g_0) u_1 \qquad (8.79)$$
$$A_{1i} = 0 \quad \text{for} \quad 3 \leq i \leq N. \qquad (8.80)$$

Continuing in this way we see that **A** has a special form: only the main diagonal and the ones immediately above it and below it have nonzero entries. Matrices of this form, which consist primarily of zero entries, are known as *sparse matrices* and **A** is in a special class of these called *tridiagonal* matrices. Since they arise so often in numerical computations, algorithms for inverting sparse matrices are very well developed. The interested reader is referred to [69, 159, 165] for a variety of discussions of different depths. Note that the tridiagonal form of **A** arose from two convenient choices: the second-order differencing (8.68) and the choice to order the rows of the matrix in the same way as the grid points r_i.

8.5.3 Solutions

In what follows we set $r_0 = 0.75$. Figure 8.11 shows the computed solutions for various values of β along with the central deflection as a function of β. Here we have plotted the numerically computed solution along with the outer solution given by (8.30). We again see that the membrane is never pulled down to the $z = 0$ plane, no matter how large β becomes. We also see that a large central portion of the membrane is pulled down very close to the $z = 0$ plane. This is in contrast to the solutions we obtained for the previous membrane model in which $r_0 \ll 1$. The deflection of the membrane in that case (Figure 8.9) happens primarily near $r = 0$. For a device like a pump this is not a very desirable property since the effectiveness of the pump increases as the volume of the chamber below the membrane is reduced. Hence it is rather a lucky break that the membrane is pulled down over such a large portion of its surface area in the small-aspect ratio model. But why does this happen?

Once the solution is computed, we can use the data and the finite difference approximation (8.68) to compute the dimensionless pressure term P given in (8.52). This is plotted in Figure 8.12 for the solutions in Figure 8.11. Now we see why the membrane is pulled down over such a large fraction of its area: the pressure increases drastically as we go outward from the center. The ratio of the pressure at the edge to the central pressure as a function of β is shown in Figure 8.12. Note the logarithmic scaling of the y-axis in this figure: for $\beta = 0.5$ the ratio is about 420. The central pressure actually reaches a maximum at $\beta \approx 0.15$ and then decreases for increasing β, while the pressure at the edge keeps increasing with β.

A moment's reflection will be enough to see why this has to be the case. The pressure on the membrane is essentially caused by the gradient of the magnetic field. Both the field and its gradient increase as we get closer to the coils so the pressure increases as we move from the center outward. This causes the membrane to experience a much stronger pressure in the region near the coil than in the center.

8.5.4 How Close Did We Get?

The results presented in the previous section all contain a certain amount of error. It would be reassuring if we could in some way verify that this error is small, something like "error $\ll |u_i|$" for every value of i. Note that we certainly cannot conclude that since $\Delta \approx 10^{-2}$ and the error scales as Δ^2, then the error is approximately 10^{-4}. The leading term of the truncation error for the difference approximations (8.68) are proportional to the third and fourth derivatives of u. Now for the solutions shown above, it certainly looks like the third and fourth derivatives of u are well behaved, but we still don't know whether the total error is small or not for the grid on which the solution was computed ($N = 100$).

To get a handle on this, consider the value of u at the origin, $u_0 = u(0)$,

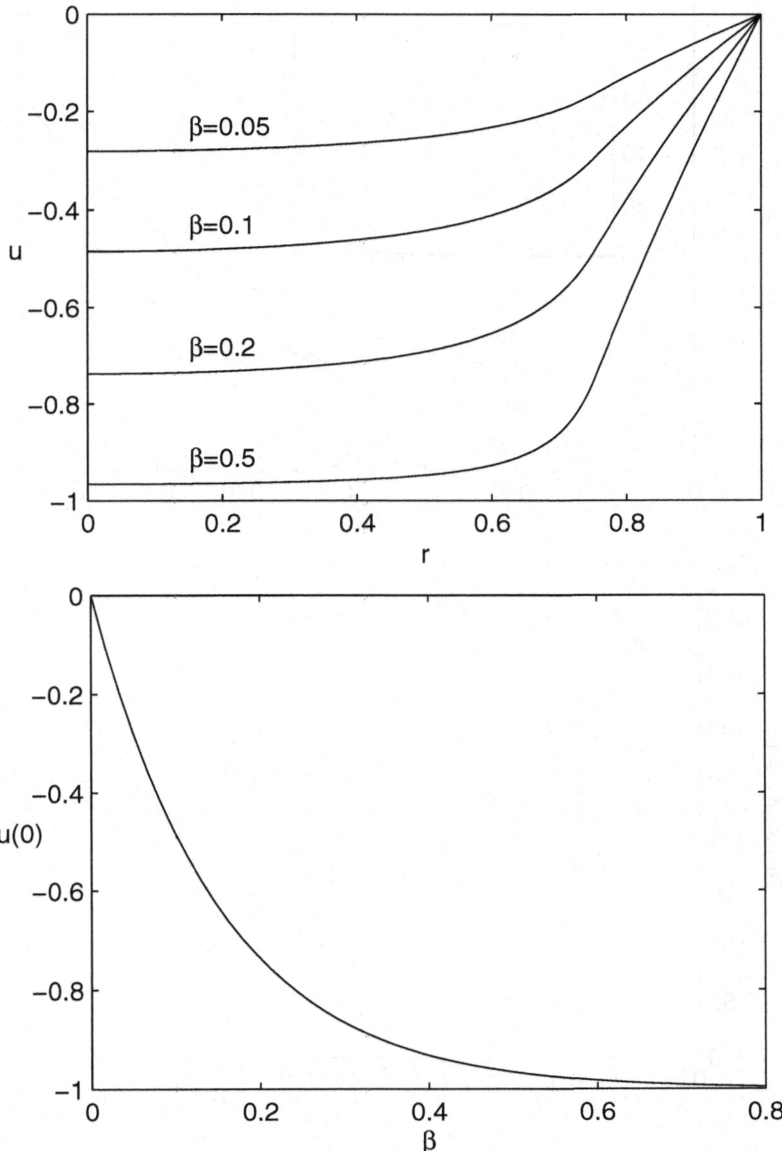

FIGURE 8.11: Numerically computed solutions for the deflection of the small aspect ratio micropump model. In the calculations in this section we have used $N = 100$.

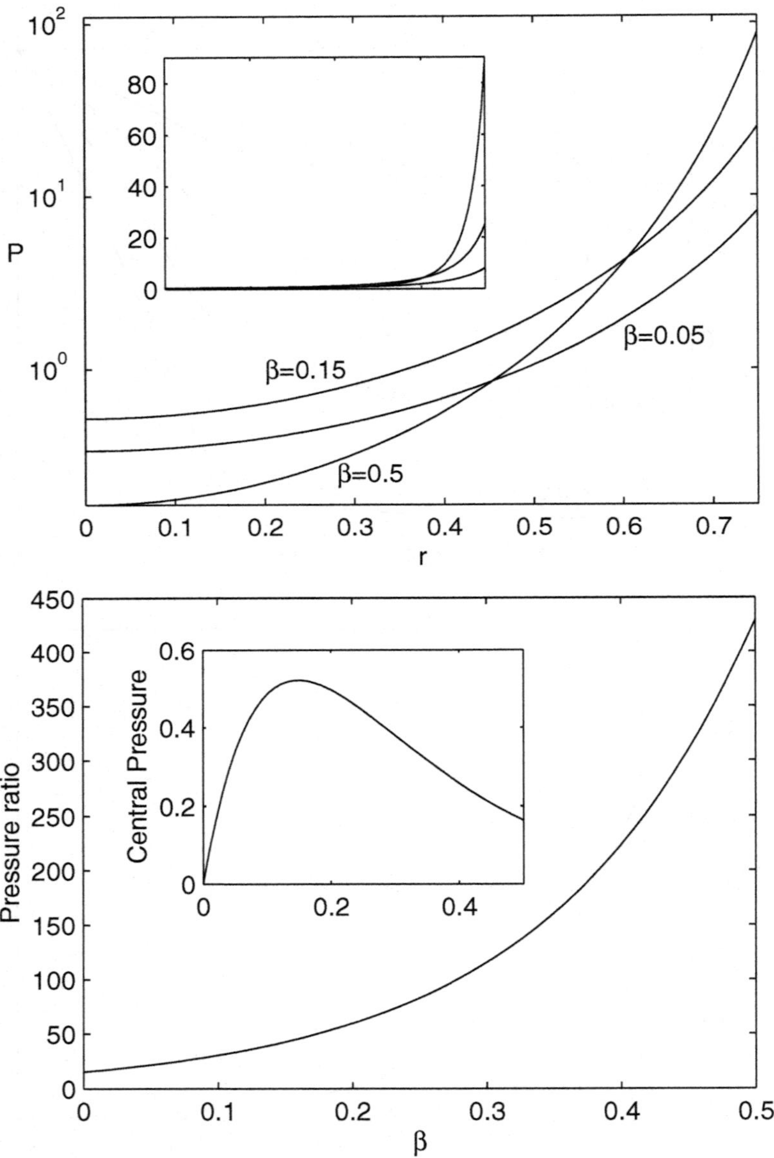

FIGURE 8.12: Pressure on the membrane. The top figure is a semilog plot of the pressure as a function of r for $\beta = 0.05, 0.15, 0.5$. The insert shows the corresponding linear plot. The bottom figure shows the ratio of the edge pressure to the central pressure for $0 \leq \beta \leq 0.5$, while the insert shows the central pressure as a function of β. Note the peak at about $\beta = 0.15$.

and suppose that u_0^* is the exact solution for $u(0)$ for a given value of r_0 and β. Assume that u_0 is related to u_0^* by

$$u_0 = u_0^* + \kappa\Delta^p, \tag{8.81}$$

where for the moment we pretend we don't know the value of p. Now we perform the same calculation but at a smaller grid spacing, $f\Delta$, with $0 < f < 1$. This will give a different result:

$$u_0' = u_0^* + \kappa f^p \Delta^p. \tag{8.82}$$

Consider a third measurement at grid spacing $f^2\Delta$. We have

$$u_0'' = u_0^* + \kappa f^{2p} \Delta^p. \tag{8.83}$$

We may solve the system consisting of (8.81), (8.82), and (8.83) for p,

$$p = \frac{1}{\log f} \log\left(\frac{u_0' - u_0''}{u_0 - u_0'}\right) \tag{8.84}$$

as well is u_0^* itself[10]

$$u_0^* = \frac{u_0 u_0'' - (u_0')^2}{u_0 - 2u_0' + u_0''}. \tag{8.85}$$

Of course, this u_0^* cannot be the exact solution itself since we have ignored many terms in the approximation (8.81). Instead it is simply another approximation to the exact solution, and it can be shown that this extrapolated approximation converges at rate Δ^{p+1}. This is technique is known as *Richardson extrapolation.*

How do we go about computing u_0^*? Notice that f cannot be chosen arbitrarily but instead must be such that $M = N/f$ is an integer. Of course, $f = 1/2$ satisfies this, but we would rather not have to double the size of the grid in order get a handle on the error. Instead we choose

$$f = \frac{\sqrt{N}}{1 + \sqrt{N}}, \tag{8.86}$$

so that $M = N + \sqrt{N}$. As long as N is a perfect square, then M will be an integer. Notice that $N/f^2 = N + 2\sqrt{N} + 1$ is also an integer. The restriction that N be a perfect square is not burdensome since a moment's reflection will verify that are more than enough of them (The reader may care to quantify this statement for him or herself.) The solution, of course, can be computed for any value of $N > 2$.

Hence the procedure is the following: for a given value of N, we compute u_0 on a grid of size N, u_0' on a grid of size $N + \sqrt{N}$, and u_0'' on a grid of size $N+2\sqrt{N}+1$. From these we may compute u_0^* and p as per the above. Table 8.1 shows the value of u_0, u_0^*, and p, for various values of N with f computed

TABLE 8.1: Convergence of the numerically computed solution

N	u_0	u_0^*	p	f	$\|u\|$
9	-0.984074	-0.965588052	2.19985	0.750	0.729733
25	-0.967433	-0.965167899	2.03068	0.833	0.707901
49	-0.965744	-0.965167899	2.00866	0.875	0.705609
100	-0.965297	-0.965156321	2.00222	0.909	0.704995
225	-0.965184	-0.965156260	2.00047	0.938	0.704840
441	-0.965163	-0.965156258	2.00006	0.955	0.704811

by (8.86). We see that for $N = 100$, the central deflection is accurate to three or four decimal places. Note also that the Richardson extrapolated value is not only more accurate (six decimal places) than the computed value of u_0, but it is also converging faster to the exact solution.

As a final note, the reader may wonder why we choose to focus solely on the convergence of the central deflection; why not choose another point of the grid? The answer is practical: when the grid size is changed, the position of most of the points r_i changes; for an arbitrary size grid there are only two fixed points, which of course are the end points, $r = 0, r_0$. Another more serious issue is that we have not looked at the convergence of the solution as a whole. In other words, we have only confirmed that $u(0)$ converges at rate $p = 2$, the convergence and accuracy of the solution at other values of r is still unknown. To address both of these issues, it is only necessary to define a scalar function of the solution, a norm in other words, and look at its convergence. For this problem we could define

$$\|u\| = -\int_0^{r_0} u dr, \tag{8.87}$$

which is always positive since $u \leq 0$ (at least for $\beta > 0$!). Table 8.1 shows that $\|u\|$ has also converged to about three decimal places at $N = 100$.

8.6 Chapter Highlights

- The pull-in phenomenon, which is ubiquitous for electrostatic actuators, is not necessarily present in magnetic systems.
- Because of the ability to distribute magnetic forces in space via coils and permanent magnets, the bifurcation diagrams of magnetic systems can be easily tailored.
- The magnetic pressure distribution on the membrane of a magnetic pump can be highly variable.

- An example of the finite difference method was used to solve a linear ordinary differential equation. The concepts of convergence, accuracy, and Richardson extrapolation were introduced.

8.7 Exercises

Section 8.2.3

1. The scaling here differs from that in [158]. Look up [158] and explain the difference.

Section 8.3.1

2. Compute the stability of the two equilibrium states of the rigid wire mass-spring model (8.9).

3. Replace the bottom wire in this model with two wires spaced a distance $2a$ apart and each carrying current $I/2$. Assume that the top wire is restricted to the plane perpendicular to and equidistant from them. Compute the equilibrium states and bifurcation diagram of this model. What happens to the pull-in current? Now add two more wires parallel to the first two, a distance $2b$ apart with $b > a$, and carrying current $J/2$ with J of the same sign as I. Come up with values for b and J that add more stationary states to the model. How many more stationary states do you get? Hypothesize about the effect of adding more pairs of wires with appropriate spacing and current on the bifurcation diagram. Come up with a pair of series $\{b_i\}$ and $\{J_i\}$ such that in the limit $i \to \infty$ there exists an arbitrarily large number of stationary states. What is the physical size of your device in the limit $i \to \infty$? If it's not finite, can you modify the series so that it is?

4. Consider the model in the previous problem but with only the first two pairs of bottom wires. For a conveniently chosen value of b, relax the restriction that $J > 0$ and examine the effect of J on the bifurcation diagram. If we consider the current J as a "control" parameter, what do we get to control? Compute as many things as you think appropriate.

5. We can create a model in which the magnetic force depends on the displacement of the wire in the following way. Replace the wires with flexible, conducting strings under tension T and carrying parallel currents I. As usual, let the strings have length L and have separation d when $I = 0$ and let the endpoints of the strings be fixed. As the current is increased, the strings will deflect toward each other, so in this model the deflection and the magnetic field are coupled in a different

way than in the rigid wire model. Given that the equation governing the deflection of a string with a load (force per unit length), F is

$$\frac{d^2u}{dx^2} = \frac{F}{T},$$

where x is a coordinate along the length of one of the strings and u is the deflection of the string perpendicular to the x direction, show that a reasonable model is

$$\frac{d^2u}{dx^2} = \frac{\beta}{d + 2u}.$$

What is the expression for β? List the assumptions you had to make in order to arrive at this model. Does this model have an analytic solution? If not, can you compute the bifurcation diagram anyway?

Section 8.3.2

6. Why is it important in this model that the current be restricted to the the wires making up the sheet? If this condition is dropped, and we consider a model consisting simply of two conducting plates carrying current, what changes?

Section 8.3.3

7. Show that (8.19) has either one or three solutions for $0 < z \leq 1$. For each value of ϵ, find the number of solutions as a function of β.

8. Compute the stability of the stationary states for all possible values of β and ϵ.

9. In this model we made the assumption that the location of the dipole is restricted to the z axis. Using the expression (8.33) for the magnetic field of the coil, determine whether or not these states are stable with respect to displacements away from the z axis.

10. Consider the same model but with a square coil instead of a circular one. Using the result of Problem 28 in Chapter 2, find the equilibrium equation for this model and put it in dimensionless form. Using any means necessary, compute the bifurcation diagram. Is it the same as in the circular loop model?

Section 8.5

11. Explain why r_0 must be strictly less than one in the small aspect ratio limit. Is the value $r_0 = 0.75$ used in the numerical computations reasonable?

12. Compute the tangential force on the membrane due to the coil in the small aspect ratio limit. What is the magnitude of this term compared to the normal component of the force? Phrase your answer in terms of a function of ϵ. Is it reasonable to neglect this term? If your answer is no, how would you modify the model to include it?

13. Even though (8.60) is linear, the existence and uniqueness of solutions is not a trivial issue. To illustrate this, consider the system consisting of a flexible string of length L, under tension T, which has been given a linear electric charge density $\sigma = Q/L$. Let the string lie on the x axis and suppose it is exposed to an external electric field E of the form

$$\mathbf{E} = E_0 \frac{y}{L} \hat{\mathbf{y}},$$

where E_0 is constant and $\hat{\mathbf{y}}$ is a unit vector in the y direction (perpendicular to the string). Under these conditions show that the equation governing small deflections of the string is

$$\frac{d^2 y}{dx^2} = -\frac{\sigma E_0 y}{TL},$$

with $y(0) = y(L) = 0$. For what values of E_0 does this equation have a unique solution? In particular, what happens when $E_0 = \pi^2 T/Q$?

14. Using the finite difference method, compute solutions to the model described in Problem 5. Note that since the differential equation is nonlinear, the set of algebraic equations you will obtain from the method will also be nonlinear. To solve them, use the *Newton-Raphson* algorithm described in [159]. Plot the bifurcation diagram for this model using the computed solution.

8.8 Notes

1. It should be noted that almost anything will exhibit some diamagnetism if the field strength is large enough. Indeed, magnetic levitation using this effect has been demonstrated on a wide variety of materials not normally considered magnetic, such as hazelnuts, tulips, strawberries, and live frogs [21]. The frogs were apparently unharmed by the 20-Tesla field.

2. Recall that most magnetic materials fall into two categories: "soft" materials, which are magnetized by an external field but lose much of their magnetization once the field is removed, and "hard" materials, which retain their

field after the external field is removed. Soft materials have slender hysteresis curves while hard ones have wide curves.

3. For example, microgear systems [181] of diameter 50μ can run at 10^5 revolutions per minute, meaning that the average velocities are on the order of a meter per second, or a million microns per second.

4. The reciprocal effect, whereby a magnetic field is produced by applying stress to a material, is know as the *Villari effect.*

5. Note that the strains referred to here are "saturation" values, that is, the maximum strain produced by a very large field. There apparently is no general definition of a "magnetostricion coefficient" since the small field behavior of the strain is highly material dependent.

6. As in previous chapters, we use notation whereby the del operator, ∇, in this equation remains unprimed even though it is with respect to primed (i.e., dimensional) variables. The context of each particular equation will make it clear which ∇ is intended.

7. Tension is one of those quantities whose units depend on the dimension of the space it acts in. The units are force $\times$ length^{1-n}, where n is the dimension of the space. Hence in three dimensions, tension is the same as pressure. The surface tension of a liquid interface can properly be thought of as a two-dimensional pressure.

8. In one dimension, discretization of the domain and the coordinate system are synonymous. In higher dimensions this is not the case and the fact that finite differencing depends on coordinate discretization rather than geometrical discretization (as in the finite element method) causes a number of complications.

9. Here the word "order" is used in a different sense than in "order of magnitude." In the latter we are referring to the approximate size of a quantity while in the former we are talking about how it changes with respect to something else. The former is slope, the latter is a size.

10. The reader may recognize this formula as the *Aitken* δ^2 *method* for the acceleration of a convergent series.

Chapter 9

Microfluidics

Meanwhile the appropriate and legitimate postulate of the physicist, in approaching the physical problems of the living body, is that with these physical phenomena no alien influence interferes. But the postulate, though it is certainly legitimate, and though it is the proper and necessary prelude to scientific enquiry, may some day be proven to be untrue; and its disproof will not be to the physicist's confusion, but will come as his reward.

D'Arcy Thompson

9.1 Introduction

Despite its youth, microfluidics is a large and rapidly growing field. So much so that almost all of the material amply merits the traditional "beyond the scope of this book" salute. Rather than trying to give a representative overview of the field, as was done in previous chapters, we will simply attempt to whet the reader's appetite.[1]

Microfluidics has been at the forefront of miniaturization technology since the beginning some 30 years ago. The field is mature enough to have had a number of commercial successes, two of the most prominent being the manifold intake pressure sensor (automotive industry), which was commercially available in 1979, and the ink jet printer head, first used in Hewlett-Packard printers in 1984. In the intervening two decades researchers have worked on a plethora of devices including microvalves, micropumps, filters, mixers, separators, reactors, microcooling systems, microturbines for power generation, and sensors of all types. These have been used in miniaturized chemical and bioanalytical tools (lab-on-a-chip systems), diagnostic microchips, miniaturized bioassays, microchannel cellular component separators, DNA amplifiers, microelectrochromatography, and many other systems. The effects used in these devices run the gamut from electrostatic, magnetic, piezoelectric, thermal, bubble, electrokinetic, and shape memory alloys. The field of micropumps alone spans many types of devices including peristaltic, rotary, membrane, and electrokinetic devices. Not surprisingly, the biomedical and pharmaceutical industries have supplied the primary motivation for the rapid explosion of microfluidic technology.

This chapter follows the format of the preceding ones. In Section 9.2 we give a short description of a small number of microfluidic devices. This is followed by a section expanding on the scaling results of Section 3.2.1, which is intended to round out our intuition about microfluidics. This is followed by the modeling and analysis of an important microfluidic effect, squeeze film damping.

9.2 Microfluidic Devices

Because of the wide range of microfluidics, the devices in this section are by no means representative of the field as a whole. Rather, we have chosen several "classic" devices (the fluistor, the ink jet printer head, and the pressure sensor), a discussion of electrokinetics, and two examples from nanotechnology: nanolithography and nanojets.

9.2.1 The Thermopneumatic "Fluistor" Valve

Figure 9.1 shows a sketch of one of the first commericially successful active micromachined valves, the thermopneumatically actuated "fluistor" (see Kovacs [108]). The valve sketched in the figure is normally open. Actuation is by a heating element that causes the thermal expansion of a fluid trapped in a cavity. The expansion in turn creates a deflection of a silicon membrane against the valve seat, which closes it. Valves of this type typically require quite a bit of power, on the order of hundreds or thousands of milliwatts to activate and the switching times are typically hundreds of milliseconds. Several variations of the original design have been investigated, including the differential flow rate valve [77] and a version utilizing silicone rubber, which has a very low Young's modulus (1 MPa), and a correspondingly larger maximum deflection (860 microns) [213].

9.2.2 The Ink Jet Printer Head

Ink jet printers are a good example of microfluidic technology which has reached the "household name" status. These printers operate by expelling single droplets of ink through very small nozzles. Following Kovacs [108], the operating principle is shown in Figure 9.2. A joule heater rapidly heats a volume of ink, which vaporizes and forms a bubble. The bubble then acts as a piston, forcing the volume of ink above it out of the nozzle at speeds up to 10 m/s. The vapor bubble formation time is about one microsecond, and its lifetime is about 20 microseconds. When the heater is turned off, the bubble

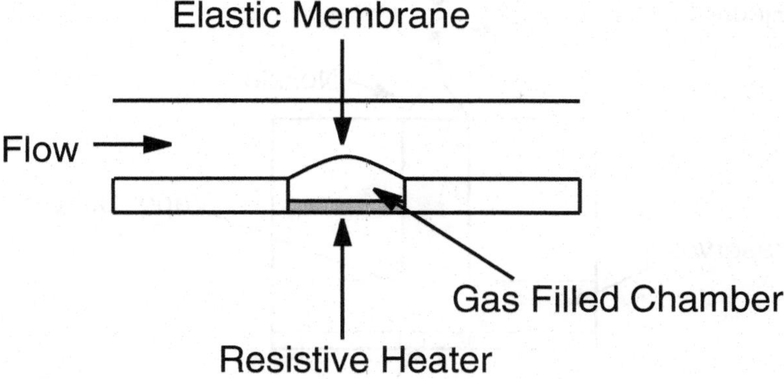

FIGURE 9.1: Sketch of a simple thermopneumatically actuated valve.

quickly cools and collapses, drawing more ink into the channel.

9.2.3 Membrane Pumps

A membrane pump is any device that use the deflection of a membrane to change the volume of a cavity, which causes an amount of fluid to be either expelled or pulled into the cavity. Membrane pumps are not widely used in macroscale engineering, but they are common in biological organisms, the human heart being perhaps the most familiar example. While the heart uses layers of muscle to control the volume of its cavities, most human-engineered membrane micropumps have used a wide variety of other actuating mechanisms. Examples include the electrostatically actuated pump of Zengerle et. al. [218], the magnetic pump of Zhang et. al. [219], and the thermally driven martensitic phase change pump of Benard et. al. [9]. Clearly there are many other alternatives: the thermopneumatic valve discussed in the previous section could double as a pump and one could easily design a thermal bimorph pump. In fact, most of the actuation mechanisms discussed in this book could be used to drive a membrane pump of one sort or another. Modeling of a simple magnetically actuated membrane pump is the central theme of Chapter 8.

9.2.4 The MAP Sensor

Mass-produced microsystems have been in use in the automotive industry since 1979 in the form of the manifold absolute pressure (MAP) sensor. MAP sensors measure the partial vacuum in an engine's intake manifold and reports the measurements to the engine control unit so that the amount of fuel

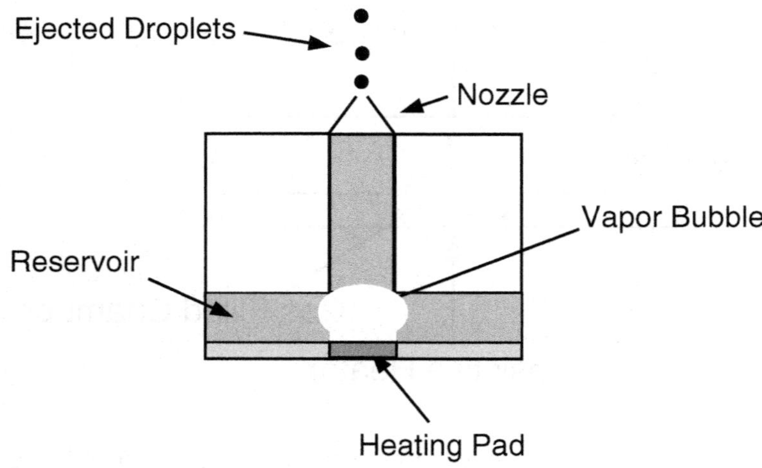

FIGURE 9.2: An ink jet printer head.

required for each engine cylinder can be computed. Descriptions of the MAP sensor and other piezoresistive pressure sensors can be found in many places, including [46] and [175].

MAP sensors typically have a flexible diaphragm that deforms in the presence of a pressure difference. They are called absolute pressure senors since they maintain a vacuum (zero pressure) on one side. The deformation is converted to an electrical signal appearing at the sensor output, in many cases the signal is generated by a series of piezoresistors patterned across the surface of the membrane. A change in ambient pressure deforms the membrane, resulting in a change in resistance of the piezoresistors.

9.2.5 Electrokinetic Flow

Membrane pumps move fluid by creating pressure differences. However, as we have seen, fluids require very large pressure drops to move through small channels (imagine if your heart were to have to pump honey through your capillaries). Membrane pumps typically do not produce very large pressures so that they are most effective if used in a distributed pumping system. However, many fluids of interest have electrical properties that can be exploited to devise new transportation mechanisms. The propulsion of ions or fluids using external electric fields is known as *electrokinetic flow.* This usually takes two forms: *electrophoresis*, which acts on a charged fluid, and *electroosmosis*, which acts on a neutral fluid. Overviews of electrokinetic flow can be found in a number of places, including [108, 128, 175].

In electrophoresis, an electric field is created along the length of a channel containing a fluid consisting in part of positive or negative ions. Each ion attracts a layer of polarized but neutral molecules around it and drags them along as it's moved by the field. If the ionic number density is large enough this will be enough to move the entire fluid body through the length of the channel. Note that conducting fluids are usually not candidates for this type of flow mechanism since they will short circuit the potential that creates the field along the length of the channel. Unfortunately this leaves out many biological fluids. Electrophoresis is often used to separate out ions of opposite charge in otherwise neutral fluids.

Elecroosmosis acts on a neutral fluid in which there are oppositely charged groups of ions and does not require an overall electric field to permeate the fluid. Suppose the channel is made of a conducting material that has been given a positive surface charge density that is not necessarily uniform. Any negative ions in the fluid in the channel will be attracted to the walls of the channel, forming a layer called the *electrochemical double layer*. The layer normally consists of two parts: a rigid part, of only about one ionic diameter in extent, which is more or less attached to the channel wall, and a diffuse part, 1 to 10nm thick, which extends out into the liquid. In the remaining volume of liquid, any polarizable molecules will arrange themselves in a series of dipole layers which can be 10 to 1000 microns in thickness, and are loosely attached to the diffuse layer. If the charge distribution on the channel wall is then given a time dependence, so that waves of charge move along the channel wall, the rigid layer will be dragged along with it, and so in turn will the fluid layers above it. Note that (by Gauss's law) that even though the channel is charged, the interior will in general not have a net electric field. This dragging effect is strong enough that in small channels the flow rate is independent of the width and geometry of the channel.

9.2.6 Nanolithography

Lithographic fabrication of nanostructures is limited by the wavelength of ultraviolet light. Most photolithographic techniques developed by the microelectronics industry rely upon UV light that limits the achievable feature size to approximately 250nm. While lithography has been demonstrated with high-energy light of shorter wavelength, such as x-rays, the challenges associated with producing masks and optics for these systems have not been overcome. As a consequence researchers are actively exploring alternative technologies to allow creation and reproduction of structures on scales smaller than 250nm. A recent review of these efforts can be found in [208].

One such effort is the "nanoscale photocopy machine" developed by Schaffer, Thurn-Albrecht, Russell and Steiner [125, 171, 172]. In this system a thin polymer film of thickness h is spin-coated from solution onto a silicon wafer. A second silicon wafer is then suspended a distance $d > h$ above this wafer. The entire assembly is heated in an oven to a temperature greater than the glass transition temperature of the polymer. This allows the polymer to flow as a liquid. An electric potential difference is then applied between the two wafers. The surface of the polymer film is unstable in this configuration and a periodic disturbance begins to grow. Figure 9.3 shows the setup. As time

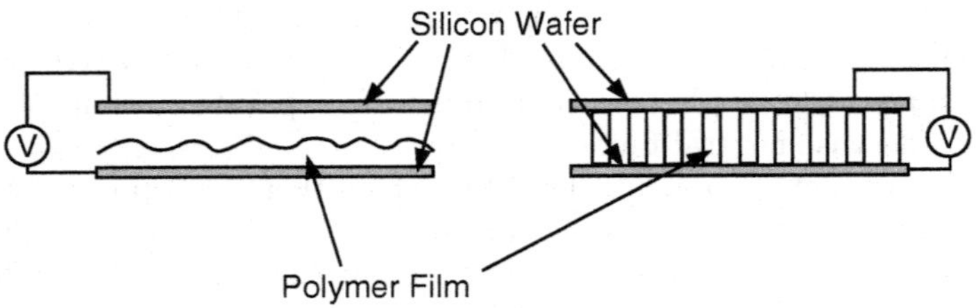

FIGURE 9.3: The basic nanoscale photocopy machine. The figure on the left shows the onset of the instability. The figure on the right shows the developed instability

.

goes on, the disturbance grows, eventually causing the columnar structure shown in Figure 9.3. The scale of this structure depends on d and h. Schaffer et. al., report the formation of patterns with lateral dimensions of 140nm, significantly better than UV photolithographic techniques. Further, the instability can be controlled by etching the surface of the upper silicon wafer. In Figure 9.4 we show an upper wafer with a periodic pattern. When this

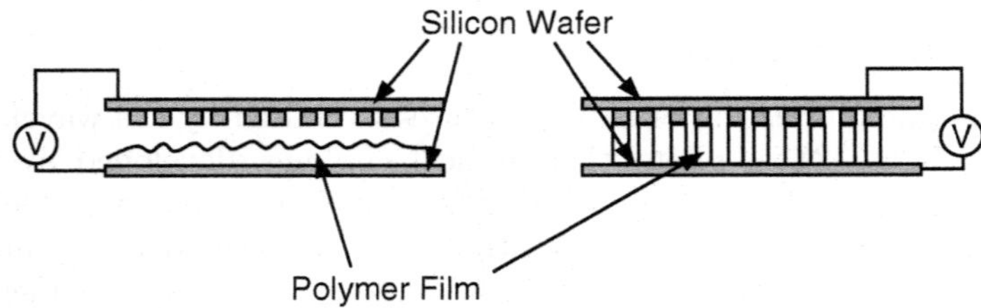

FIGURE 9.4: The nanoscale pattern replicator. The figure on the left shows the onset of the instability with a pattern dictated by the topography of the upper electrode. The figure on the right shows the developed instability.

wafer is used in the polymer-silicon system described above, the instability in the polymer film develops first at the locations where the wafers are closest. As time progresses, the disturbance grows in the induced pattern, resulting in a columnar structure dictated by the topography of the upper wafer. Effectively, this allows replication of the topography of the upper silicon wafer, a nanoscale Photocopy machine.

9.2.7 Nano-Jets

Molecular dynamics simulations performed by Moseler and Landman [135] indicate that the formation of nanoscale liquid jets is possible. Their simulations of the pressurized injection of propane into a gold nozzle with a nanoscale orifice revealed behavior both similar to and different from macroscopic fluid jets. In their initial simulations, with an orifice only a few tens of molecular diameters wide, they discovered that the propane molecules tended to stick to the orifice, quickly creating a nanoscale cork and shutting off further flow. When heating of the nozzle was introduced into the model, the cork effect was destroyed. A simulated coating of the nozzle surface also prevented cork formation. After establishing that jet formation was possible, Moseler and Landman turned their attention to the dynamics of the nanoscale jet. In the macroworld the Rayleigh instability leads to undulations and breakup of liquid jets. In nanojets a similar phenomenon was observed. However, the breakup occurs in about one half the distance predicted by the macroscopic hydrodynamic theory. Moseler and Landman attributed the difference to the effect of thermal fluctuations. This effect, while negligible in the macroworld, is appreciable for nanojets. The stochastic lubrication equation derived by Moseler and Landman, which included the effect of thermal fluctuations, is a beautiful example of the limits and possible extensions of continuum theory. While demonstrating that a continuum description of nanojets is indeed possible, their study shows that unexpected phenomena, such as thermal fluctuations, need to be accounted for in a modified continuum theory.

9.3 More Fluidic Scaling

We saw in Chapter 3 that on a small enough length scale we expect fluid flow to be laminar and that the motion of particles through fluids is dominated by viscosity. This picture was obtained by simple scaling arguments using standard fluidic examples. In this section we'll use more of those types of calculations in order to round out our picture of microfluidics.

9.3.1 Liquids and Gases

The Navier-Stokes equations described in Section 2.6 are equally valid for both liquids and gases and the two are often treated on an equal footing, with the exception that gases are generally compressible while liquids are not. However, as length scales decrease several crucial differences emerge, which require that gases and liquids be treated differently. The first of these is the existence of interfaces. Liquids support interfaces, that is, there can exist a sharp (on the order of Angstroms) boundary between different liquid species which is dynamically stable. Such interfaces have their own sets of equations, as described in Sections 2.6 and 3.2, and a corresponding set of characteristic phenomena, e.g., the capillary effect discussed in Section 3.2. Gases, on the other hand, do not support interfaces; they mix readily while liquids do not. On small length scales, gases in contact will mix very rapidly by diffusion alone.

The second important difference becomes apparent when we examine where the assumptions behind the Navier-Stokes equations begin to break down. For gases the crucial number is called the *Knudsen number* and is defined as

$$Kn = \frac{\lambda}{L}, \tag{9.1}$$

where λ is the mean free path between collisions of the molecules of the gas and L is again a characteristic length of the system. A small Knudsen number ($Kn < 0.01$) indicates that the equations of continuum theory ought to be a good approximation, while a Knudsen number approaching unity means that the gas must be treated as a collection of individual particles rather than a continuum.[2] A good example is air, which consists primarily of the molecules N_2 and O_2, which are each a few Angstroms (say 0.3nm) in diameter. At standard temperature and pressure, the average intermolecular spacing is on the order of 3nm while the mean free path is about 100nm or about 30 times greater. Hence λ can be as great as 0.1 microns, which is considerable given that even with current micromachining technology it is easily possible to make channels a micron or smaller in width.

On the other hand, the predictions of the continuum Navier-Stokes equations seem to hold well in liquid flow even in the smallest microchannels in which measurements have been made, about one micron. It is speculated that the limit might be as small as 100Å [128].

9.3.2 The Boundary Layer

Fluids flowing through microchannels are constrained by walls. On the surface of the wall, a commonly used boundary condition for the velocity of the fluid is the *no-slip* condition discussed in Section 2.6. Physically, the reasoning behind this is that fluid molecules adjacent to the walls will stick to them and not move. Since the bulk of the fluid is moving, there will be a transition region between the surface of the wall and the bulk of the fluid where the velocity increases from zero to the bulk velocity. This transition region is called the *boundary layer* and it plays a very important role in fluid dynamics.

It can be shown [46] that if L is a characteristic length of the system (the diameter of a pipe for instance) and the flow is laminar, then the magnitude of the boundary layer is approximately

$$\delta = \frac{L}{\sqrt{Re}}. \tag{9.2}$$

Hence in the limit of a very small Reynold's number, the boundary layer thickness becomes larger than the characteristic size of the system. Since many microsystems operate in a regime in which $Re \ll 1$, it is a common situation that the entire fluidic system is a boundary layer. This is not very surprising, since the boundary layer is a surface effect and all we are doing is again demonstrating the dominance of surface over body effects in the limit of small length scales.

9.3.3 Flow Rates and Pressure

Consider a pipe of length l and circular cross-section with radius r. For the steady flow of an incompressible fluid, the Navier-Stokes equations can be solved exactly and the result is called *circular Poiseuille flow*. The volumetric flow rate (volume of fluid passing through a given cross-section perpendicular to the pipe per unit time) is given by the *Hagen-Poiseuille law*,

$$Q = \frac{\pi r^4 \Delta P}{8\nu l}, \tag{9.3}$$

where ΔP is the pressure drop over the length of the pipe and ν is the dynamic viscosity of the liquid. Note that the no-slip boundary condition has been used here. The average fluid velocity is

$$v = \frac{Q}{\pi r^2} = \frac{r^2 \Delta P}{8\nu l}; \tag{9.4}$$

hence, the pressure drop over the length l is

$$\Delta P = \frac{8\nu v l}{r^2}. \tag{9.5}$$

Now let $r = \epsilon\, l$ and suppose that we scale the velocity v such that

$$v = \alpha r^n, \tag{9.6}$$

where $n = 0$ is the case of constant velocity and $n = 1$ is the case of constant time for a particle of the fluid to traverse the pipe. Then

$$\Delta P = \frac{8\nu\alpha r^{n-1}}{\epsilon}, \quad Re = \frac{\rho\alpha r^{n+1}}{\nu}, \quad Q = \pi\alpha r^{n+2}. \tag{9.7}$$

We see that to maintain a constant velocity of the fluid as the length scale is decreased will require very large pressures. The more sensible constant time of traversal results in a constant pressure drop across the pipe. However this case requires extremely high pressure gradients since

$$\nabla P \approx 8\nu\alpha r^{n-2}. \tag{9.8}$$

Note that in both cases the Reynold's number decreases with decreasing r. If we desire to avoid both large pressures and large pressure gradients, then we must have $r \geq 2$, which means that Q scales as at least the fourth power of r, resulting in a very small flow rate. Hence either large pressures or large pressure gradients are unavoidable in microsystems.

The *fluidic resistance* is defined as the pressure drop over the flow rate. In this case this is independent of the average velocity of the fluid,

$$R = \frac{\Delta P}{Q} = \frac{8\nu l}{\pi r^4} = \frac{8\nu}{\pi \epsilon r^3}, \tag{9.9}$$

so that R always scales as r^{-3}, i.e., small pipes have very high resistance.

9.3.4 Entrance Length

In the example of the circular pipe above, the velocity distribution across the diameter of the pipe is given by

$$v(r) = \frac{\Delta P}{4\nu}\left(R^2 - r^2\right). \tag{9.10}$$

The parabolic profile is typical of laminar flow in pipes and is fundamentally caused by the existence of the boundary layer. When the fluid first enters the pipe, however, the velocity profile will be inherited from the flow outside of the pipe and will not in general be the parabolic profile given in (9.10). Instead, this profile will develop over a characteristic distance called the *entrance length.* For circular Poiseuille flow this can be estimated [175] to be

$$L_e = \beta r Re, \tag{9.11}$$

where $\beta \approx 0.06$. Defining the fractional entrance length as

$$L_f = \frac{L_d}{l} \tag{9.12}$$

gives

$$L_f = \frac{\rho\epsilon\alpha\beta r^{n+1}}{\nu}. \tag{9.13}$$

Hence even with the constant velocity scaling, $n = 0$, we see that L_f scales as r. With the more reasonable constant time scaling, L_f scales as r^2. The conclusion is that entrance lengths are generally negligible in microsystems.

9.3.5 Mixing

When you pour milk in your coffee and stir it, you mix the two fluids by creating turbulence. Even without computing a Reynolds number, this is readily apparent from the lovely vortices in the turbulent wake behind the spoon. Now suppose you always like a nice big bowl of peanut butter and honey with your morning coffee. You take a bowl, pour in some peanut butter, then honey, and start stirring. But instead of one or two deft moves of the spoon, to mix these two liquids sufficiently requires quite a bit of elbow grease. This is becauce the flow in the bowl is laminar and no amount of human stirring will get the Reynolds number up to anywhere near 4000. This is what happens when you try to mix two liquids under the conditions of creeping flow: they don't, but rather become more and more twisted up in each other. Eventually the twisting becomes so convoluted and the layers of each fluid so thin that the two simply diffuse into one another, at which point they can be said to be mixed.

Now, just as in the above example, turbulence cannot be used for mixing in microsystems because the Reynold's numbers are far too low. Even though diffusion is fast, the length scales of many microsystems are large enough for diffusion to be ineffective on short enough scales. This is because the diffusion coefficient of liquids can be very small. If we denote this coefficient as D, then the time scale for the liquids to diffuse a distance L into each other is[3]

$$\tau \approx \frac{L^2}{D}. \tag{9.14}$$

For many liquids of interest the diffusion coefficient can be less than $10^{-5}\text{cm}^2\text{s}^{-1}$, which means $T > 10$ seconds for a 100 micron gap.[4] Mixing in microsystems is an active area of current research; the reader is directed to [108, 128, 130, 175] for discussions.

9.4 Modeling Squeeze Film Damping

In this section we focus on modeling a particularly important microfluidic phenomenon, squeeze film damping. This phenomenon occurs in microsystems when two components of the system lie close to one another. If one or both of the components are in motion, the fluid filling the region between them contributes significantly to the drag on their motion. For example, in the parallel plate system shown in Figure 9.5 if the upper plate is set in motion, it will encounter a resistance. When it moves downward it must compress or force the fluid in the gap outward. When it moves upward it creates a pressure drop sucking fluid back into the gap. The fluid flow dissipates energy creating a drag on the plate.

This phenomenon plays an important role in the functioning of many devices. Examples include the GLV of Chapter 7 [74], electrostatic torsion mirrors [201], and resonant sensors [80]. The reader is directed to the literature for many other examples of squeeze film damping in MEMS and NEMS. In addition to MEMS and NEMS, squeeze film effects arise in the study of other systems. Perhaps the most important of these is the study of hard disk drives. The read-write head on a hard disk drive, called an air bearing slider, typically hovers about 50nm above the disk. The behavior of the fluid in the gap between the slider and disk surface affects the motion and stability of the slider. Due to its commercial importance, this system has been studied extensively. References [59, 148, 205] provide a first look at this literature. In this section we will study the system shown in Figure 9.5 and compute the effect of the fluid in the gap on the motion of the mass-spring system.

9.4.1 Lubrication Theory

Throughout this text we have exploited geometry and the size of dimensionless parameters in the simplification of various models. For the isothermal flow of fluids through a thin gap, the Navier-Stokes equations can be similarly simplified. In particular, in the limit of small Reynolds number, the Navier-Stokes equations can be reduced to the one-dimensional compressible Reynolds equation,

$$\frac{\partial(ph)}{\partial t'} + \frac{U}{2}\frac{\partial(ph)}{\partial x'} = \frac{1}{12\mu}\frac{\partial}{\partial x'}\left(ph^3\frac{\partial p}{\partial x'}\right). \tag{9.15}$$

Here p is the pressure in the fluid, U is the velocity of the background flow, h is the height of the fluid layer, and μ is the fluid viscosity. Equation (9.15) is also referred to as the lubrication approximation. The reader is referred to [31] for details of the reduction of the Navier-Stokes equations to the Reynolds equation.

9.4.2 The Model

We consider the system shown in Figure 9.5. We assume the system is one-dimensional. In particular, we assume that the pressure, p, is a function of x' and time only. We assume no background flow of the fluid. This implies that $U = 0$ and the Reynold's equation reduces to

$$\frac{\partial(ph)}{\partial t'} = \frac{1}{12\mu}\frac{\partial}{\partial x'}\left(ph^3\frac{\partial p}{\partial x'}\right). \tag{9.16}$$

Note that h in equation (9.16) is function of time only. Bear in mind that h is the size of the gap between the mass of Figure 9.5 and the stationary plate. At the ends of the gap, we assume that fluid pressure is some known constant and impose

$$p(-L/2, t') = p(L/2, t') = p_A. \tag{9.17}$$

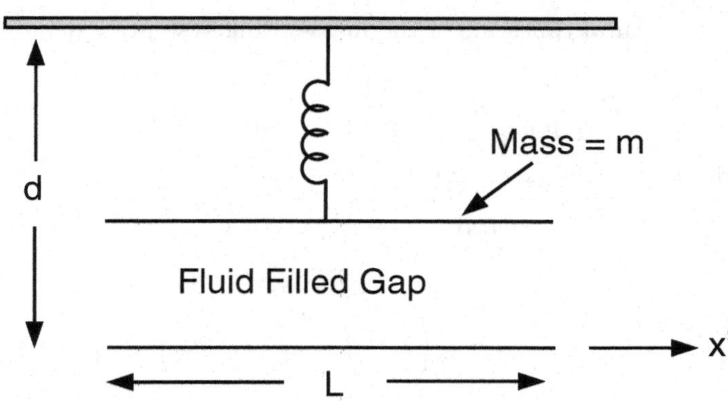

FIGURE 9.5: The mass-spring system for squeeze film damping study.

Next, we must write down an equation governing the motion of the mass-spring system. This equation of motion follows directly from Newton's second law, that is

$$m\frac{d^2u'}{dt'^2} = \sum \text{forces}. \tag{9.18}$$

Here the forces acting on the mass are the spring force and the pressure force exerted by the fluid. Assuming the spring obeys Hooke's law and that the plate has area A, we have

$$m\frac{d^2u'}{dt'^2} + k(u' - l) = Ap_A - w\int_{-L/2}^{L/2} p(x', t')dx', \tag{9.19}$$

where l is the rest length of the spring and w is the width of the plate. Note that h is related to u' via

$$h(t') = d - u'(t') \tag{9.20}$$

and when the system is at rest, the gap between the mass and bottom plate is

$$g = d - l. \tag{9.21}$$

Note that we have not explicitly included a damping term in our mass-spring equation. Any damping in the system will be due to squeeze film damping alone. Equations (9.16)–(9.20) together with appropriate initial conditions represent a closed system for the displacement of the mass, u', and the pressure in the fluid, p.

9.4.3 Scaling

We scale x' with the length of the plates, the pressure with the ambient pressure of the external environment p_A, u' with the rest gap g, h with g, and time with the natural timescale of the mass-spring system by defining

$$x = \frac{x'}{L}, \quad P = \frac{p}{p_A}, \quad u = \frac{u' - l}{g}, \quad s = \frac{h}{g}, \quad t = \sqrt{\frac{k}{m}} t'. \tag{9.22}$$

Inserting into equations (9.16)–(9.20) we obtain

$$\frac{\partial (Ps)}{\partial t} = \gamma \frac{\partial}{\partial x}\left(P s^3 \frac{\partial P}{\partial x} \right) \tag{9.23}$$

$$\frac{d^2 u}{dt^2} + u = \alpha \left(1 - \int_{-1/2}^{1/2} P(x,t) dx \right) \tag{9.24}$$

$$P(-1/2, t) = P(1/2, t) = 1 \tag{9.25}$$

$$s(t) = 1 - u(t). \tag{9.26}$$

The dimensionless parameters α and γ are defined as

$$\alpha = \frac{A p_A}{kg}, \quad \gamma = \frac{p_A g^2}{12 \mu L^2} \sqrt{\frac{m}{k}}. \tag{9.27}$$

The parameter α is simply a ratio of a representative pressure force to a representative spring force. We should think of α as measuring the relative strengths of the spring and pressure forces. The parameter γ measures the relative effects of pressure forces and viscous forces in the system.

9.4.4 A Linear Theory

Our mass-spring lubrication theory model of squeeze film damping, equations (9.23)–(9.26), is a nonlinear partial differential equation coupled to an ordinary differential equation. Note that the coupling is two-way. The displacement of the mass appears in the equation for the pressure through s, while the integrated pressure appears as a force in the equation of motion for the mass. Our first step in analyzing solutions of this system is to compute steady-state solutions by setting time derivatives to zero. Doing so yields

$$u = 0, \quad P = 1, \quad s = 1. \tag{9.28}$$

This is unsurprising. It simply says in the steady-state the pressure in the gap is the same as atmospheric pressure and the mass-spring has come to rest at the equilibrium length of the spring.

For this system, our primary interest is not in the steady state, but in the approach to the steady state. For general initial conditions, this approach is difficult to determine. We can however, perturb about our steady state, linearize equations (9.23)–(9.26), and study the linear system to get some understanding of the decay rate. To do so we set

$$P(x,t) = 1 + \epsilon P_1(x,t) \tag{9.29}$$

$$u(t) = \epsilon v(t) \tag{9.30}$$

$$s(t) = 1 - \epsilon v(t) \tag{9.31}$$

in equations (9.23)–(9.26), let $\epsilon \to 0$, and obtain

$$\frac{\partial P_1}{\partial t} = \gamma \frac{\partial^2 P_1}{\partial x^2} - \frac{dv}{dt} \tag{9.32}$$

$$\frac{d^2 v}{dt^2} + v = -\alpha \int_{-1/2}^{1/2} P_1(x,t)dx \tag{9.33}$$

$$P_1(-1/2, t) = P_1(1/2, t) = 0. \tag{9.34}$$

Equations (9.32)–(9.34) are a pair of linear coupled equations. The equation for the perturbation to the mass position, v, is a linear second-order ordinary differential equation. The equation for the perturbation to the pressure, P_1, is a linear partial differential equation. Note that the equation for P_1 is a diffusion equation for P_1 with v as a source term. This source depends on the velocity of the plate.

A crude first approximation to solving the linear system, equations (9.32)-(9.34), is to make a so-called *quasi-steady* approximation. In this approximation we set the time derivative on P_1 in equation (9.32) to zero so that the time change of the pressure tracks the velocity of the mass. With this approximation, we can solve for P_1 to find

$$P_1(x,t) = \frac{1}{2\gamma} \frac{dv}{dt} \left(\frac{1}{4} - x^2 \right). \tag{9.35}$$

Using this expression in equation (9.33) we obtain

$$\frac{d^2 v}{dt^2} + \frac{\alpha}{12\gamma} \frac{dv}{dt} + v = 0. \tag{9.36}$$

From this we see that the motion of the mass, $v(t)$, is that of a damped harmonic oscillator. The damping has arisen naturally as a consequence of the coupling to the Reynolds equation rather than having been imposed *a priori.* The dimensionless damping constant sits in front of the first derivative of v. We note that

$$\frac{\alpha}{12\gamma} = \frac{AL^2 \mu}{g^3 \sqrt{mk}}. \tag{9.37}$$

From our mass spring analysis of Chapter 5, we know that this quantity is the inverse Q of the system. Consequently, we can solve for the dimensional damping constant, say a, as

$$a = \frac{AL^2 \mu}{g^3}. \tag{9.38}$$

Notice that the size of this damping coefficient increases with viscosity, increases with plate size, and decreases with equilibrium gap size. This is in accord with our intuition.

A more accurate analysis of the linear system, (9.32)–(9.34), can be performed by solving the equation for P_1 via the method of eigenfunction expansion. Seeking a solution of the form,

$$P_1(x,t) = \sum_{n=1}^{\infty} A_n(t) \sin(n\pi(x - 1/2)), \tag{9.39}$$

and using orthogonality of the eigenfunctions yields

$$A_n(t) = A_n(0)e^{-n^2\pi^2\gamma t} + \frac{2}{n\pi}(\cos(n\pi) - 1) \int_0^t \frac{dv}{d\tau} e^{-n^2\pi^2\gamma(t-\tau)} d\tau. \tag{9.40}$$

Choosing an initial condition for P_1 fixes the $A_n(0)$ in the usual way. If $P_1(x,0) = 0$, then $A_n(0) = 0$ for each n, which implies

$$A_n(t) = \frac{2}{n\pi}(\cos(n\pi) - 1) \int_0^t \frac{dv}{d\tau} e^{-n^2\pi^2\gamma(t-\tau)} d\tau. \tag{9.41}$$

Using this solution for P_1 in the equation for v yields

$$\frac{d^2v}{dt^2} + v = -\alpha \sum_{n=1}^{\infty} \frac{2}{n^2\pi^2}(\cos(n\pi) - 1)^2 \int_0^t \frac{dv}{d\tau} e^{-n^2\pi^2\gamma(t-\tau)} d\tau. \tag{9.42}$$

This equation, while complicated, is still linear. Further progress in its solution is possible via the Laplace transform. The reader is directed to the exercises.

9.5 Chapter Highlights

- Although both are described by the Navier-Stokes equations, liquids and gases exhibit markedly different behavior in very small systems. Liquids support interfaces while gases do not and the validity of the continuum equations for gas flows generally breaks down at much larger length scales then it does for liquids.

- Fluid boundary layers will usually be large compared to the dimensions of a microsystem. Hence boundary layer effects will be dominant in such systems.

- Microfluidic systems using pressure (i.e., hydraulics) to transport fluids will have very large pressures and/or pressure gradients. In these systems distributed pumping will be much more effective than centralized pumping.

- Mixing of fluids is problematic, despite the fact that diffusion is fast. This is because of the very low diffusion coefficients of many liquids.

- A model for squeeze film damping, an important effect in many microsystems, is presented.

9.6 Exercises

Section 9.3

1. Consider a closed fluidic system with a single centralized pump (e.g., the human circulatory system). Assume that the pump has a single outlet and a single inlet each of radius r_0, which are connected via a branching system of pipes. Assume that when a pipe of radius r splits that it splits into two pipes each of radius αr. After n branchings, what is the velocity of the fluid in the pipes as a function of the pipe radius and outlet velocity? Relate your answer to the constant time and constant velocity scalings discussed in the text. In the human circulatory system the blood velocity drops from about 30 cm/sec in the largest arteries to about 0.07 cm/sec in the smallest capillaries. Given that a typical capillary is about 10 microns in diameter, or about 1000 times smaller than a large artery, what can you say about α for this system?

Section 9.4

2. Consider squeeze film damping of the system in Figure 9.5 with the background flow velocity, U, not equal to zero. Scale the system. You should encounter a new dimensionless parameter. Give a physical interpretation of this parameter.

3. Again consider the system of Figure 9.5, but remove the spring and allow the mass to fall freely from some initial height. If fluid effects are ignored, compute the time for the mass to hit the bottom plate. Now, include fluid effects and compute the time for the mass to hit the bottom plate. Compare.

4. Assume the plate in the model of this section is composed of silicon with density 2300kg/m^3. Assume the plate is 10μm on a side, 0.01μm thick, and the gap at equilibrium is 0.1μm. Compute the approximate damping constant from the quasi-steady theory.

5. Assume the spring in the model of this section is a microbeam of length 1500μm, width 5μm, and height 10μm. Compute an effective spring constant for this spring. Using the damping coefficient of problem 4 and

assuming the plate dimensions of problem 4 compute an approximate quality factor for this system from the quasi-steady theory.

6. Use the method of Laplace transforms to solve equation (9.42). (Hint: You will need to recall the convolution theorem for Laplace transforms.)

9.7 Notes

1. In our defense we point out that this is only the second bad pun in 297 pages.

2. This is usually broken down into four regimes: the continuum regime in which $Kn < 0.01$, the "slip-flow" regime in which $0.01 < Kn < 0.1$ where the Navier-Stokes equations still hold but the normal no-slip boundary condition has to be modified, the "transition" regime in which $0.1 < Kn < 3$, and finally the "free-molecular regime," for which $Kn > 3$.

3. Diffusion of one particle species into another is of course governed by the heat equation. In this context it is often referred to as *Fick's second law of diffusion*,

$$\frac{\partial C}{\partial t} = D\nabla^2 C, \tag{9.43}$$

where C is the concentration of the diffusing particle and D is the diffusion coefficient. Fick's first law is just the diffusion counterpart of Fourier's law (2.8).

4. For instance, the diffusion coefficient of glucose in water is about $7 \times 10^{-6}\text{cm}^2\text{s}^{-1}$ while that of hemoglobin is about 10 times smaller. The self-diffusion coeffient of water is $2.5 \times 10^{-5}\text{cm}^2\text{s}^{-1}$ at 25°C.

Chapter 10

Beyond Continuum Theory

Now is the psychological time for the creative mind to work out the many, many new uses from the inexhaustible deposits of our fine Southern clay; vegetable dye stuffs; mineral deposits, new and old; various and varied mineral waters; Southern fiber plants; mineral for paper pulp, and many, many other things too numerous to mention in an article of this kind.

George Washington Carver

10.1 Introduction

The premise of this book is that modeling microsystems using continuum theories is not only useful but offers a wealth of new insights about the nature of the microscale world. However, it seems that the reader who has gotten this far and agrees with the above statement must be well practiced in the art of suspension of disbelief. We are balanced on the knife edge of a ridiculous oxymoron: reduce the length scale and learn new and amazing things, reduce it a bit more, and the entire basis of the analysis disappears amid a busy cloud of atoms, molecules, photons, and wave functions.

The purpose of this chapter is to attempt to delineate this length scale boundary as well as we can. To give the prospective microworld explorer yet another tool for her toolbox: this one says when to give up and go get another tool. In Section 10.2 we will attempt to roughly probe the boundaries of the continuum approximation. We follow this by a description of some recent nanotechnology systems that cannot be modeled by the methods used in the rest of this book.

10.2 Limits of Continuum Mechanics

The continuum theories outlined in Chapter 2 rest on two basic types of observations about how the world works:

- Physical laws that are deduced from empirical observations. These include Newton's laws of motion, Maxwell's equations, conservation laws, and so forth.

- Statistical laws about the behavior of systems containing many particles. This includes the law of large numbers and the empirical law of averages. The latter states that if a random experiment is repeated many times, independently and under identical conditions, the fraction of trials that result in a given outcome converges to a limit as the number of trials grows without bound. In other words, averaging things makes sense. Note that this is not a mathematical theorem, but essentially a physical law in the sense of the first item in that it is an observation about the universe.

The validity of these two items in any particular system is independent. The first asks the question: are you using the right physical theory *on a microscopic level* to describe the system? While the second asks: is the system large enough to reliably use averages? We explore each of these assumptions in turn.

10.2.1 The Limit of Newtonian Mechanics

Newtonian mechanics is roughly bounded by special relativity in one limit and quantum mechanics in another. In both cases, Newton's second law of motion fails and the laws governing the dynamics of particles must be replaced.

The relativistic case is simpler. Even in a very high temperature gas, the molecular velocities are only on the order of a few thousand meters per second, which is a small fraction of the speed of light. Hence for the practicing MEMS and NEMS researcher, relativistic mechanics is not a consideration.[1] The sole exception is a system involving free electrons. Electrons in vacuum can be quickly accelerated close to the speed of light, even by fairly modest electric fields. A potential drop of only few thousand volts is enough to accelerate an electron to $0.1c$.

Quantum effects become important when a particle's *De Broglie wavelength*,

$$\lambda = \frac{h}{p}, \tag{10.1}$$

is comparable to a typical length scale in the system. In this formula p is the momentum of a free particle and h is Plank's constant (given in appendix B). For a nonrelativistic gas we have $p = mv$ and typical values of the velocity for room temperature gases are on the order of 500 m/s. This gives $\lambda = 0.2\text{Å}$, which is only a little smaller than the smallest atomic diameter. An electron accelerated across a 100-volt gap has a velocity of about 6×10^6m/s and a De Broglie wavelength of about 12Å, which is the same order of magnitude as a very small microsystem. Overall the length scales are in the ballpark of micro- or nanosystems. In these cases free particles will diffract through structures and matter waves will interfere with each other. Hence if one has a system containing *free* atoms or electrons, then quantum effects are often important.[2]

What about bound atoms? In this case the atoms interact with their neighbors through their electron clouds. Since electrons bound to atoms are intrinsically quantum in nature, this interaction must be as well. However, unlike the electrons themselves, this interaction can very effectively be *modeled* and encapsulated in a Newtonian potential well. A typical example is the *Lennard Jones* potential between certain types of molecules,

$$U = 4\epsilon\left[\left(\frac{\sigma}{r}\right)^6 - \left(\frac{\sigma}{r}\right)^{12}\right], \tag{10.2}$$

where r is a characteristic "radial" coordinate. A good example is the interaction between water molecules, for which $\sigma = 0.317\text{nm}$ and $\epsilon = 0.65\text{kJ/mol}$. Note that this is an empirical law and values for σ and ϵ are usually experimentally determined. The derivation from the equations of nonrelativistic quantum mechanics for any but the simplest atoms is difficult. This type of modeling makes it possible to model molecular dyamics using Newtonian mechanics alone.

There is an additional limit in which quantum effects become very important: ultralow temperatures. In this case, even bound atomic systems will exhibit quantum effects such as Bose-Einstein condensation, quantized thermal and electric conductance, and superfluidity. Superconductivity is another example, but as is by now well known, it is no longer confined to low-temperature systems.

10.2.2 The Limit of Large Numbers

Continuum mechanics describes matter in terms of continuous fields. For instance, in the Navier-Stokes equations the velocity field $\mathbf{v}$ is a function that returns a vector given a point in the domain. If we accept the currently popular view that matter is fundamentally discontinuous, then we must admit that no such field exists.[3] Instead, these fields are only meaningful in the *continuum approximation.*

Following Section 2.2, one way to think of this approximation is the following: the value corresponding to the velocity field at a point is actually an average of the velocities of all the particles in a small volume, $V = \delta^3$, centered around the point. The continuum equations will be a reasonable approximation as long as one condition holds: the number of particles in V is still large when δ is small compared to a typical length scale of the system. In other words, we think of V as being a volume small compared to the domain as a whole, but large compared to a typical microscopic scale volume. The key point here is that the continuum equations will fail to hold when no such volume exists. This condition is inherently vague, but evidently hinges on the words "small" and "large."

Consider an ordinary, macroscopic volume of air at room temperature and pressure with a net 10 cm/s flow; a nice breeze in other words. Under these conditions the number density of air molecules is on the order of 10^{19} per cubic centimeter. Suppose we take "large" to mean that the small volume, V, contains a million particles. Then δ is on the order of a micron. This seems more than reasonable considering our room is several meters across.

One might think we've been overly conservative here. But consider the following facts: The average thermal velocity of an air molecule at room temperature and pressure is about 500 meters per second, which is *5000 times* the flow velocity. In other words, we are averaging out a lot of "noise" to get the relatively small overall flow velocity. Suddenly a million doesn't seem like such a big number. The second worrisome fact is a certain corollary of the law of large numbers in statistics. This law states that the arithmetic mean actually does converge to the expectation value in the limit of very many independent observations. In other words, averaging does what you intuitively expect it should do. However, the corollary states that *rate* of convergence is very slow, roughly

$$\text{error} \propto N^{-1/2}. \tag{10.3}$$

This is unfortunate. Suppose we decide to back off to ten million molecules in our small volume, then (10.3) states that the improvement in the error of the average velocity is a mere factor of three.

But still, ten million particles and $\delta/L \approx 10^{-7}$ seems like a good basis for trusting the continuum approximation. And of course it is, otherwise atmospheric science and aerodynamics wouldn't be the fields they are today.

Shrink the room down to one cubic micron. Now we are justifiably nervous. We have only one million total particles to work with and have yet to even think about defining a suitable δ. Taking this into account and considering the fact that any velocity field we define will have large averaging errors due to (10.3) makes it questionable that the Navier- Stokes equations can be used to describe the flow of gas in this room with any accuracy at all. This is more or less in accord with Section 9.3 where the Knudsen number was used to delineate the regime of validity of the Navier-Stokes equations for gas flows. Following the example in that section, all we need is a channel width of approximately 0.03 microns in order to create a flow of air at standard temperature and pressure with Knudsen number greater than 3, which puts it squarely in the free-molecular regime. In this regime the mean free path of the air molecules is greater than three times the width of the channel, so that the molecules interact with the channel walls much more than they do with each other. For this example the intermolecular spacing is still only 3nm, about ten times smaller than the channel width. Of course, if we lower the pressure or increase the temperature of the air in this example, the Knudsen number increases even further.

The particles of a liquid are perpetually in contact with one another (which accounts for the generally very low compressibility of most liquids), so the mean free path is much smaller, about the size of the particle size itself. In this case the Knudsen number is approximately D/L, where D is a typical particle diameter. Based on the above, we would expect Navier-Stokes flow to hold for approximately $D/L < 0.01$. If we extend this to include the transition region where the slip boundary condition (2.141) holds, then we might expect the Navier-Stokes equations for a liquid to be useful all the way up to $D/L \approx 0.1$. As one can imagine, the case in which the channel is only 10 times larger than the fluid particles it contains would have to be considered a lower limit on the applicability of a continuum equation.

Even in this rather generous case, many situations of practical importance cannot be considered in the continuum regime. For highly heterogeneous fluids, such as those encountered in biological applications, the average size and molecular weight of the components of the fluid will be much larger. For instance, the glucose molecule, which is a very small molecule by biological standards, is about 15Å in diameter, while an average size protein molecule can easily be 100Å across. Hence a biggish protein molecule wouldn't even fit through a 100Å channel in a cell membrane, not to mention obey the Navier-Stokes equation inside it. A fluid like blood is even more extreme, consisting of cells which are seven to eight *microns* in size flowing through a 10 micron capillary.

Of course, the velocity field in a gas at room temperature and pressure was specifically chosen because the number density is low and the variance of particle velocities is large. For other fields under other conditions, this will not be the case. For instance the density field of a homogeneous fluid will have a much smaller variance so that smaller numbers of particles and a smaller δ will be required. Careful judgement and intuition are required when applying continuum equations in micro- and nanoscale systems.

10.3 Devices and Systems Beyond Continuum Theory

In this section we take a brief look at several systems whose modeling requires the researcher to go beyond continuum theory. As the development of NEMS progresses and researchers seek to understand the behavior of smaller and smaller systems, the explosion of research in this area is sure to continue. The reader is referred to the literature for further information on this rapidly developing area.

10.3.1 Nanostructured Materials

It is likely that the first true success for nanoscale technology will be in the area of nanostructured materials. During the later half of the 20th century the field of materials science came of age. Investigation of the relationship between microstructure and macroscopic properties proceeded apace from the theoretical, physical, and engineering communities. Nanotechnology opens the door to the possibility of controlling the structure of materials on the atomic level. Just as tailoring of microstructure has led to stronger, lighter, more durable materials, tailoring of the atomic structure promises materials with other desirable advanced properties. One interesting recent advance in this area is the collaboration between Nano-Tex of Greensboro, North Carolina, and Burlington Industries. Nano-Tex, under the direction of David Soane, has developed new nanostructured textiles with remarkable properties [53]. For example, by impregnating denim with billions of "nanowhiskers," each only 10nm long, Nano-Tex has succeeded in making waterproof denim with a feel indistinguishable from ordinary denim. Further advances in the field of nanostructured materials are likely to impact optical materials, medical materials, and many others.

10.3.2 Nanobiological Systems

Nanotechnology can take inspiration and comfort from biology. After all, the proof of concept behind the idea of self-assembled nanoscale machines lies all around us in the form of biological cells. Moreover, nanotechnology may be able to take direct advantage of naturally occurring nanosystems. One example of this is the "nanopropeller" built at Cornell University and powered by a natural ATP based motor [183]. This naturally occurring motor, which provides a motive force in living organisms, was harnessed to actuate a synthetic propeller. The system rotated at approximately ten revolutions per second for a period of several hours. Hybrid systems such as the nanopropeller are a first step toward the construction of complex nanosystems capable of performing useful tasks.

10.3.3 Nanobots and Other Dream Machines

In 1999 a team led by William Goddard at the California Institute of Technology won the Feynman Prize in Nanotechnology.[4] In this instance the prize was awarded for groundbreaking work in the modeling and simulation of molecular machines. Such machines represent the penultimate goal of the nanotechnology endeavor. The dream of nanoscale robots coursing through veins repairing damaged cells, the idea of a utility fog surrounding each individual creating objects on demand, the promise of a universal assembler rendering the modern economics system moot, all rely on the mundane first step of fabricating nanoscale gears, bearings, and motors. And, before they can be fabricated, they must be simulated. Understood. Proved to work. Goddard's molecular dynamics simulations represent a first step in this direction. There are many more steps yet to be taken. The role of mathematical modeling is sure to be important; the form such models will take remain to be discovered.

10.4 Chapter Highlights

- Continuum mechanics rests on two assumptions about how the world works: physical laws that govern the interaction of particles and statistical laws that govern the overall nature of systems of large numbers of particles.
- In many cases, MEMS and NEMS can be modeled using Newtonian mechanics. Relativistic effects are only important in systems involving free electrons. Quantum effects are important when analyzing systems involving single, free particles, and systems of particles at very low temperatures.
- Often quantum effects, particularly the interaction of atoms and molecules, can be modeled empirically and Newtonian mechanics can be used.
- The error in averaging a quantity over N independent observations scales as $N^{-1/2}$. Thus it is often the case that a very large N will be required to make such averages meaningful.

10.5 Exercises

Section 10.2

1. The electric field inside a dielectric medium obeys

$$\mathbf{D} = \epsilon \mathbf{E},$$

where $\mathbf{E}$ is the external field. Assuming that the electric field really is a field, discuss what this equation means in terms of the continuum limit. In other words, how exactly is the field, $\mathbf{D}$, inside the medium produced? What values of N and δ would you feel comfortable with in a device involving a dielectric?

2. Now consider the same situation except for magnetics. That is,

$$\mathbf{H} = \frac{\mathbf{B}}{\mu},$$

Now what is your opinion? Consider how the **H** field is produced inside the medium.

3. Consider a thin flexible beam made of a bundle of single-walled carbon nanotubes. Suppose each tube is 1 nm in diameter and 100 microns long and the bundle is one micron in diameter. Suppose we make a cantilever beam out of the bundle and measure the deflection of the tip of the beam given an applied force at the tip F. For small deflections, elementary beam theory gives

$$u = \frac{4FL^3}{3\pi E a^4}, \tag{10.4}$$

where a is the radius of the beam, L its length, and E is the Young's modulus of the beam material. Now this relationship can be used to experimentally determine the Young's modulus of the beam via $E = 4FL^3/(3\pi u a^4)$. So far so good. Now suppose we perform the same experiment on a beam consisting of a *single nanotube.* Do you expect a linear relationship between F and u in this case? If so, the experiment will give us a value for the Young's modulus of the tube. Does it makes sense to interpret the result of the experiment in this way?

4. Consider the same nanotube bundle, but this time we perform a thermal experiment. The ends of the bundle are attached to two heat reservoirs at temperatures T_0 and T_1. The experiment is performed in vacuum so that the bundle does not lose heat through its surface (except where attached to the walls). The temperature along the bundle is given by

$$T(x) = T_0 + (T_1 - T_0)(x/L), \tag{10.5}$$

where L is the length of the bundle and x a coordinate along the bundle. At time $t = 0$ the bundle is disconnected from both walls. Compute the time required for it to reach thermal equilibrium. A measurement of this time will enable you to experimentally determine the thermal diffusivity, κ, of the bundle. Now perform the same experiment on a single nanotube. What do you expect to conclude? What is the difference, if any, between this situation and that of the previous problem?

10.6 Notes

1. We are, of course, ignoring the fact that Maxwell's equations are intrinsically relativistic.

2. The diffraction and interference of atomic matter waves was first observed in the 1980s, and nowadays this effect is commonly observed even for relatively large molecules like buckeyballs [17].

3. The one exception to this is, of course, electromagnetics. For all intents and purposes, low intensity electric and magentic fields are continuous functions of space.

4. Several annual awards in nanotechnology named in honor of Richard P. Feynman are administered by the Foresight Institute. See www.foresight.org.

References

[1] R.C. Ackerberg, *On a Nonlinear Differential Equation of Electrohydrodynamics*, Proc. Roy. Soc. A, 312 (1969), pp. 129-140.

[2] C.H. Ahn, Y.J. Kim, and M.G. Allen, *A Planar Variable Reluctance Magnetic Micromotor with Fully Integrated Stator and Coils*, J. Microelectromech. Sys., 2 (1993), pp. 165-173.

[3] N.R. Aluru and J. White, *Direct Newton Finite-Element/Boundary-Element Technique for Micro-Electro-Mechanical-Analysis*, Solid State Sensor and Actuator Workshop, Hilton Head, South Carolina (1996), pp. 54-57.

[4] M.J. Anderson, J.A. Hill, C.M. Fortunko, N.S. Dogan, and R.D. Moore, *BroadBand Electrostatic Transducers: Modeling and Experiments*, J. Acoust. Soc. Am., 97 (1995), pp. 262–272.

[5] A.A. Andronov, A.A. Vitt, and S.E. Khaikin, *Theory of Oscillators* (1966), Dover, New York.

[6] M. Ataka, A. Omodaka, N. Takeshima, and H. Fujita, *Fabrication and Operation of Polymide Bimorph Actuators for a Ciliary Motion System*, J. Microelectromech. Sys., 2 (1993), pp. 146-150.

[7] M. Bao and W. Wang, *Future of Microelectromechanical Systems (MEMS)*, Sensors and Actuators A, 56 (1996), pp. 135-141.

[8] J. Bebernes and D. Eberly, *Mathematical Problems from Combustion Theory* (1989), Springer-Verlag, New York.

[9] W.L. Benard, H. Kahn, A.H. Heuer, and M.A. Huff, *Thin-Film Shape-Memory Alloy Actuated Micropumps*, J. Microelectromech. Sys., 7 (1998), pp. 245-251.

[10] D. Bernstein, P. Guidotti, and J.A. Pelesko, *Analytical and Numerical Analysis of Electrostatically Actuated MEMS Devices*, Proc. MSM 2000 (2000), pp. 489-492.

[11] A. Beskok, G. Karniadakis, and W. Trimmer, *Rarefaction and Compressibility Effects in Gas Microflows*, Trans. ASME, 118 (1996), pp. 448-456.

[12] S. Bhansali, A.L. Zhang, R.B. Zmood, P. Jones, and D.K. Sood, *Prototype Feedback Controlled Bi-Directional Actuation System for MEMS Applications*, J. Microelectromech. Syst., 9 (2000), pp. 245-251.

[13] J. Bienstman, J. Vandewalle, and R. Puers, *The Autonomous Impact Resonator: A New Operating Principle for a Silicon Resonant Strain Gauge*, Sensors and Actutators A, 66 (1998), pp. 40-49.

[14] G. Birkhoff and G.-C. Rota, *Ordinary Differential Equations* (1978), John Wiley, New York.

[15] R.H. Blick, A. Erbe, H. Krommer, A. Kraus, and J.P. Kotthaus, *Charge Detection with Nanomechanical Resonators*, Physica E, 6 (2000), pp. 821-827.

[16] W.E. Boyce and R.C. DiPrima, *Elementary Differential Equations and Boundary Value Problems* (1992), John Wiley, New York.

[17] B. Brezger, L. Hackermuller, S. Uttenthaler, J. Petschinka, and A. Zeilinger, *Matter-Wave Interferometer for Large Molecules*, Phys. Rev. Kett., 88 (2002), pp. 1-4.

[18] D. Briand, M.A. Gretillat, B. van der Schoot, and N.F. de Rooij, *Thermal Management of Micro-Hotplates Using MEMCAD as Simulation Tool*, Proc. MSM 2000 (2000) pp. 640-643.

[19] D. De Bruyker, A. Cozma, and R. Puers *A Combined Piezoresistive/Capacitive Pressure Sensor with Self-Test Function Based on Thermal Actuation*, Sensors and Actuators A (1998), pp. 70-75.

[20] J. Bryzek, K. Peterson and W. McCulley, *Micromachines on the March*, IEEE Spectrum, 31 (1994), pp. 20-31.

[21] M. Buchanan, *And God said - let there be levitating strawberries, flying frogs and humans that hover over Seattle*, New Scientist, 155 (1997), pp. 42.

[22] J. Buhler, J. Funk, J.G. Korvink, F. Steiner, P.M. Sarro, and H. Baltes, *Electrostatic Aluminum Micromirrors Using Double-Pass Metallization*, J. Microelectromech. Sys., 6 (1997), pp. 126-135.

[23] J.T. Butler, V.M. Bright, and W.D. Cowan, *Average Power Control and Positioning of Polysilicon Thermal Actuators* Sensors and Actuators, (1999), pp. 88-97.

[24] A. Cabal and D.SṘoss, *Snap-Through Bilayer Microbeam*, Proc. MSM 2002 (2002), pp. 230-233.

[25] H. Camon, F. Larnaudie, F. Rivoirard, and B. Jammes, *Analytical Simulation of a 1D Single Crystal Electrostatic Micromirror*, Proc. MSM 1999, (1999).

[26] E.K. Chan, E.C. Kan, R.W. Dutton, and P.M. Pinsky, *Nonlinear Dynamic Modeling of Micromachined Microwave Switches*, IEEE MTT-S Digest, (1997), pp. 1511-1514.

[27] E.K. Chan and R.W. Dutton, *Effects of Capacitors, Resistors and Residual Charge on the Static and Dynamic Performance of Electrostatically Actuated Devices*, Proceedings of SPIE Symposium on Design, Test and Microfabrication of MEMS and MOEMS (1999), pp. 120-130.

[28] S. Chatzandroulis, D. Tsoukalas, and P.A. Neukomm, *A Miniature Pressure System with a Capacitive Sensor and a Passive Telemetry Link for Use in Implantable Applications*, J. Microelectromech. Sys., 9 (2000), pp. 18-23.

[29] A.V. Chavan and K.D. Wise, *Batch-Processed Vacuum-Sealed Capacitive Pressure Sensors*, J. Microelectromech. Sys., 10 (2001), pp. 580-588.

[30] T.-C. Chin, *Permanent Magnet Films for Application in Microlectromechanical Systems*, J. Magnetism Magnetic Materials, 209 (2000), pp. 75-79.

[31] M. Chipot and M. Luskin, *The Compressible Reynolds Lubrication Equation* in *Metastability and Incompletely Posed Problems* (1987), Springer-Verlag, pp. 61-75.

[32] P.B. Chu and K.S.J. Pister, *Analysis of Closed-Loop Control of Parallel-Plate Electrostatic Microgrippers*, Proc. IEEE Int. Conf. Robotics and Automation (1994), pp. 820-825.

[33] B.W. Chui, T.D. Stowe, T.W. Kenny, H.J. Mamin, B.D. Terris, and D. Rugar, *Low-Stiffness Silicon Cantilevers for Thermal Writing and Piezoresistive Readback with the Atomic Force Microscope*, App. Phys. Lett. (1996), pp. 2767-2769.

[34] D.S. Cohen and A. Roger, *Chemical Reactor Theory and Problems in Diffusion*, Phys. D., 20 (1986), pp. 122-141.

[35] D.S. Cohen and T.W. Laetsch, *Nonlinear Boundary Value Problems Suggested by Chemical Reactor Theory*, J. Differential Equations, 7 (1970) pp. 217-226.

[36] D.S. Cohen, *Multiple Stable Solutions of Nonlinear Boundary Value Problems Arising in Chemical Reactor Theory*, SIAM J. Appl. Math., 20 (1971) pp. 1-13.

[37] D.S. Cohen and B.J. Matkowsky, *On Inhibiting Runaway in Catalytic Reactors*, SIAM J. Appl. Math., 35 (1978) pp. 307-314.

[38] J.H. Comtois and V.M. Bright, *Applications for Surface-Micromachined Polysilicon Thermal Actuators and Arrays*, Sensors and Actuators A (1997), pp. 19-25.

[39] J.H. Comtois, V.M. Bright, and M.W. Phipps, *Thermal Microactuators for Surface-Micromachining Processes*, Proc. SPIE (1995), pp. 10-21.

[40] J.H. Daniel, S. Iqbal, R.B. Millington, D.F. Moore, C.R. Lowe, D.L. Leslie, M.A. Lee, and M.J. Pearce, *Silicon Microchambers for DNA Amplification*, Sensors and Actuators, A71 (1998), pp. 81-88.

[41] P. Dario, M.C. Carrozza, A. Benvenuto, and A. Menciassi, *Micro-Systems in Biomedical Applications*, J. Micromech. Microeng., 10 (2000), pp. 235-244.

[42] A. Dec and K. Suyama, *Micromachined Varactor with Wide Tuning Range*, Electronics Lett., 33 (1997), pp. 22-25.

[43] O. Degani, E. Socher, A. Lipson, T. Leitner, D.J. Setter, S. Kaldor, and Y. Nemirovsky, *Pull-In Study of an Electrostatic Torsion Microactuator*, J. Microelectromech. Sys., 7 (1998), pp. 373-379.

[44] A. Duwel, M. Weinstein, J. Gorman, J. Borenstein, and P. Ward, *Quality Factors of MEMS Gyros and the Role of Thermoelastic Damping*, Proc. MEMS 2002 (2002). pp. 214-219.

[45] W.P. Eaton and J.H. Smith, *Micromachined Pressure Sensors: Review and Recent Developments*, Smart Mater. Struct., 6 (1997), pp. 530-539.

[46] M. Elwenspoek and R. Wiegerink, *Micromechanical Sensors* (2001), Springer-Verlag, Berlin.

[47] G. Emanuel *Analytical Fluid Dynamics* (2001), CRC Press, Boca Raton.

[48] A. Erbe, R.H. Blick, A. Tilke, A. Kriele, and J.P. Kotthaus, *A Mechanically Flexible Tunneling Contact Operating at Radio Frequencies*, App. Phys. Lett., 73 (1998), pp. 3751-3753.

[49] A. Erbe, C. Weiss, W. Zwerger, and R.H. Blick, *Nanomechanical Resonator Shuttling Single Electrons at Radio Frequencies*, Phys. Rev. Lett., 87 (2001), pp. 1-4.

[50] A. Erbe, H. Krommer, A. Kraus, and R.H. Blick, *Mechanical Properties of Suspended Structures at Radio Frequencies*, Physica B, 280 (2000), pp. 553-554.

[51] B.J. Feder, *A Big Step Forward in Tiny Technology*, New York Times, May 8 (2000), pp. C1-C4.

[52] Z. Feng, W. Zhang, B. Su, K.F. Harsh, K.C. Gupta, V. Bright, and Y.C. Lee, *Design and Modeling of RF MEMS Tunable Capacitors Using Electro-Thermal Actuators*, IEEE MTT-S Digest (1999), pp. 1507-1510.

[53] M. Fitzgerald, *The Next Wave*, Business 2.0, July (2002).

[54] M. Flynn and J.P. Gleeson, *A Study of the Behavior of Magnetic Microactuators*, Proc. MSM 2002 (2002), pp. 322-325.

[55] J.A. Folta, N.F. Raley, and E.W. Hee, *Design, Fabrication and Testing of a Miniature Peristaltic Membrane Pump*, Solid-State Sensor and Actuator Workshop (1992), pp. 186-189.

[56] O. Francais and I. Dufour, *Enhancement of Elementary Displaced Volume with Electrostatically Actuated Diaphragms: Application to Electrostatic Micropumps*, J. Micromech. Microeng., 10 (2000), pp. 282-286.

[57] O. Francais and I. Dufour, *Dynamic Simulation of an Electrostatic Micropump with Pull-in and Hysteresis Phenomena*, Sensors and Actuators A, 70 (1998), pp. 56-60.

[58] B. Franklin, *Letters and Papers of Benjamin Franklin and Richard Jackson, 1753-1785* (1947), Reprint Services Corporation.

[59] S. Fukui and R. Kaneko, *Dynamic Analysis of Flying Head Sliders with Ultra-Thin Spacing Based on the Boltzmann Equation*, JSME Int. J., Series III, 33 (1990), pp. 76.

[60] J. Funk, J. Buhler, J.G. Korvink, and H. Baltes, *Thermomechanical Modeling of and Actuated Micromirror*, Sensors and Actuators A, 46 (1995), pp. 632-636.

[61] J. Funk, G. Korvink, M. Bachtold, J. Buhler, and H. Baltes, *Coupled 3D Thermo-Electro-Mechanical Simulations of Microactuators*, Proc. MEMS 1996 (1996), pp. 133-138.

[62] J.M. Funk, J.G. Korvink, J. Buhler, M. Bachtold, and H. Baltes, *Solidis: A Tool for Microactuator Simulation in 3-D*, J. Microelectromech. Sys., 6 (1997), pp. 70-82.

[63] G. Galilei, *Dialogues Concerning Two New Sciences*, (1954) Dover, New York.

[64] M.K. Ghantasala, L. Qin, D.K. Sood, and R.B. Zmood, *Design and Fabrication of a Micro Magnetic Bearing*, J. Smart Mat. Struc., 9 (2000), pp. 1-6.

[65] J.R. Gilbert, G.K. Ananthasuresh, and S.D. Senturia, *3D Modeling of Contact Problems and Hysteresis in Coupled Electro-Mechanics*, Proceedings of the 9th Annual International Workshop on Micro Electro Mechanical Systems (1996), pp. 127-132.

[66] J.R. Gilbert and S.D. Senturia, *Two-Phase Actuators: Stable Zipping Devices without Fabrication of Curved Structures*, Solid-State Sensor and Actuator Workshop (1996), pp. 98-100.

[67] J.J. Gill, K. Ho, and G.P. Carman, *Three-Dimensional Thin-Film Shape Memory Alloy Microactuator with Two-Way Effect*, J. Microelectromech. Sys., 11 (2002), pp. 68-77.

[68] C. Goldsmith, T.H. Lin, B. Powers, W.R. Wu, and B. Norvell, *Micromechanical Membrane Switches for Microwave Applications*, IEEE MTT-S Digest (1995), pp. 91-94.

[69] G.H. Golub and C.F. Van Loan, *Matrix Computations* (1983), Johns Hopkins University Press, Baltimore.

[70] C. Grosjean, X. Yang, and Y.C. Tai, *A Thermopneumatic Microfluidic System*, Proc. MEMS 2002 (2002), pp. 24-27.

[71] C. Grosjean, X. Yang, and Y.C. Tai, *A Practical Thermopneumatic Valve*, Proc. MEMS 1999 (1999), pp. 147-152.

[72] H. Guckel, T. Earles, J. Klein, J.D. Zook, and T. Ohnstein, *Electromagnetic Linear Actuators with Inductive Position Sensing*, Sensors and Actuators A, 53 (1996), pp. 386-391.

[73] H. Guckel, K.J. Skrobis, T.R. Christenson, J. Klein, S.I. Han, B. Choi, E.G. Lovell, and T.W. Chapman, *Fabrication and Testing of the Planar Magnetic Micromotor*, J. Micromech. Microeng. 1 (1991), pp. 135-138.

[74] C.S. Gudeman, B. Staker, and M. Daneman, *Squeeze Film Damping of Doubly Supported Ribbons in Noble Gas Atmospheres*, Technical Digest Solid-State Sensor and Actuator Workshop (1998), pp. 288-291.

[75] E. Hashimoto, Y. Uenishi, and A. Watabe, *Thermally Controlled Magnetization Micro Relay*, Transducers '95 (1995), pp. 361-364.

[76] E. Hashimoto, H. Tanka, Y. Suzuki, Y. Uenishi, and A. Watabe, *Thermally Controlled Magnetization Actuator for Microrelays*, IEICE Trans. Electron, Vol. E80-C, No. 2 (1997), pp. 239-245.

[77] A.K. Henning, J. Fitch, D. Hopkins, L. Lilly, R. Faeth, E. Falsken, and M. Zdeblick, *A Thermopneumatically Actuated Microvalve for Liquid Expansion and Proportional Control*, Transducers '97 (1997), pp. 825-828.

[78] A.E. Herr, J.I. Molho, J.G. Santiago, M.G. Mungal, and T.W. Kenny, *Electroosmotic Capillary Flow with Nonuniform Zeta Potential*, Analytical Chemistry, 72 (2000), pp. 1053-1057.

[79] R.C. Hibbeler, *Mechanics of Materials* (2000), Prentice Hall, Upper Saddle River.

[80] J. Hietanan, J. Bomer, J. Jonsmann, W. Olthuis, P. Bergveld, K. Kaski, *Damping of a Vibrating Beam*, Sensors and Actuators A (2000), pp. 39-44.

[81] T. Honda, K.I. Arai, and M. Yamaguchi, *Fabrication of Magnetorestrictive Actuators Using Rare-Earth (Tb,Sm)-F3 Thin Films*, J. Appl. Phys. 76 (1994), pp. 6994-6999.

[82] M. Hoffmann, P. Kopka, and E. Voges, *All-Silicon Bistable Micromechanical Fiber Switch Based on Advanced Bulk Micromachining*, Selected Topics in Quantum Electronics, IEEE Journal, 5 (1999), pp. 46-51.

[83] M.H. Holmes, *Introduction to Perturbation Methods* (1991), Springer-Verlag, New York.

[84] M.N. Horenstein, S. Pappas, A. Fishov, and T. Bifano, *Electrostatic Micromirrors for Subaperturing in an Adaptive Optics System*, J. Electrostatics, 54 (2002), pp. 3-4.

[85] M. Horenstein, T.G. Bifano, S. Pappas, J. Perreault, and R. Krishnamoorthy-Mali, *Real Time Optical Correction Using Electrostatically Actuated MEMS Devices*, J. Electrostatics, 46 (1999), pp. 91-101.

[86] B.H. Houston, D.M. Photiadis, M.H. Marcus, J.A. Bucaro, and X. Liu, *Thermoelastic Loss in Microscale Oscillators*, Appl. Phys. Lett., 80 (2002), pp. 1300-1302.

[87] C. Hsu and W. Hsu, *A Two-Way Membrane-Type Micro-Actuator with Continuous Deflections*, J. Micromech. Microeng., 10 (2000), pp. 387-394.

[88] Q.A. Huang and N.K.S. Lee, *A Simple Approach to Characterizing the Driving Force of Polysilicon Laterally Driven Thermal Microactuators*, Sensors and Actuators, 80 (2000), pp. 267-272.

[89] M. Huja and M. Husak, *Thermal Microactuators for Optical Purpose*, Coding and Computing 2001, (2001), pp. 137-142.

[90] I.W. Hunter, S. Lafontaine, J.M. Hollerbach, and P.J. Hunter, *Fast Reversible NiTi Fibers for Use in Microrobotics*, Proc. MEMS '91 (1991), pp. 166-170.

[91] K. Ikuta and H. Shimizu, *Two Dimensional Mathematical Model of Shape Memory Alloy and Intelligent SMA-CAD*, Proc. MEMS '93, (1993) pp. 87-92.

[92] F.P. Incropera and D.P. DeWitt, *Introduction to Heat Transfer* (1996), John Wiley, New York.

[93] J.D. Jackson, *Classical Electrodynamics* (1975), John Wiley, New York.

[94] H. Jerman, *Electrically-Activated Micromachined Diaphragm Valves*, Technical Digest IEEE Solid-State Sensor and Actuator Workshop (1990), pp. 65-69.

[95] M. Jianqiang, C. Shixin, L. Yi, and L.B. Buan, *Thermal Stability Evaluation of MEMS Microactuator for Hard Disk Drives*, Proc. MSM 2002 (2002), pp. 222-225.

[96] B.P. Johnson, S. Kim, J.K. White, and S.D. Senturia, *MEMCAD Capacitance Calculations of Mechanically Deformed Square Diaphragm and Beam Microstructures*, Proceedings of the 6th International Conference on Solid State Sensors and Actuators (1991), pp. 494-497.

[97] D.D. Joseph and T.S. Lundgren, *Quasilinear Dirichlet Problems Driven by Positive Sources*, Arch. Rational Mech. Anal., 49 (1973), pp. 241-268.

[98] H. Kahn, M.A. Huff, and A.H. Heuer, *The TiNi Shape-Memory Alloy and its Applications for MEMS*, J. Micromech. Microeng., 8 (1998), pp. 213-221.

[99] H. Kapels, R. Aigner, and J. Binder, *Fracture Strength and Fatigue of Polysilicon Determined by a Novel Thermal Actuator*, IEEE Tran. on Elec. Dev., 47 (2000), pp. 1522- 1528.

[100] H.B. Keller, *Numerical Methods in Bifurcation Problems* (1987), Tata Institute of Fundamental Research, Bombay.

[101] H.B. Keller and J.M. Fier, *Equilibrium Chaos and Related Parameter Sequences*, Analyse Mathematique et Applications (1988), pp. 235-244.

[102] M. Khoo and C. Liu, *Development of a Novel Micromachined Magnetostatic Membrane Actuator*, 58th Device Research Conference, Conference Digest (2000), pp. 109-110.

[103] P. Kim and C.M. Lieber, *Nanotube Nanotweezers*, Science, 286 (1999), pp. 2148-2150.

[104] W.P. King, T.W. Kenny, K.E. Goodson, G. Cross, M. Despont, U. Durig, H. Rothuizen, G.K. Binnig, and P. Vettiger, *Atomic Force Microscope Cantilevers for Combined Thermomechanical Data Writing and Reading*, App. Phys. Lett. (2001), pp. 1300-1302.

[105] P.E. Kladitis, V.M. Bright, K.F. Harsh, and Y.C. Lee, *Prototype Microrobots for Micro Positioning in a Manufacturing Process and Micro Unmanned Vehicles*, Proceedings of IEEE Conference on MEMS (1999), pp. 570-575.

[106] P.E. Kladitis and V.M. Bright, *Prototype Microrobots for Micropositioning and Micro-Unmanned vehicles*, Sensors and Actuators (2000), pp. 132-137.

[107] M.U. Kopp, A.J. de Mello, and A. Manz, *Chemical Amplification: Continuous Flow PCR on a Chip*, Science, 280 (1998), pp. 1046-1048.

[108] G.T.A. Kovacs, *Micromachined Transducers Sourcebook* (1998), McGraw-Hill, Boston.

[109] G.A. Kriegsmann, M.E. Brodwin, and D.G. Watters, *Microwave Heating of a Ceramic Halfspace*, SIAM J. Appl. Math., 50 (1990), pp. 1088-1098.

[110] G.A. Kriegsmann, *Microwave Heating of Dispersive Media*, SIAM J. Appl. Math., 53 (1993), pp. 655-669.

[111] M.R. Booty and G.A. Kriegsmann, *Microwave Heating and Joining of Ceramic Cylinders: A Mathematical Model*, Methods Appl. Anal., 1 (1994), pp. 403-414.

[112] G.A. Kriegsmann, *Hot Spot Formation in Microwave Heated Ceramic Fibers*, IMA J. Appl. Math., 59 (1997), pp. 123-148.

[113] J.A. Pelesko and G.A. Kriegsmann, *Microwave Heating of Ceramic Laminates*, J. Engrg. Math., 41 (2001), pp. 345-366.

[114] P. Krulevitch, A.P. Lee, P.B. Ramsey, J.C. Trevino, J. Hamilton, and M.A. Northrup, *Thin Film Shape Memory Alloy Microactuators*, J. Microelectromech. Sys., 5 (1996), pp. 270-282.

[115] J. van Kuijk, T.S.J. Lammerink, H.E. de Bree, M. Elwenspoek, and J.H.J. Fluitman, *Multi-Parameter Detection in Fluid Flows*, Sensors and Actuators A (1995), pp. 369-372.

[116] J.T. Kung and H.S. Lee, *An Integrated Air-Gap-Capacitor Pressure Sensor and Digital Readout with Sub-100 Attofarad Resolution*, J. Microelectromech. Sys., 1 (1992), pp. 121-129.

[117] I. Kuzin and S. Pohozaev, *Entire Solutions of Semilinear Elliptic Equations* (1997), Birkhauser, Basel.

[118] A.A. Lacey, *Thermal Runaway in a Non-Local Problem Modeling Ohmic Heating. I. Model Derivation and Some Special Cases*, European J. Appl. Math., 6 (1995), pp. 127-144.

[119] A.A. Lacey, *Thermal Runaway in a Non-Local Problem Modelling Ohmic Heating. II. General Proof of Blow-Up and Asymptotics of Runaway*, European J. Appl. Math., 6 (1995), pp. 201-224.

[120] W.E. Leary, *Giant Hopes for Tiny Satellites*, New York Times, November 9 (1999), pp. D2.

[121] L.A. Liew, V.M. Bright, M.L. Dunn, J.W. Daily, and R. Raj, *Development of SiCN Ceramic Thermal Actuators*, Proc. MEMS 2002 (2002), pp. 590-593.

[122] L.A. Liew, A. Tuantranont, and V.M. Bright, *Modeling of Thermal Actuation in a Bulk-Micromachined CMOS Micromirror*, Microelectron. J., 31 (2000), pp. 791-801.

[123] R. Lifshitz and M.L. Roukes, *Thermoelastic Damping in Micro and Nano-Mechanical Systems*, Phys. Rev. B, 61 (2000), pp. 5600-5609.

[124] C.C. Lin and L.A. Segel, *Mathematics Applied to Deterministic Problems in the Natural Sciences* (1988), SIAM, Philadelphia.

[125] Z. Lin, T. Kerle, S.M. Baker, D.A. Hoagland, E. Schaffer, U. Steiner, and T.P. Russell, *Electric Field Induced Instabilities at Liquid/Liquid Interfaces*, J. Chem. Phys., 114 (2001), pp. 2377-2381.

[126] C.D. Lott, T.W. McLain, J.N. Harb, and L.L. Howell, *Thermal Modeling of a Surface-Micromachined Linear Thermomechanical Actuator*, Proc. MSM 2001 (2001), pp. 370-373.

[127] A. Ludwig and E. Quandt, *Giant Magnetostrictive Thin Films for Applications in Microelectromechanical Systems*, J. App. Phys., 87 (2000), pp. 4691-4695.

[128] M. Madou, *Fundamentals of Microfabrication* (1997), CRC Press, Boca Raton.

[129] J.M. Maloney, D.L. DeVoe, and D.S. Schreiber, *Analysis and Design of Electrothermal Actuators Fabricated From Single Crystal Silicon*, Proc. MEMS 2000 (2000), pp. 233-240.

[130] N. Maluf, *An Introduction to Microelectromechanical Systems Engineering* (2000), Artech House, Boston.

[131] N. Mankame and G.K. Ananthasuresh, *The Effect of Thermal Boundary Conditions and Scaling on Electro-Thermal-Compliant Micro Devices*, Proc. MSM 2000 (2000), pp. 609-612.

[132] C.H. Mastrangelo, M.A. Burns, and D.T. Burke, *Microfabricated Devices for Genetic Diagnostics*, Proc. IEEE, 86 (1998), pp. 1769-1787.

[133] P. Melvas, E. Kalvesten, and G. Stemme, *A Surface-Micromachined Resonant-Beam Pressure-Sensing Structure*, J. Microelectromech. Sys., 10 (2001), pp. 498-502.

[134] S. Moon, S.H. Lim, S.B. Lee, H.K. Kang, M.C. Kim, S.H. Han, and O.H. Oh, *Optical Switch Driven by Giant Magnetorestrictive Thin Films*, Symposium on Desgin, Test, and Microfabrication of MEMS and MOEMS (1999), pp. 854-862.

[135] M. Moseler and U. Landman, *Formation, Stability and Breakup of Nanojets*, Science, 289 (2000), pp. 1165-1169.

[136] J.B. Muldavin and G.M. Rebeiz, *30 GHz Tuned MEMS Switches*, IEEE MTT-S Digest (1999), pp. 1511-1518.

[137] S.J. Murray, M. Marioni, S.M. Allen, R.C. O'Handley and T.A. Lograsso, *6% Magnetic-Field-Induced Strain by Twin-Boundary Motion in Ferromagnetic Ni-Mn-Ga*, Appl. Phys. Lett. 87 (2000), pp. 886.

[138] J.D. Murray, *Mathematical Biology* (1989), Springer-Verlag, New York.

[139] D. Myers, *Surfaces, Interfaces, and Colloids*, (1999), John Wiley, New York.

[140] T.Y. Na, *Computational Methods in Engineering Boundary Value Problems* (1979), Academic Press, New York.

[141] N. Nakajima, K. Ogawa, and I. Fujimasa, *Study on Micro Engines—Miniaturizing Stirling Engines for Actuators and Heatpumps*, Proc. IEEE Micro Electro Mechanical Systems (1989), pp. 145-148.

[142] H.J. Nam, S.M. Cho, Y.S. Kim, D.C. Kim, and J.U. Bu, *Micromirror Actuated by PZT Film for Optical Application*, Integr. Ferroelectr., 41 (2001), pp. 1715-1724.

[143] H.C. Nathanson, W.E. Newell, R.A. Wickstrom and J.R. Davis, *The Resonant Gate Transistor*, IEEE Trans. on Electron Devices, 14 (1967), pp. 117-133.

[144] N.T. Nguyen, S. Schubert, S. Richter, and W. Dotzel, *Hybrid-Assembled Micro Dosing System Using Silicon Based Micropump/Valve and Mass Flow Sensor*, Sensors and Actuators A (1998), pp. 85-91.

[145] N.T. Nguyen, D. Bochnia, R. Kiehnscherf, and W. Dotzel, *Investigation of Forced Convection in Microfluid Systems*, Sensors and Actuators A (1996), pp. 49-55.

[146] B. Ni, S.B. Sinnott, P.T. Mikulski, and J.A. Harrison, *Compression of Carbon Nanotubes Filled with* C_60*,* CH_4*, or* Ne*: Predictions from Molecular Dynamics Simulations*, Phys. Rev. Lett., 88 (2002), pp. 1-4.

[147] Y. Okano and Y. Hirabayashi, *Magnetically Actuated Micromirror and Measurement System for Motion Characteristics Using Specular Reflection*, IEEE J. Sel. Top. Quantum Electron., 8 (2002), pp. 19-25.

[148] K. Ono, *Dynamic Characteristics of Air-Lubricated Slider Bearing for Non-Contact Magnetic Recording*, ASME J. Lub. Technol., 97 (1975), pp. 250.

[149] V.K. Pamula, A. Jog, and R.B. Fair, *Mechanical Property Measurement of Thin-Film Gold using Thermally Actuated Bimetallic Cantilever Beams*, Proc. MSM 2001 (2001), pp. 410-413.

[150] C.V. Pao, *Nonlinear Parabolic and Elliptic Equations* (1992), Plenum Press, New York.

[151] T.C. Papanastasiou, G.C. Georgiou, and A.N. Alexandrov, *Viscous Fluid Flow* (2000), CRC Press, Boca Raton.

[152] M. Parameswaran, Lj. Ristic, K. Chau, A.M. Robinson, and W. Allegretto, *CMOS Electrothermal Microactuators*, Proc. IEEE Micro Electro Mechanical Systems (1990), pp. 128-131.

[153] J.A. Pelesko and A.A. Triolo, *Nonlocal Problems in MEMS Device Control*, Proceedings of Modeling and Simulation of Microsystems 2000 (2000), pp. 509-512.

[154] J.A. Pelesko and A.A. Triolo, *Nonlocal Problems in MEMS Device Control*, J. Eng. Math., 41 (2001), pp. 345-366.

[155] J.A. Pelesko and X.Y. Chen, *Electrostatically Deflected Circular Elastic Membranes*, J. Electrostatics, in press.

[156] J.A. Pelesko, *Nonlinear Stability Considerations in Thermoelastic Contact*, ASME J. Appl. Mech., 66 (1999), pp. 109-116.

[157] J.A. Pelesko, *Diffusive and Wavelike Phenomena in Thermal Processing of Materials*, Ph.D. Thesis, New Jersey Institute of Technology (1997).

[158] R. Pelrine and I. Busch-Vishniac, *Magnetically Levitated Micro-Machines*, IEEE Micro Robots and Teleoperators (1987).

[159] W.H. Press, S.A. Teukolsky, W.T. Vetterling, and B.P. Flannery, *Numerical Recipes in C* (1993), Cambridge University Press, Cambridge.

[160] R.H. Price, J.E. Wood, and S.C. Jacobsen, *The Modelling of Electrostatic Forces in Small Electrostatic Actuators*, Technical Digest IEEE Solid-State Sensor and Actuator Workshop (1988), pp. 131-135.

[161] A. Rasmussen, C. Mavriplis, M.E. Zaghloul, O. Mikulchenko, and K. Mayaram, *Simulation and Optimization of a Microfluidic Flow Sensor*, Sensors and Actuators A (2001), pp. 121-132.

[162] J.R. Reid, V.M. Bright, and J.T. Butler *Automated Assembly of Flip-up Micromirrors*, Sensors and Actuators A (1998), pp. 292-298.

[163] W. Riethmuller, W. Benecke, U. Schnakenberg, and A. Heuberger, *Micromechanical Silicon Actuators Based on Thermal Expansion Effects*, Transducers '87 (1987), pp. 834-837.

[164] M. Gongora-Rubio, L.-.M. Sola-Laguna, P.J. Moffett, and J.J. Santiago-Aviles, *The Utilization of Low Temperature Co-Fired Ceramics Technology for Meso-Scale EMS, A Simple Thermistor Based Flow Sensor*, Sensors and Actuators A (1999), pp. 215-221.

[165] Y. Saad, *Iterative Methods for Sparse Linear Systems* (1996), PWS Publishing Company, Boston.

[166] M.T.A. Saif, B.E. Alaca and H. Sehitoglu, *Analytical Modeling of Electrostatic Membrane Actuator Micro Pumps*, IEEE J. Microelectromech. Sys., 8 (1999), pp. 335-344.

[167] D.J. Sadler, T.M. Liakopoulos, and C.H. Ahn, *A New Electromagnetic Actuator Using Through Hole Plating of Nickel/Iron Permalloy*, Proceedings of the Fifth International Symposium on Magnetic Materials, Processes, and Devices Applications to Storage and Microelectromechanical Systems (MEMS) (1999), pp. 377-388.

[168] D.J. Sadler, T.M. Liakopoulos, and C.H. Ahn, *A Universal Electromagnetic Microactuator Using Magnetic Interconnection Concepts*, J. Microelectromech. Sys., 9 (2000), pp. 460-468.

[169] D.H. Sattinger, *Topics in Stability and Bifurcation Theory* (1973), Springer-Verlag, New York.

[170] S.A. Schaaf and P.L. Chambre, *Flow of Rarefied Gases* (1961), Princeton University Press, Princeton.

[171] E. Schaffer, T. Thurn-Albrecht, T.P. Russell, and U. Steiner, *Electrically Induced Structure Formation and Pattern Transfer*, Nature, 403 (2000), pp. 874-876.

[172] E. Schaffer, T. Thurn-Albrecht, T.P. Russell, and U. Steiner, *Electrohydrodynamic Instabilities in Polymer Films*, Europhys. Lett., 53 (2001), pp. 518-524.

[173] J.I. Seeger and S.B. Crary, *Stabilization of Electrostatically Acutated Mechanical Devices*, Proceedings of the 1997 International Conference on Solid-State Sensors and Actuators (1997), pp. 1133-1136.

[174] J.J. Seeger and S.B. Crary, *Analysis and Simulation of MOS Capacitor Feedback for Stabilizing Electrostatically Actuated Mechanical Devices*, Second International Confernce on the Simulation and Design of Microsystems and Microstructures—MICROSIM97 (1997), pp. 199-208.

[175] S.D. Senturia, *Microsystem Design* (2001), Kluwer Academic Publishers, Boston.

[176] F. Shi, P. Ramesh, and S. Mukherjee, *Simulation Methods for Micro-Electro-Mechanical Structures (MEMS) with Applications to a Microtweezer*, Computers and Structures, 56 (1995), pp. 769-782.

[177] F. Shi, P. Ramesh, and S. Mukherjee, *Dynamic Analysis of Micro-Electro-Mechanical Systems*, Int. J. Num. Meth. Eng., 39 (1996), pp. 4119-4139.

[178] W. Sim, B. Kim, B. Choi, and J. Park, *Thermal and Load-Deflection FE Analysis of Parylene Diaphragms*, Proc. MSM 2002 (2002), pp. 210-213.

[179] I. Shimoyama, O. Kano, and H. Miura, *3D Micro-Structures Folded by Lorentz Force*, Proc. MEMS '98 (1998), pp. 24-28.

[180] S.A. Smee, M. Gaitan, D.B. Novotny, Y. Joshi, and D.L. Blackburn, *IC Test Structures for MultiLayer Interconnect Stress Determination*, IEEE Electron Dev. Lett., 21 (2000), pp. 12-14.

[181] J.H. Smith, *Micromachined Sensor and Actuator Research at Sandia's Microelectronics Development Laboratory*, Sensors Expo (1996), pp. 119-123.

[182] W.R. Smythe, *Static and Dynamic Electricity* (1950), Hemisphere Publishing Corporation, New York.

[183] R.K. Soong, G.D. Bachand, H.P. Neves, A.G. Olkhovets, and H.G. Craighead, Science, 290 (2000), pp. 1555-1558.

[184] I. Stackgold, *Green's Functions and Boundary Value Problems* (1998), Wiley Interscience, New York.

[185] J. Steurer and F. Kohl, *Adaptive Controlled Thermal Sensor for Measuring Gas Flow*, Sensors and Actuators A (1998), pp. 116-122.

[186] S. Takeuchi and I. Shimoyama, *A Three-Dimensional Shape Memory Alloy Microelectrode with Clipping Structure for Insect Neural Recording*, J. Microelectromech. Sys., 9 (2000), pp. 24-31.

[187] N. Tas, J. Wissink, L. Sander, T. Lammerink, and M. Elwenspoek, *Modeling, Design and Testing of an Electrostatic Shuffle Motor*, Sensors and Actuators A (1998), pp. 171-178.

[188] G.I. Taylor, *The Coalescence of Closely Spaced Drops When They are at Different Electric Potentials*, Proc. Roy. Soc. A, 306 (1968), pp. 423-434.

[189] W.P. Taylor, M.G. Allen, and C.R. Dauwalter, *A Fully Integrated Magnetically Actuated Micromachined Relay*, Solid State Sensor and Actuator Workshop (1996), pp. 231-234.

[190] W.P. Taylor, O. Brand, and M.G. Allen, *Fully Integrated Magnetically Actuated Micromachined Relays*, J. Microelectromech. Sys., 7 (1998), pp. 181-191.

[191] M. Terry, J. Reiter, K.F. Bohringer, J.W. Suh, and G.T.A. Kovacs, *A Docking System for Microsatellites Based on MEMS Actuator Arrays*, Smart Mater. Struct., 10 (2001), pp. 1176-1184.

[192] D.W. Thompson, *On Growth and Form* (1992), Dover, New York.

[193] H.A.C. Tilmans, M. Elwenspoek, and J.H.J. Fluitman, *Micro Resonant Force Gauges*, Sensors and Actuators A, 30 (1992), pp. 35-53.

[194] H.A.C. Tilmans and R. Legtenberg, *Electrostatically Driven Vacuum-Encapsulated Polysilicon Resonators. Part II. Theory and Performance*, Sensors and Acutators A, 45 (1994), pp. 67-84.

[195] S. Timoshenko, *Theory of Plates and Shells*, (1940), McGraw-Hill, New York.

[196] S. Timoshenko, *Bending and Buckling of Bimetallic Strips*, J. Optical Soc. Am., 11 (1925), pp. 233.

[197] K. Toda, Y. Maeda, I. Sanemasa, K. Ishikawa, and N. Kimura, *Characteristics of a Thermal Mass-flow Sensor in Vacuum Systems*, Sensors and Actuators A (1998), pp. 62-67.

[198] K. Toepfer, *Grenzschichten in Flussigkeiten mit kleiner Reibung*, Z. Math. Phys., 60 (1912), pp. 397-398.

[199] W.S.N. Trimmer, *Microrobots and Micromechanical Systems*, Sensors and Actuators A, 19 (1989), pp. 267-287.

[200] A. Tuantranont and V.M. Bright, *Segmented Silicon-Micromachined Microelectromechanical Deformable Mirrors for Adaptive Optics*, IEEE J. Sel. Top. Quantum Electron., 8 (2002), pp. 33-45.

[201] N. Uchida, K. Uchimaru, M. Yonezawa, and M. Sekimura, *Damping of Micro Electrostatic Torsion Mirror Caused by Air-Film Viscosity*, J. Japan Society of Precision Engineering, 65 (1999), pp. 1301-1305.

[202] J.R. Vig and Y. Kim, *Noise in Microelectromechanical System Resonators*, IEEE Tran. on Ultrasonic, Ferroelectrics, and Frequency Control, 46 (1999), pp. 1558-1565.

[203] C. Wang, B.P. Gogoi, D.J. Monk, and C.H. Mastrangelo, *Contamination-Insensitive Differential Capacitive Pressure Sensors*, J. Microelectromech. Sys., 9 (2000), pp. 538-543.

[204] R. Weinstock, *Calculus of Variations* (1952), McGraw-Hill, New York.

[205] T.P. Witelski, *Dynamics of Air Bearing Sliders*, Physics of Fluids, 10 (1998), pp. 698-708.

[206] R.H. Wolf and A.H. Heuer, *TiNi (Shape Memory) Films on Silicon for MEMS Applications*, J. Microelectromech. Sys., 4 (1995), pp. 206-212.

[207] D.R. Wur, J.L. Davidson, W.P. Kang, and D.L. Kinser, *Polycrystalline Diamond Pressure Sensor*, J. Microelectromech. Sys., 4 (1995), pp. 34-41.

[208] Y. Xia, J.A. Rogers, K.E. Paul, and G.M. Whitesides, *Unconventional Methods for Fabricating and Patterning Nanostructures*, Chem. Rev., 99 (1999), pp. 1823-1848.

[209] Z.Y. Xue, M.T.A. Saif, and Y.G. Huang, *The Strain Gradient Effect in Microelectromechanical Systems (MEMS)*, J. Microelectromech. Sys., 11 (2002), pp. 27-35.

[210] N. Yao and V. Lordi, *Carbon Nanotube Caps as Springs: Molecular Dynamics Simulations*, Phys. Rev. B, 58 (1998), pp. 649-651.

[211] X. Yang, C. Grosjean, and Y.C. Tai, *Design, Fabrication, and Testing of Micromachined Silicone Rubber Membrane Valves*, J. Microelectromech. Sys., 8 (1999), pp. 393-402.

[212] X. Yang, Y.C. Tai, and C.M. Ho, *Micro Bellow Actuators*, Transducers '97 (1997), pp. 45-48.

[213] X. Yang, C. Grosjean, Y.C. Tai, and C.M. Ho, *A MEMS Thermopneumatic Silicone Membrane Valve*, Proc. MEMS 1997 (1997), pp. 114-118.

[214] Y. Yi and C. Liu, *Parallel Assembly of Hinged Microstructures Using Magnetic Actuation*, Solid State Sensor and Actuator Workshop (1998), pp. 269-272.

[215] Y. Yi and C. Liu, *Parallel Assembly of Hinged Microstructures Using Magnetic Actuation*, Solid State Sensors and Actuators Workshop (1996), pp. 269-272.

[216] L. Yin and G.K. Ananthasuresh, *A New Material Interpolation Scheme for the Topology Optimization of Thermally Actuated Compliant Micromechanisms*, Proc. MSM 2001 (2001), pp. 124-127.

[217] Y. Zhang, Y. Zhang, and R.B. Marcus, *Thermally Actuated Microprobes for a New Wafer Probe Card*, J. Microelectromech. Sys., 8 (1999), pp. 43-49.

[218] R. Zengerle, S. Kluge, M. Richter, and A. Richter, *A Bi-Directional Silicon Micropump*, Proc. MEMS '95 (1995), pp. 19-24.

[219] W. Zhang and C.H. Ahn, *A Bi-Directional Magnetic Micropump on a Silicon Wafer*, Solid State Sensor and Actuator Workshop (1996), pp. 94-97.

[220] X.M. Zhang, F.S. Chau, C. Quan, Y.L. Lam, and A.Q. Liu, *A Study of the Static Characteristics of a Torsional Micromirror*, Sens. Actuator A-Phys., 90 (2001), pp. 73-81.

[221] J. Zou, J. Chen, C. Liu, and J.E. Schutt-Aine, *Plastic Deformation Magnetic Assembly (PDMA) of Out-of-Plane Microstructures: Technology and Application*, J. Microelectromech. Sys., 10 (2001), pp. 302-309.

[222] R.B. Zmood, L. Qin, D.K. Sood, T. Vinay, and D. Meyrick, *Magnetic MEMS Used in Smart Structures Which Exploit Magnetic Materials and Properties*, Smart Structures and Devices (SPIE vol. 4235) (2001), pp. 173-187.

Appendix A

Some Useful Mathematical Results

A.1 The Divergence Theorem

The divergence theorem relates the surface integral over a surface S to the volume integral over the volume V contained within the surface

$$\int_S \mathbf{v} \cdot \mathbf{n}\, dS = \int_V \nabla \cdot \mathbf{v}\, dV \tag{A.1}$$

where $\mathbf{n}$ is an outward pointing unit normal.

A.2 Stoke's Theorem

Stoke's theorem relates the surface integral over a surface S to the line integral over the perimeter C

$$\int_S (\nabla \times \mathbf{F}) \cdot dS = \int_C \mathbf{F} \cdot d\mathbf{l}. \tag{A.2}$$

A.3 The Laplacian

In rectangular Cartesian coordinates the Laplacian is

$$\nabla^2 = \frac{\partial^2}{\partial x^2} + \frac{\partial^2}{\partial y^2} + \frac{\partial^2}{\partial z^2}, \tag{A.3}$$

while in cylindrical coordinates it is

$$\nabla^2 = \frac{1}{r}\frac{\partial}{\partial r}\left(r\frac{\partial}{\partial r}\right) + \frac{1}{r^2}\frac{\partial^2}{\partial \theta^2} + \frac{\partial^2}{\partial z^2}, \tag{A.4}$$

and in spherical coordinates it takes the form

$$\nabla^2 = \frac{1}{\rho^2}\frac{\partial}{\partial \rho}\left(\rho^2 \frac{\partial}{\partial \rho}\right) + \frac{1}{\rho^2 \sin\theta}\frac{\partial}{\partial \theta}\left(\sin\theta \frac{\partial}{\partial \theta}\right) + \frac{1}{\rho^2 \sin^2\theta}\frac{\partial^2}{\partial \phi^2}. \tag{A.5}$$

A.4 The Bi-Laplacian

In rectangular Cartesian coordinates the bi-Laplacian or biharmonic operator takes the form

$$\nabla^4 = \frac{\partial^4}{\partial x^4} + \frac{\partial^4}{\partial y^4} + \frac{\partial^4}{\partial z^4} + 2\left(\frac{\partial^4}{\partial x^2 \partial y^2} + \frac{\partial^4}{\partial y^2 \partial z^2} + \frac{\partial^4}{\partial x^2 \partial z^2}\right). \tag{A.6}$$

A.5 The Mean Value Theorem for Integrals

The mean value theorem for integrals is

THEOREM A.1

Let $f(x)$ be continuous on $[a, b]$, then there exists at least one $c \in [a, b]$ such that

$$\int_a^b f(x)dx = f(c)(b-a). \tag{A.7}$$

A.6 The Dubois-Reymond Lemma

The Dubois-Reymond Lemma says that given a continuous function f defined over a domain D and if

$$\int_V f dV = 0 \tag{A.8}$$

for every region V contained in D, then $f \equiv 0$. This deceptively simply but powerful lemma is used repeatedly in deriving the equations of continuum mechanics via the integral method. Note it is necessary that f be *continuous.* To convince yourself of this, consider the function $f(x)$ defined on the interval $[0, 1]$ as $f(x) = 0$ if x is irrational and $f(x) = 1$ if x is rational. Since the rational numbers in the interval $[0, 1]$ are a set of measure zero, the integral of f over any subinterval is identically zero, yet f itself is not.

A.7 A Brief Look at Asymptotics, Big O and Little o

Hopefully, you're convinced that asymptotics is something worth knowing. Let's make a few things precise. We'll begin by thinking about limits.

$$\lim_{n\to\infty} f(n) = L. \tag{A.9}$$

We all know from calculus what such a statement means. But, in applications, we usually don't care what value $f(n)$ takes at "∞," but rather we'd like to know how f behaves when n is large. That is how does f get to L? Our calculus limit notation hides this information! How? Well, what we really mean when we write A.9 is that for all $\epsilon > 0$, there exists an $N(\epsilon)$ such that if $n \geq N(\epsilon)$, then

$$|f(n) - L| < \epsilon. \tag{A.10}$$

From the point of view of asymptotics, $N(\epsilon)$ is the key! It gives explicit numerical information about how $f(n)$ behaves. Let's look at an example. Consider

$$\lim_{n\to\infty} \frac{n+1}{n} = 1. \tag{A.11}$$

This means that given any $\epsilon > 0$, we must choose $N(\epsilon)$ so that

$$\left|\frac{n+1}{n} - 1\right| < \epsilon \tag{A.12}$$

or

$$\left|\frac{1}{n}\right| < \epsilon. \tag{A.13}$$

Now, it's clear that we should choose $N(\epsilon) = 1/\epsilon$. This means that

$$|f(n) - 1| \leq 1/n. \tag{A.14}$$

In asymptotics, we introduce the symbol O and write

$$f(n) - 1 = O(1/n) \tag{A.15}$$

or

$$f(n) = 1 + O(1/n). \tag{A.16}$$

Note, the "big O" notation makes it clear how f approaches 1 as n tends to infinity.

Let's make this precise with a definition.

DEFINITION A.1 *Let S be any set of real or complex numbers and f, ϕ functions defined on S. Then, we say $f = O(\phi)$ on S iff there exists a* *number $A > 0$ which is independent of s such that $|f(s)| \leq A|\phi(s)|$ is true for all $s \in S$.*

Consider some examples:

$$x^2 = O(x) \;\; |x| < 2, \tag{A.17}$$

$$\sin(x) = O(1) \;\; (-\infty, \infty) \tag{A.18}$$

$$\sin(x) = O(x) \;\; (-\infty, \infty). \tag{A.19}$$

Often in practice, we're not so interested in the behavior over a specific interval, but in how f behaves over some subinterval. That is, we say things like

$$f(x) = O(\phi(x)) \tag{A.20}$$

as $x \to \infty$ or

$$f(x) = O(\phi(x)) \tag{A.21}$$

as $x \to 0$. We'll modify our notation accordingly. That is, when we say

$$f(x) = O(\phi(x)), \tag{A.22}$$

as $x \to \infty$, we mean that $f(x) = O(\phi(x))$ on (a, ∞) for some a. Often, we don't care what a is, just that it exists. We'll usually be sloppy and not even talk about a.

Next, we need two other pieces of notation

DEFINITION A.2 *We say that $f(x) = o(\phi(x))$ as $x \to \infty$ iff*

$$\lim_{x \to \infty} \frac{f}{\phi} = 0. \tag{A.23}$$

We often write $f \ll \phi$ as $x \to \infty$ in place of the "little o" notation. It's also standard to consider $f = o(1)$ and in perturbation theory, we'll write $\epsilon \ll 1$ to mean $\epsilon = o(1)$ as $\epsilon \to 0$. The other definition we need is

DEFINITION A.3 *We say $f(x)$ is asymptotic to $g(x)$ and write $f(x) \sim g(x)$ as $x \to x_0$ iff,*

$$\lim_{x \to x_0} \frac{f}{g} = 1. \tag{A.24}$$

Note that it's equivalent to write $f \sim g$, $f - g \ll g$ or $f = g(1 + o(1))$. You should convince yourself of this.

Finally, we have enough defined to get to the heart of the matter. This notion of *asymptotic equivalence* is what we are really after. For example, we'd like to say things like

$$x + 1 \sim x \;\; \text{as} \;\; x \to \infty \tag{A.25}$$

or

$$\sinh(x) \sim e^x/2 \quad \text{as} \quad x \to \infty \tag{A.26}$$

or

$$\Pi(n) \sim \frac{n}{\log(n)} \quad \text{as} \quad x \to \infty. \tag{A.27}$$

Now, consider the following three statements:

$$e^x \sim 1 \quad \text{as} \quad x \to 0, \tag{A.28}$$

$$e^x \sim 1 + x \quad \text{as} \quad x \to 0, \tag{A.29}$$

$$e^x \sim 1 + x + \frac{x^2}{2} \quad \text{as} \quad x \to 0. \tag{A.30}$$

Are these all true? Of course they are; we're simply including more terms in the Taylor series expansion of e^x! Suppose we wanted to compute $e^{1/10}$. Using Matlab we obtain $e^{1/10} \approx 1.1052$. Now, using our asymptotic approximations, we can write

$$e^{1/10} \sim 1 + \frac{1}{10} + O\left(\frac{1}{10^2}\right) = 1.1 + O\left(\frac{1}{100}\right) \tag{A.31}$$

from our best approximation. The others would be cruder. This gives us the idea of better asymptotic approximation and of pinning down the error. Often, we'll write statements like

$$e^x \sim 1 + x + O(x^2) \quad \text{as} \quad x \to 0 \tag{A.32}$$

to mean

$$e^x \sim 1 + x, \tag{A.33}$$

with

$$e^x - (1 + x) = O(x^2). \tag{A.34}$$

The introduction to asymptotics provided here has been brief. The reader is directed to the related reading at the end of Chapter 4 for more information.

A.8 Fundamental Lemma of Calculus of Variations

The fundamental lemma of the calculus of variations may be stated for one-dimensional integrals as

LEMMA A.1
Let $f(x)$ be a continuous function on the interval $[a, b]$. If

$$\int_a^b f(x)\eta(x)dx = 0 \tag{A.35}$$

for every choice of the continuously differentiable function $\eta(x)$ *for which* $\eta(a) = \eta(b) = 0$ *then* $f(x)$ *is identically zero in* $[a, b]$.

A.9 Linear Spaces

Recall the definition of a vector space.

DEFINITION A.4 *A* vector space X *is a collection of elements (vectors)* $u, v, w, \ldots$*, together with a binary operation called addition, which maps the space to itself. That is* $+ \colon X \to X$ *or if* $u, v \in X$ *then* $u + v \in X$*. Also, we have an operation called scalar multiplication between our vectors and the elements* $\alpha, \beta, \ldots$ *of some field. These two operations must satisfy*

- $u + v = v + u$
- $u + (v + w) = (u + v) + w$
- *There exists a unique vector* 0 *such that* $u + 0 = u$
- *For all* u*, there exists a unique vector* $-u$ *such that* $u + (-u) = 0$
- $\alpha(\beta u) = (\alpha\beta)u$
- $1u = u$
- $(\alpha + \beta)u = \alpha u + \beta u$
- $\alpha(u + v) = \alpha u + \alpha v$.

Working within a vector space X we also want to define the notion of a basis and the notion of dimension.

DEFINITION A.5 *Let* $u_1, u_2, \ldots, u_k \in X$*, then*

- $\alpha_1 u_1 + \cdots \alpha_k u_k$ *is a linear combination*
- $\alpha_1 u_1 + \cdots \alpha_k u_k$ *is nontrivial if not all the* α_i*'s are zero*
- $u_1, u_2, \ldots, u_k$ *are linearly dependent if there exists a nontrivial linear combination equal to the zero vector.*

DEFINITION A.6 X *is* n*-dimensional if there exists an* n *element set of independent vectors, but every* $n + 1$ *element set is linearly dependent.* X *is infinite dimensional if for all* n*, there exists a* n *element independent set.*

DEFINITION A.7 *$h_1, h_2, ..., h_k \in X$ is said to be a basis if for all $x \in X$ there exists a unique linear combination of the h_i's such that $\alpha_1 h_1 + \cdots \alpha_k h_k = x$.*

Some easy results that you should recall are

THEOREM A.2
Let X be finite dimensional with dimension n, then

- *The vectors in a basis are independent*
- *Any n independent vectors form a basis*

Recall that sets of functions can form vector spaces. Some examples are

- $P_n = \{$all polynomials of degree less than n$\}$; note that $\{1, x, x^2, ..., x^{n-1}\}$ is a basis
- The set of all real valued continuous functions on $[a, b]$
- The set of all square integrable functions on $[a, b]$ (If you recall the Schwartz inequality, you can easily prove this.)

Notice that the second and third examples are infinite dimensional according to our definition above.

So, thus far we have set up sets with an algebraic structure. That is, we have sets with addition and scalar multiplication defined on them. But, we'd also like to introduce a notion of distance and length, i.e., geometric concepts. To do this we need the notion of a metric.

DEFINITION A.8 *A set X of elements is said to be a metric space iff $\forall u, v \in X$ there is an associated real number, $d(u, v)$, such that*

- *$d(u, v) > 0$ for $u \neq v$*
- *$d(u, u) = 0$*
- *$d(u, v) = d(v, u)$*
- *$d(u, w) \leq d(u, v) + d(v, w)$.*

We call d a metric.

You should notice that a metric space need not be a vector or linear space. Next, we introduce spaces with a notion of length defined on them.

DEFINITION A.9 *A normed linear space is a linear space in which a norm $||u||$ is defined. The norm must satisfy*

- $||u|| > 0$ *if* $u \neq 0$
- $||0|| = 0$
- $||\alpha u|| = |\alpha|||u||$
- $||u + v|| \leq ||u|| + ||v||$.

Now, a normed linear space is a metric space with $d(u, v) = ||u - v||$. Once we have a metric, we can do calculus. Of course, doing calculus means we can take limits, i.e., we can talk about convergence of sequences.

DEFINITION A.10 *In a metric space the sequence* $\{u_k\} \to u$ *iff* $\forall \epsilon > 0$, $\exists N$ *such that* $d(u, u_k) < \epsilon \ \forall k > N$.

In a normed linear space we can even discuss summation. (Exercise: Why can't we talk about summation in a general metric space?)

DEFINITION A.11 *In a normed linear space* $\sum u_k$ *converges to* u *iff* $\forall \epsilon > 0$, $\exists N > 0$ *such that* $|| \sum u_k - u|| < \epsilon \ \forall n > N$.

As stated our notions of convergence rely upon knowledge of the limit. That is, we need to know the limit *a priori*. Of course, we can redo the notion of convergence in the Cauchy sense and remove this restriction.

DEFINITION A.12 *A sequence* $\{u_k\}$ *is Cauchy iff* $\forall \epsilon > 0$, *there exists* $N > 0$ *such that* $d(u_m, u_p) \leq \epsilon$ *for all* $m, p > N$.

As with sequences of numbers, a convergent sequence in a metric space is convergent if and only if it is Cauchy.

Finally, we can state a few new definitions.

DEFINITION A.13 *A metric space* X *is complete iff every Cauchy sequence in* X *converges to a point in* X.

DEFINITION A.14 *A normed linear space which is complete in its natural metric* $d(u, v) = ||u - v||$ *is called a Banach space.*

In this text, we'll find that Banach spaces are sufficient for our purposes. You've probably heard of Hilbert spaces and perhaps even Sobolev spaces. As an aside, these are just Banach spaces with even more structure. For example, a Hilbert space is a Banach space with an inner product as well as a norm. This means there is a notion of angle as well as distance and length. For those of you who are interested, some further reading on this topic is [184]. But, let's return to Banach space. Here's some examples

- The set of rational numbers is not complete and hence is not a Banach space
- $[a, b]$ is a Banach space
- (a, b) is not a Banach space
- The space of continuous functions on $[a, b]$ denoted $C[a, b]$ is a Banach space with the infinity norm $||u||_\infty = \max_{[a,b]} |u|$

The first three examples are clear, but what about the last example? Why is $C[a, b]$ a Banach space? The only question we really need to think about is the question of completeness. Suppose $\{u_k\}$ is a Cauchy sequence in our space. Then, for any $\epsilon > 0$ there exists an $N > 0$ such that

$$\max_{[a,b]} |u_k(x) - u_m(x)| < \epsilon \tag{A.36}$$

for any $m, k > N$. But, this means that the sequence converges uniformly! Recalling that the uniform limit of continuous functions is a continuous function, we see that $C[a, b]$ is complete in the maximum norm.

Appendix B

Physical Constants

The constants tabulated in Table B.1 are in the International System of units (SI), which consists of seven fundamental units: length (*meter*, m), mass (*kilogram*, kg), time (*second*, s), electric current (*ampere*, A), thermodynamic temperature (*kelvin*, K), amount of substance (*mole*, mol), and luminous intensity (*candela*, cd). These are joined by any number of derived units, some of which are given in Table B.2

TABLE B.1: Useful physical constants in SI units

Quantity	Symbol	Values	Units
Speed of light	c	2.99792458×10^8	$\mathrm{m\,s^{-1}}$
Plank's constant	h	6.63×10^{-34}	$\mathrm{J\,s}$
Permittivity constant	ϵ_0	$8.854187817 \times 10^{-12}$	$\mathrm{F\,m^{-1}}$
Permeability constant	μ_0	$4\pi \times 10^{-7}$	$\mathrm{N\,A^{-1}}$
Gravitational constant	G	6.673×10^{-11}	$\mathrm{m^3\,s^{-2}\,kg^{-1}}$
Stefan-Boltzmann constant	σ	5.670400×10^{-8}	$\mathrm{W\,m^{-2}\,K^{-4}}$
Avogadro constant	N_A, L	$6.02214199 \times 10^{23}$	$\mathrm{mol^{-1}}$
Elementary charge	e	$1.602176462 \times 10^{-19}$	C
Boltzmann constant	k	$1.3806503 \times 10^{-23}$	$\mathrm{J\,K^{-1}}$

TABLE B.2: Selected derived SI units

Quantity	Symbol	Measures	Derived	Fundamental
Newton	N	Force	—	$\mathrm{m\,kg\,s^{-2}}$
Pascal	Pa	Pressure	$\mathrm{N\,m^{-2}}$	$\mathrm{m^{-1}\,kg\,s^{-2}}$
Joule	J	Energy	$\mathrm{N\,m}$	$\mathrm{m^2\,kg\,s^{-2}}$
Watt	W	Power	$\mathrm{J\,s^{-1}}$	$\mathrm{m^2\,kg\,s^{-3}}$
Poiseuille	PI	Dynamic viscosity	$\mathrm{N\,s\,m^{-2}}$	$\mathrm{kg\,m^{-1}\,s^{-1}}$
Hertz	Hz	Frequency	—	$\mathrm{s^{-1}}$
Coulomb	C	Electric charge	—	$\mathrm{s\,A}$
Volt	V	Electrical potential	$\mathrm{W\,A^{-1}}$	$\mathrm{m^2\,kg\,s^{-3}\,A^{-1}}$
Farad	F	Capacitance	$\mathrm{C\,V^{-1}}$	$\mathrm{m^{-2}\,kg^{-1}\,s^4\,A^2}$
Weber	Wb	Magnetic flux	$\mathrm{V\,s}$	$\mathrm{m^2\,kg\,s^{-2}\,A^{-1}}$
Henry	H	Inductance	$\mathrm{Wb\,A^{-1}}$	$\mathrm{m^2\,kg\,s^{-2}\,A^{-2}}$
Tesla	T	Magnetic flux density	$\mathrm{Wb\,m^{-2}}$	$\mathrm{kg\,s^{-2}\,A^{-1}}$
Ohm	Ω	Electrical resistance	$\mathrm{V\,A^{-1}}$	$\mathrm{m^2\,kg\,s^{-3}\,A^{-2}}$
Siemens	S	Electrical conductance	$\mathrm{A\,V^{-1}}$	$\mathrm{m^{-2}\,kg^{-1}\,s^3\,A^2}$

Index

Printed in the United States
by Baker & Taylor Publisher Services

Printed in the United States
by Baker & Taylor Publisher Services